OXFORD MATHEMATICAL MONOGRAPHS

Series Editors

OXFORD MATHEMATICAL MONOGRAPHS

The Mathematical Theory of Thermodynamic Limits: Thomas–Fermi type models

ISABELLE CATTO

CNRS URA 749, Université Paris Dauphine, Paris, France

CLAUDE LE BRIS

Ecole Nationale des Ponts et Chaussées, Marne La Vallée, France

PIERRE-LOUIS LIONS

CNRS URA 749, Université Paris Dauphine and Ecole Polytechnique, Paris, France

CLARENDON PRESS · OXFORD
1998

Oxford University Press, Great Clarendon Street, Oxford OX2 6DP
Oxford New York
Athens Auckland Bangkok Bogota Buenos Aires Calcutta
Cape Town Chennai Dar es Salaam Delhi Florence Hong Kong Istanbul
Karachi Kuala Lumpur Madrid Melbourne Mexico City Mumbai
Nairobi Paris São Paolo Singapore Taipei Tokyo Toronto Warsaw
and associated companies in
Berlin Ibadan

Oxford is a registered trade mark of Oxford University Press

Published in the United States by
Oxford University Press Inc., New York

A catalogue record for this book is available from the British Library

Library of Congress Cataloging in Publication Data
(Data applied for)

ISBN 0 19 850161 7 (Hbk.)

Typeset by The Author
Printed in Great Britain by
Bookcraft Ltd.,
Midsomer Norton, Avon

PREFACE

We begin with this book a long-term study of the thermodynamic limit for various models in quantum molecular chemistry; namely, the Thomas–Fermi–von Weizsäcker and Hartree–Fock type models.

Generally speaking, the problem under consideration may be stated as follows (more sophisticated forms, involving the most important physically relevant cases, will be given in the course of our work, and we only define it here in a somewhat academic fashion). Take a model of quantum chemistry for a molecule of N electrons and N nuclei of unit charge. Denote by E_N the ground-state energy of this molecule, and by ρ_N the minimizing density. Let N now go to infinity. Does the energy per unit volume $\frac{1}{N}E_N$ and does the density ρ_N go to a limit?

Such questions have been studied by many authors in various frameworks: among others, Ruelle and Fisher [48], Lieb and Lebowitz [29, 36, 37], Lieb and Narnhofer [38], Lieb [32, 35], Fefferman [21, 22], and Gregg [25]. However, for the models we are interested in, the ground-breaking study is undoubtedly due to Lieb and Simon, in their pioneering work [40] on the Thomas–Fermi model.

Proving the existence of such limits and identifying them may be seen as a first step in the enterprise of going (mathematically) from the microscopic scale to the macroscopic scale. In particular, apart from helping to understand how thermodynamics arises, it can give some insight into the elaboration of models for the condensed phase. Our motivation primarily lies in this second aspect and, more particularly, in the derivation of rigorous models for the solid phase. Since at low temperature the nuclei in a solid mostly arrange themselves in regular structures, it seems rather natural, at least from an academic viewpoint, to construct a crystal step by step through the limiting process described above. For this purpose, we have chosen to deal first with the Thomas–Fermi–von Weizsäcker model, which is the simplest model in which this fundamental issue may be meaningfully discussed: indeed, as opposed to the Thomas–Fermi model, the Thomas–Fermi–von Weizsäcker model allows binding of molecules. Even if the challenging question of explaining why heavy nuclei with enough electrons bind themselves to form periodic structures is still completely open, it seems interesting to lay some groundwork in this direction, and we hope this book will contribute to such a preparation.

Let us now emphasize that our purpose here is not to give a complete description of the state of the art, but rather to present new results on models that are already well known in quantum molecular chemistry, but have not yet been considered from this standpoint. It is to be mentioned that our work requires a good familiarity with the theory of elliptic partial differential equations and variational problems. Therefore, it is at first designed for mathematicians and mathematical physicists. However, we hope that the first chapter and the outlines of our work will also interest physicists and chemists who are curious to see the mathematical difficulties raised by such problems. Let us also underline that we are aware that the model we focus on in this book, namely the Thomas–Fermi–von Weizsäcker model, is not the ultimate model in quantum molecular chemistry, and that it is broadly overtaken by Hartree–Fock type models or modern models issuing from density functional theory. However, the Thomas–Fermi–von Weizsäcker model, as will be detailed below, provides a convenient framework for the new mathematical tools we introduce and that we shall need in the future for the study of more physically relevant models.

The book is organized as follows.

The first chapter is a wide introduction to all the questions we shall address not only in this book but also in a following work [17]. We define the so-called thermodynamic limit problem first in a very general setting (in Section 1.2 from the mathematical standpoint, and in Section 1.3 from the physical standpoint, in a way that of course does not pretend to compete with textbooks by physicists, our purpose being only to give the flavour of what is important in this field). Then we briefly survey in Section 1.4 the works devoted so far to this problem. In Section 1.5, we particularize the thermodynamic limit problem to Thomas–Fermi–von Weizsäcker type settings, which will be the topic of this book. In this section, we state the main results we shall obtain in our work; namely, the fact that in Thomas–Fermi–von Weizsäcker type settings, both the energy per unit volume and the electronic density go to a limit, that can be identified and recast as a periodic problem. Thinking of other questions of great interest that will not be addressed in this work but in a subsequent one [17], we present in Section 1.6 the thermodynamic limit problem in Hartree–Fock type settings, and also connect our work to related questions from other fields.

In Chapter 2, we get to the heart of the matter. This chapter and the following one are devoted to the question of finding the limit for the energy per unit volume. It presents the proof of the existence of the thermodynamic limit for the energy per unit volume for the Thomas–Fermi–von Weizsäcker problem with a Yukawa potential ($\frac{e^{-a|x|}}{|x|}$, $a > 0$) modelling the interaction between particles (the electrons and the nuclei, the latter being fixed at the points of a lattice of $\mathbf{R}^3$). We show (Sections 2.2–2.5) that the energy per unit volume converges to the ground-state energy of some periodic minimization problem of the same family as the original one, but set on the unit cell of the crystal and involving a periodic potential for the modelling of the interaction between the particles. Sections 2.6 and 2.7 give various comments and extensions.

Chapter 3 deals with the same question of the existence of the limit for the energy per unit volume for the Thomas–Fermi–von Weizsäcker problem with a Coulomb potential, which is the physically relevant model. In this chapter, we shall provide many proofs of the convergence, some of them holding true under the same assumptions, some others holding true for different situations. Whatever the strategy of proof is, we begin by proving *a priori* estimates on the energy and the density. Some of them are borrowed from the work by Solovej [53] on the asymptotics of the Thomas–Fermi–von Weizsäcker model for large molecules, a problem that is in fact closely related to ours. Next, we present the various proofs. It is to be remarked that when the nuclei are smeared nuclei, the proofs are often simpler than when the nuclei are point nuclei; in the latter case especially, the approximation by the Yukawa potential is often (but not always) needed. The reason why we consider a Yukawa potential at first, before tackling the physically relevant case of the Coulomb potential, is that the decay of the potential that models the interactions between particles at infinity plays a significant role in the analysis of the thermodynamic limit problem, as it will become clear in the course of Chapter 1. In other words, the fact that the Coulomb potential is a long-ranged potential creates serious mathematical difficulties that we can circumvent through the approximation by Yukawa potentials. When the exponent of the Yukawa potential goes to zero, one recovers the model with Coulomb potential from the model with Yukawa potential. This is useful for the comments of Section 2.6 of Chapter 2, but also provides a strategy (amongst others) to find the lower limit in the Coulomb case (Section 3.4). On the other hand, the upper limit is obtained in Section 3.5 by some direct argument, which requires some additional technical assumptions on the geometry of the lattice. It is worth noticing already here that a proof of the convergence of the energy per unit volume that does not make use of the approximation by the Yukawa potential and foremost that holds for more general situations than the situations considered in this third chapter will be given in Chapter 6. The last section of the chapter (Section 3.6) is devoted to some extensions, and in particular the Thomas–Fermi–Dirac–von Weizsäcker model. Basically, the main result on the convergence of the energy that we obtain in this book is summarized in a somewhat vague way in the following:

Theorem *In the setting of the Thomas–Fermi–von Weizsäcker theory, the ground-state energy of a neutral molecule whose nuclei are located on a finite subset Λ of $\mathbf{Z}^3$, divided by the volume enclosed in Λ, converges, when Λ tends to cover the entire periodic lattice $\mathbf{Z}^3$, to the ground-state energy of a periodic minimization problem of the Thomas–Fermi–von Weizsäcker type set on the unit cell of the lattice and involving a periodic interaction potential of the same period as the lattice.*

After these questions on the convergence of the energy, Chapters 4 and 5 are devoted to the more difficult question of the existence of a limit for the minimizing electronic density, respectively, in the Yukawa case and in the Coulomb case. In both chapters, we first prove (in Section 4.2, and respectively Section 5.2) preliminary results of convergence of the densities. These results are somewhat weak, but, as we shall explain then, they are in some sense the best results one may prove, using arguments only based upon energetic estimations. In order to strengthen these results of convergence and to identify the limit density, we next focus on the Euler–Lagrange equations obtained in the thermodynamic limit, and prove the uniqueness of their solution. (The result we obtain in Chapter 5 will be improved in Chapter 6, and leads to a general result of existence and uniqueness for a system of coupled partial differential equations.) This will consequently give the periodicity of the limit density and conclude the proof of the convergence of the sequence of electronic densities to the periodic density minimizing the periodic energy obtained (in Chapters 2 and 3) by passing to the thermodynamic limit in the energy per unit volume. The strategy of proof for the uniqueness in the Coulomb case (Section 5.3) is slightly different from the Yukawa case (Section 4.3), but also allows us to conclude in this latter case (see Subsection 5.3.3.1). Finally, we prove in Section 5.5 the convergence of the electronic density to the limit periodic density in a very strong sense, thereby completing the 'weak' results of Section 5.2.

Chapter 6 has two goals. First, we give in Section 6.2 another proof of the convergence of the energy per unit volume in the Coulomb case. It is based upon the convergence of the density in the strong sense shown in Section 5.5. Therefore, it does not appeal at all to the technical assumption needed in Chapter 3.

Next, in Section 6.3 we prove the following general existence and uniqueness result:

Theorem *Let $c > 0$, let m be a non-negative measure on $\mathbf{R}^3$ satisfying*

$$(H1) \qquad \sup_{x \in \mathbf{R}^3} m(x + B_1) < \infty,$$

$$(H2) \qquad \lim_{R \longrightarrow +\infty} \inf_{x \in \mathbf{R}^3} \frac{1}{R} \, m(x + B_R) = +\infty,$$

where B_R denotes the ball of radius R centered at 0.

Then, there is one and only one solution (u, Φ) on $\mathbf{R}^3$ of the system

$$\begin{cases} -\Delta u + c u^{7/3} - \Phi u = 0, \\ u \geq 0, \\ -\Delta \Phi = 4\pi \left[m - u^2 \right], \end{cases}$$

with $u \in L^{7/3}_{\text{loc}} \cap L^2_{\text{unif}}(\mathbf{R}^3)$ and $\Phi \in L^1_{\text{unif}}(\mathbf{R}^3)$.

In addition, $\Phi \in L^{3,\infty}_{\text{unif}}(\mathbf{R}^3)$, $\inf_{\mathbf{R}^3} u > 0$, and $u \in L^\infty(\mathbf{R}^3) \cap C^{0,\alpha}(\mathbf{R}^3) \cap W^{2,p}_{\text{unif}}(\mathbf{R}^3)$ for all $0 < \alpha < 1$, $1 \leq p < 3$.

This result is then used to prove the existence of the thermodynamic limit in more general cases than the rather academic ones studied so far. In particular, we shall deal with quasicrystals.

Let us mention that the main results presented in this book have been announced in [16].

In a following work ([17]), we shall be concerned with the questions analogous to those of this book, this time in Hartree and Hartree–Fock type settings, and also to questions related to the optimized geometries. Indeed, whereas we assume here that the nuclei are located at fixed points of a periodic lattice of $\mathbf{R}^3$, it is possible to consider the situation when, for each number of nuclei, both the electronic state of the molecule and the positions of the nuclei are optimized, and to ask the question of the existence of the thermodynamic limit. We shall present in the near future some preliminary results towards the solution of this difficult issue.

Finally, we would like to thank Professor Roger Balian, who made many valuable comments on a preliminary version of this work and in particular on the physical considerations developed in Chapter 1.

Paris   I.C.
Marne La Vallée   P.-L.L.
June 1997   C. Le B.

CONTENTS

1

GENERAL PRESENTATION

1.1 Introduction

This chapter is a global presentation of the problems we are going to consider
in this book, and also in a forthcoming work [17]. We therefore successively
define the problem of investigating the thermodynamic limit of a given model
in quantum molecular chemistry and survey the main mathematical difficulties
encountered (Section 1.2), give some clues on the physical background of this
mathematical problem (Section 1.3), and briefly review the state of the art of the
mathematical knowledge on this type of problem (Section 1.4). Next, we introduce the thermodynamic limit problem for the Thomas–Fermi–von Weizsäcker
model for a crystal and some related models, detail the main results we have
obtained in this setting, and indicate some extensions of our work to particular
situations (Section 1.5). Section 1.6 presents some future prospects, in particular
about the thermodynamic limit in the Hartree–Fock setting; we also list a few
open problems that we have isolated as interesting and difficult, and we connect these problems, and more generally all our work, to some related questions
arising in other fields of mathematical analysis and mathematical physics.

1.2 The thermodynamic limit from the mathematical viewpoint

In this section, we formulate in a somewhat vague way a general problem of
existence of the thermodynamic limit for a molecular model in the framework of
quantum molecular chemistry and emphasize the main mathematical difficulties.
In the next section we shall see that this thermodynamic limit problem can be set
in a far more general physical context. In Section 1.5 below, the formal setting
of this section will be made precise and detailed for the Thomas–Fermi–von
Weizsäcker model for a crystal. Likewise, the problem for Hartree and Hartree–
Fock type models will be considered in a following work [17].

Consider a finite number of nuclei, each nucleus being of unit charge and
being located at a point $k = (k_1, k_2, k_3)$ of integral coordinates in $\mathbf{R}^3$, which is the
centre of a cubic unit cell $\Gamma_k = \{(x_1, x_2, x_3) \in \mathbf{R}^3; -\frac{1}{2} < x_i - k_i \leq \frac{1}{2}, \ i = 1, 2, 3\}$.

The set of the positions of these nuclei is then a finite subset Λ of the set of
all points of integral coordinates; that is, $\mathbf{Z}^3 \subset \mathbf{R}^3$. The union of all cubic cells
whose centre is a point of Λ is denoted by $\Gamma(\Lambda)$; its volume is denoted by $|\Lambda|$ (see
Figure 1.1 below). Since each cell has unit volume and each nucleus is of unit
charge, $|\Lambda|$ is also the number of nuclei and the total nuclear charge. (All the
arguments we present hereafter also hold *mutatis mutandis* when each nucleus
is of charge z–the number of electrons that have then to be considered below to

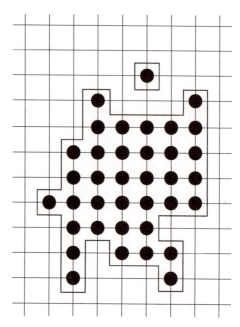

FIG. 1.1. **The sets Λ and $\Gamma(\Lambda)$ (in two dimensions).** The circles show the nuclei, and the solid curve shows the boundary $\partial\Gamma(\Lambda)$.

ensure neutrality is $z|\Lambda|$–or when the elementary cell is not cubic but is a fixed parallelepiped (see in particular Section 1.5).)

Suppose that for $\Lambda \subset \mathbf{Z}^3$ fixed, we have a 'good' model for the ground state of the neutral molecule consisting of $|\Lambda|$ electrons and $|\Lambda|$ nuclei located at the points of Λ; that is to say, we have the following:

1. A potential $\mathcal{V}$ modelling the interaction between particles.

2. A function $\mathcal{E}_\Lambda(\rho)$ of the electronic density ρ modelling the energy of the molecule, such that the following minimization problem

$$\mathcal{I}_\Lambda = \inf \left\{ \mathcal{E}_\Lambda(\rho) + \tfrac{1}{2} \sum_{a \neq b \in \Lambda} \mathcal{V}(a,b); \rho \in \mathcal{D}(\mathcal{E}_\Lambda), \int_{\mathbf{R}^3} \rho = |\Lambda| \right\} \qquad (1.1)$$

has a minimum; that is, there exists a density ρ_Λ in $\mathcal{D}(\mathcal{E}_\Lambda)$ satisfying $\int_{\mathbf{R}^3} \rho_\Lambda = |\Lambda|$ and $\mathcal{E}_\Lambda(\rho_\Lambda) = \mathcal{I}_\Lambda$, where $\mathcal{D}(\mathcal{E}_\Lambda)$ is the definition domain of $\mathcal{E}_\Lambda$. Note that, for the sake of simplicity and while other frameworks are convenient (such as the Hartree and Hartree–Fock type models which involve $|\Lambda|$ wave functions; see Section 1.6 and [17]), we formulate (1.1) in the framework of a density functional theory. In other words, we assume that the electronic state is modelled through a single quantity, namely the electronic density ρ. Such a functional $\mathcal{E}_\Lambda$ is known to exist in view of the work [26] by Hohenberg and Kohn. Unfortunately, their abstract result does not provide any explicit form of the functional to consider. Many approximated functionals have been proposed (see, for instance, Dreizler and Gross [19] and Parr and Yang [45]). A typical example of such an energy $\mathcal{E}_\Lambda$ is a Thomas–Fermi type energy. The fact that the Thomas–Fermi type energies give 'good' models in the sense defined above is shown in Lieb [33] and Lions [42].

Then the question of the existence of the thermodynamic (or bulk) limit for the model (1.1) considered above may be stated as follows:

(i) *Does there exist a limit for the energy per unit volume* $\frac{1}{|\Lambda|}\mathcal{I}_\Lambda$ *when* $|\Lambda|$ *goes to infinity?*

(ii) *Does the minimizing density* ρ_Λ *approach a limit* ρ_∞ *(in a sense to be made precise later) when* $|\Lambda|$ *goes to infinity?*

(iii) *Does the limit density* ρ_∞ *have the same periodicity as the assumed periodicity of the nuclei?*

For the first two questions, a subsidiary question is the following: *Is the limit independent of the particular sequence of domains and densities used in passing to the limit?*

Since we aim in particular at putting the models used in solid state physics on a sound mathematical ground, we restrict ourselves until Chapter 5 to the study of the case where the nuclei are fixed on a periodic lattice, but leaving the positions of the nuclei free is also a case of great physical and mathematical interest (see Chapter 6 for some non-periodic structures and also [17] for other developments).

What are the main mathematical difficulties one has to overcome in order to answer the above questions?

Let us begin with question (i). In fact, we shall present in our work a way to tackle the three questions (i)–(iii) by solving first questions (ii) and (iii) and then question (i); but for the rather formal argument we make here, we prefer to begin with question (i), because it is the simplest one. Before addressing the question of the convergence, a preliminary question about the energy per unit volume is to be solved: Do there exist bounds for it? The most difficult part turns out to be finding a bound from below. This important question is actually related to the so-called stability of matter problem in physics (a problem we shall return to in Sections 1.3 and 1.4). In order to be more specific, let us assume that the molecular energy reads

$$\mathcal{E}_\Lambda(\rho) = T(\rho) - \sum_{a\in\Lambda} \int \mathcal{V}(a,x)\rho(x)\mathrm{d}x + X(\rho) + \tfrac{1}{2} \sum_{a\neq b\in\Lambda} \mathcal{V}(a,b),$$

where T is the kinetic energy and X the inter-electronic repulsion. If the potential $\mathcal{V}$ is not bounded (think, for instance, of the Coulomb potential and its singularity at 0), it is not clear whether $\left(\int \rho\right)^{-1}\mathcal{E}_\Lambda(\rho)$ is bounded from below or not, independently of Λ. It requires some work to prove that, in the good cases, it is actually bounded. On the other hand, the existence of a bound from above is related to the behaviour at infinity of the potential. The immediate consequence of such bounds on the energy is that some bounds on the densities are available, which will be particularly useful for the sequel of the study.

Once such bounds are obtained, the next difficulty is to find a limit for the potential created by charges located at the points of Λ. In the framework presented above, a good candidate for this limit is the potential created by an infinite set of charges located at all points of $\mathbf{Z}^3$, but it remains to make sense of such a potential. Note that this limit potential will be used to model not only the attractive potential of the infinite lattice of nuclei on a given electron, but also the repulsive interaction potential between electrons, since we hope that the electronic density will become periodic. A severe difficulty arises when the initial potential does not decay fast enough at infinity (say, not faster than $\frac{1}{|x|^3}$ in $\mathbf{R}^3$; think again of the Coulomb potential $\frac{1}{|x|}$). This difficulty is also identified from the physical viewpoint and is recurrent in all available proofs of thermodynamic limits (see Section 1.4).

The limit potential being imagined, the periodic limit energy may be built, say $E_{\mathrm{per}}(\rho)$, on the unit cell of the lattice, and the next mathematical challenge is to determine the correct constraint associated with the limit minimization problem (and thus identify the limit in (i); of course, there will remain afterwards the task of proving that the sequence of the energy per unit volume does indeed converge to this limit energy). In other words, the question is to discover whether or not some charge has escaped at infinity while Λ was expanding to infinity. When Λ is fixed, the molecule has $|\Lambda|$ electrons and a set of $|\Lambda|$ nuclei. A natural guess is to obtain asymptotically one electron per unit cell (and therefore to keep the global electrical neutrality), so that the periodic minimization problem one obtains formally reads

$$I_{\mathrm{per}} = \left\{ E_{\mathrm{per}}(\rho), \int_{\Gamma_0} \rho = 1 \right\}.$$

However if some charge has escaped, it rather reads

$$I_{\mathrm{per}} = \left\{ E_{\mathrm{per}}(\rho), \int_{\Gamma_0} \rho = \mu \right\},$$

with $\mu < 1$ (which corresponds to the physical case of an 'ionized crystal'). This question is closely related to what we may call the 'size' of the molecule. When Λ becomes larger and larger, the volume of the positions occupied by the nuclei grows like $|\Lambda|$: Does the volume occupied by the electronic cloud grow at the same rate, in which case the electrons stay in the surrounding of the nuclei, or does it grow faster? In the latter case, some charge may be lost while taking the thermodynamic limit. We shall see that such a situation where some charge is lost does occur when we initially put too many electrons in view of the charge of the nuclei in the setting of the Thomas–Fermi–von Weizsäcker theory with a Yukawa potential (see the details in Chapter 2 below). On the other hand, we shall prove in Chapter 3 that with the Coulomb potential and a sequence of neutral systems, no charge escapes at infinity.

Stability, determination of the periodic interaction potential (together with the possible asymptotic periodicity of the density), and proof of the compactness are the three main mathematical difficulties one has to overcome in order to answer question (i).

Let us now turn to questions (ii) and (iii), which we treat here together.

The fact that the density ρ_Λ approaches a limit when Λ goes to infinity and that this limit is a periodic function seems a much more difficult point than the convergence of the energy per unit volume.

Actually, in the proof of the convergence of the energy, some convergence and periodicity of the density in some average sense is already needed, but no *sensu stricto* periodicity is needed. However, even if the known bounds on the densities allow one to pass to the limit (at least locally) in the Euler–Lagrange equation of the minimization problem when Λ goes to infinity (which is *per se* not a straightforward fact), one then faces the following question: Does the obtained partial differential equation have a unique, periodic solution?

Take our simple formal example again. Define $u_\Lambda = \sqrt{\rho_\Lambda}$, where ρ_Λ is the minimizing density. Then u_Λ is a solution to an elliptic PDE of the kind

$$-\Delta u_\Lambda - \sum_{a \in \Lambda} \mathcal{V}(a, x) u_\Lambda + F(u_\Lambda, x, ...) u_\Lambda = 0,$$

where the term F often exhibits non-linear and non-local features. Consider a 'simple' case like the Yukawa case for a crystal where $\sum_{a \in \mathbf{Z}^3} \mathcal{V}(a, x)$ exists and defines a periodic function, and where we may pass to the limit in the above equation. The limit u_∞ of u_Λ satisfies

$$-\Delta u_\infty - \sum_{a \in \mathbf{Z}^3} \mathcal{V}(a, x) u_\infty + F(u_\infty, x, ...) u_\infty = 0.$$

Does this imply that u_∞ is periodic? We shall see that this question can be solved positively (see Chapter 4 for the Thomas–Fermi–von Weizsäcker theory with a Yukawa potential, and Chapter 5 for the analogous theory with a Coulomb potential).

The difficulties are as follows. First, even without the somewhat complicated term F, we need some spectral result on the operator $-\Delta - \sum_{a \in \mathbf{Z}^3} \mathcal{V}(a, x)$. Secondly, the term F being non-linear, we have to prove the uniqueness of the non-negative solution to

$$-\Delta u_\infty - \sum_{a \in \mathbf{Z}^3} \mathcal{V}(a, x) u_\infty + F(u_\infty, x, ...) u_\infty = 0,$$

without prescribing any boundary conditions. Thirdly, the fact that the term F is non-local (it contains the inter-electronic repulsion) is known to increase the difficulty. The uniqueness and thus the periodicity of the solution is shown in Chapters 4 and 5 by new arguments which use in a fundamental way the strict convexity of the Thomas–Fermi–von Weizsäcker functional. The Yukawa case (Chapter 4) corresponds to the situation where the non-local feature of the term that we have denoted by F above is not too wide (i.e. when in fact the interaction potential decays fast enough at infinity, again). Then, the situation we have is not too far from a local non-linear elliptic PDE, on which we shall develop comparison techniques based upon the maximum principle in order to obtain a uniqueness result (the maximum principle holding true because we know convenient spectral properties on the linearized operator). For the Coulomb case (Chapter 5), another type of argument has to be used. Using arguments dealing only with the energy, and not exploiting the Euler–Lagrange equation, a complete proof of the periodicity of the limit density is beyond one's reach, but it is, however, possible to obtain partial information such as periodicity on average (which we define as follows: $\frac{1}{\Lambda}\sum_{y\in\Lambda}\rho_\Lambda(\cdot + y)$ goes to a periodic function), or periodicity up to a translation (that is, $\rho_\Lambda(\cdot + y_\Lambda)$ goes to a periodic function for a sequence $y_\Lambda \in \mathbf{R}^3$ possibly escaping at infinity). In addition, both limit periodic densities are the expected minimizing density of the periodic problem posed on the unit cell that is the thermodynamic limit of the energy per unit volume. It is worth noticing that arguments dealing only with the energy can only give such 'weak' results and not the 'strong' result that we mentioned above. Indeed, since the asymptotic process consists of dividing by Λ a quantity that grows like Λ, it is not surprising that such pure energetic arguments may only provide such averaged or vague information, while precise information like exact periodicity requires us to go deeply in the nature of the minimization problems and work directly with the Euler–Lagrange equation.

Obtaining convenient bounds in order to pass to the local limit, and proving uniqueness of the non-negative solution to a non-local non-linear PDE, are the main two difficulties encountered in trying to answer questions (ii) and (iii).

Having emphasized the main mathematical features of the proof of the existence of the thermodynamic limit, we devote the next section to the physical aspects of this problem. Let us point out at this stage that a reader who may wish to skip the physical background can proceed directly to Section 1.4, skipping altogether Section 1.3, which has been designed to be essentially independent of the rest of this work. It is thus possible to go directly to Section 1.4, where we briefly explain how the mathematical difficulties outlined above have been solved in the various situations that have been studied so far by mathematical works dealing with the thermodynamic limit problem.

1.3   Physical background

The family of problems we study may be viewed as a meeting point of (at least) two general classical physical theories. The first one is the thermodynamic limit problem for microscopic models, and the second one is the quantum theory of the solid state.

1.3.1 *The thermodynamic limit of microscopic systems*

It is certainly an ambitious task to try to connect the macroscopic properties of matter with its microscopic nature. In the long chain that runs from the microscopic scale to the macroscopic one, the derivation of the postulates of thermodynamics with statistical mechanics applied to microscopic systems is a decisive link. By taking the so-called thermodynamic limit of statistical mechanics quantities, i.e. by considering in the limit a system whose size is infinitely as large as the typical distance between two of its microscopic constituents, one not only recovers the usual thermodynamic functions but also checks their intensivity or extensivity, that were already well known from experiment. Before going further into the details, let us mention that this section draws its inspiration from many articles and textbooks, the first of them being the book [4] by Balian. For further information, we also refer the interested reader to Ruelle [48], Tolman [55], Lieb and Lebowitz [36], Lieb [32, 35], and Fefferman [21, 22].

Macroscopic experiment confirms that the characteristic quantities whose existence is postulated by thermodynamics, namely the thermodynamic functions, may be classified into two categories, depending upon whether they are intensive, like temperature, or extensive, like entropy. While temperature does not change when the system under consideration is replaced by one of its subsystems, entropy varies linearly with respect to its three natural arguments: energy, number of particles, and volume (at zero temperature, when the entropy is zero, this yields the linear variation of the energy with respect to the number of particles). The linear variation for the energy is closely connected to the stability of matter: if the energy $E(N)$ of a system of N particles was proportional to N^p with $p > 1$ instead of being linear with respect to N, the energy released when putting two subsystems of N particles together would be of the order of $(2^p - 2)E(N)$, and thus of the order of $E(N)$, which can lead to dramatically large release of energy (see Lieb [35], from where we have borrowed the above remark). In the current century, statistical mechanics has provided a derivation of the postulates of thermodynamics (see Balian [4]). Indeed, as the principles of thermodynamics aim at describing the state of a system through the knowledge of only a few variables, the techniques and the methods of statistical mechanics give the typical behaviour of a system that is not known in its fullest details. One can thus expect the global thermodynamic behaviour of a given macroscopic system to be determined by the typical behaviour of a (possibly microscopic) representative system. For instance, the energy, whose existence is postulated in the first principle of thermodynamics, turns out to be a convenient limit (when the statistical system becomes large) of statistical quantities defined on representative ensembles. By the way, with such a limit process, it is also possible to check in a rigorous way the extensivity of the thermodynamic quantities that are known to be extensive from experiment. Consequently, this also confirms in turn the assumptions that statistical mechanics is based upon.

Efficient as they are, these limiting processes need some sound mathematical foundation and it has been a challenge in recent decades to find rigorous arguments to derivate the thermodynamics postulates, and some other well known experimental observations, from the statistical framework and related theories. We shall return to these mathematical works in more detail in Section 1.4.

Apart from the pure question of the existence of the thermodynamic limit for the statistical quantities, the related issue of consistency has been solved for the so-called extensive systems. The first concern is to verify that the same limit is obtained no matter what kind of statistical ensemble is used (canonical, microcanonical, grand canonical, and so on). A second concern is to obtain the same limit independently of the shape of the region that contains the system. These two questions are closely related to each other.

Beyond this, there is also the problem of the existence of a thermodynamic limit for the correlation functions or the reduced density matrix. This problem is a difficult one, the existence and the uniqueness of the limit thermodynamic state being far from obvious, and we refer the interested reader to Ruelle [48].

When the temperature goes to zero, the physical system we consider becomes frozen in its ground state. For the molecular systems we are interested in, we therefore enter the framework of quantum molecular chemistry. All the questions of existence of the thermodynamic limit we have examined so far may be stated in this framework. It is worth mentioning a point of terminology: we retain the term *thermodynamic limit*, while it may sound strange in such a zero temperature setting. From the physical viewpoint, the difficulties are similar. These difficulties also have their mathematical counterparts. Our work is aimed at solving the latter.

1.3.2 *Some basics of solid state theory*

By taking the thermodynamic limit of some given molecular model dealing with electrons and nuclei, the latter being fixed at the points of a periodic lattice, we aim at elaborating, in a rigorous mathematical way, a quantum model for a crystal. The setting of Section 1.2 is that of a density functional theory, but analogous work may be done with some other model of quantum chemistry involving the reduced density matrix or the wave functions of the electrons, e.g. Hartree–Fock type models (see Section 1.6 and [17]). Of course, this extension will bring us close to the models used in solid state physics.

As the typical example of quantum macroscopic matter, the solid state provides indeed the most comfortable framework in which to put a quantum model obtained by thermodynamic limit. The quantum theory of the solid state makes extensive use of quantum models of electrons and nuclei located at the points of a periodic lattice. For this purpose, it has developed a powerful machinery, which is in fact close to the kind of questions and the type of mathematical tools we use in our theoretical arguments. Various models are used, various hypotheses are made, and many techniques of numerical analysis are needed for the solution to the equations (we refer the reader, amongst other references, to Ashcroft and Mermin [2], Callaway [14], Kittel [27], Madelung [43], Pisani [46], Quinn [47], Slater [51, 52], and Ziman [58]).

We do not want to enter the details of solid state theory; however, in order to give the reader a flavour of the physical background of our work, and now that we have seen in the above subsection the physical implications of the limit process *per se*, we wish to deal with the output of this process, and briefly describe some basics of the state of the art on the following question: Which model should we use for the solid state? In doing this, we shall give also some insight into the numerical practice. Let us emphasize that, so doing, we go far beyond the scope of this book. However, our purpose here is to illustrate the link between our theoretical investigation and some very practical questions.

The starting point of the building of a model of the solid state is the Hamiltonian for the entire problem (electrons and nuclei, or ions), which contains, at least, the kinetic energy of both species and all the interaction potentials between particles. In view of the mass ratio between electrons and nuclei, this Hamiltonian may, however, be simplified by making the Born–Oppenheimer approximation and assuming the nuclei are fixed. We are aware that a lot of interesting phenomena in solid state theory arise when the motion of the nuclei around their equilibrium position is taken into account (lattice vibration, phonon interactions, and so on), but we shall not consider them in this simplified presentation. In addition, we assume that the nuclei are fixed at the points of a periodic lattice. It is of course a mathematical challenge to prove whether the minimum energy is actually achieved for such a regular periodic lattice of nuclei (crystals) or not (amorphous solids, quasicrystals).

Despite these simplifications, the model is too heavy to be used directly. Therefore one proceeds by successive steps.

Suppose $V_{e-n}(x_i, \bar{x}_k)$ models the interaction between the electron in x_i and the nucleus located at the point $\bar{x}_k$ of the lattice. The potential $\sum_{k \in \text{lattice}} V_{e-n}$ $(\cdot, \bar{x}_k)$ is a periodic potential. Thinking, for instance, of a Hartree–Fock model for the crystal, it is therefore natural to consider an operator of the form

$$H = -\Delta + V \tag{1.2}$$

where the potential V has the periodicity of the lattice (we refer the reader to [49] for a mathematical approach to the spectral properties of such an operator with a periodic potential). Now take a box (L_1, L_2, L_3) containing a large number of unit cells, and search the eigenstates of H that have the periodicity of this large box. The eigenstates of the free Laplacian on the same box with periodic boundary conditions being the plane waves

$$\varphi = e^{i\mathbf{k} \cdot \mathbf{x}},$$

where the wave vector $\mathbf{k}$, quantified by the periodic boundary condition, is

$$k_i = \frac{2\pi}{L_i} n_i, \qquad n_i \in \mathbf{Z}, \qquad i = 1, 2, 3,$$

it turns out (see, e.g., Madelung [43]) that the eigenstates of H are the same plane waves modulated by the periodicity of the lattice

$$\varphi_{n,\mathbf{k}} = e^{i\mathbf{k}\cdot\mathbf{x}} u_{n,\mathbf{k}}, \tag{1.3}$$

where the function u has the periodicity of the lattice. The subscript $n, \mathbf{k}$ denotes the nth eigenstate with wave vector $\mathbf{k}$. Functions of the form (1.3) are called Bloch functions. In order to define all the eigenstates, we may in addition restrict ourselves to a finite set of convenient $\mathbf{k}$, among those belonging to a particular region of the space called the first Brillouin zone (see Madelung [43]). For each $\mathbf{k}$ of this zone, the eigenstates of H are the $\varphi_{n,\mathbf{k}}$, $n \geq 0$, with energy $E_{n,\mathbf{k}}$. n being fixed, $E_{n,\mathbf{k}}$ varies in the nth band of energy, when $\mathbf{k}$ varies in the Brillouin zone (asymptotically, when the size of the large box is infinite, $\mathbf{k}$ varies in a continuum). This is the so-called band structure of a solid, a very powerful model that has a wide range of applications and allows one to recover many of the macroscopic properties of the solid phase (in particular, it explains why a solid is a conductor, a semi-conductor, or an insulator). Of course, the electronic eigenstates one computes depend on the periodic potential V that is used in (1.2). It is part of the theory and of the numerical *savoir-faire* to choose a suitable V (note that our mathematical work may be seen as a search for the exact potential V to consider in (1.2)). V includes the potential created by the nuclei, but also some interaction term between electrons, provided it is periodic. For instance, if the electrons are represented by Bloch functions, the density, and thus any interaction term involving only the density, is periodic. Various approximations of V are possible: muffin-tin potentials, pseudopotentials, optimized effective potentials, and so on. In one of the simplest cases, V may, for instance, come from the Jellium model. This model consists of replacing the lattice of nuclei by a uniform charged background in a domain Λ and thus we consider the Hamiltonian

$$H = T_{el} - ze \sum_i \int_\Lambda \frac{dy}{|x_i - y|} + \frac{1}{2} \sum_{i \neq j} \frac{e^2}{|x_i - x_j|}$$

in this background (T_{el} denotes the kinetic energy of the electrons). The interaction with some of the other electrons may also be incorporated in the uniform background.

Numerically, in some convenient cases, the potential V may itself be treated as a weak perturbation of the Hamiltonian $-\Delta$. The wave function is decomposed on the basis of plane waves or augmented plane waves, or on the basis of atomic orbitals as in the molecular case (these atomic orbitals may be Slater-type orbitals, Gaussians, or numerical orbitals), or is itself a fully numerical wave function.

In addition, more interaction between the electrons may be incorporated. One treats the interaction of electrons as a perturbation of the Hamiltonian and uses the machinery of perturbation theory. For this purpose, the long-range nature of the Coulomb potential is an obstacle, and one often replaces it by a short-range potential such as the Yukawa potential (or some analogous potential), taking into account the long-range effect by particular techniques (see Madelung [43]).

Going beyond this approach consists of solving directly the complete Hartree–Fock equations or Kohn–Sham equations if one works in the setting of the density functional theory (possibly with a modified exchange term; local density approximation, non-local corrections, averaged exchange term) in a self-consistent way, on a large box with periodic boundary conditions. The wave functions appearing in the Slater determinants may be Bloch functions.

Let us also mention that an alternative (and also complementary, depending on the electrons one deals with) route to the above 'global' approaches is to solve an approximate model on the unit cell of the lattice, using simplified boundary conditions, and then to try to match the local solutions together (see Ashcroft and Mermin [2], Quinn [47], and Callaway [14]). The method was originally known as the cellular method of Wigner and Seitz. The periodic potential is replaced inside the cell by a spherically symmetric potential. Various improvements of this method are possible. Note that, in view of this idea of solving a problem on the unit cell of the lattice, even in a very crude way, it is tempting to try to build a rigorous periodic model that is in some way equivalent to the initial global model on the whole crystal. This might provide a practical and numerical application of the theoretical work we present below.

1.4 Historical survey of the thermodynamic limit problem in mathematical physics

1.4.1 *Statistical mechanics works*

Although the first result on the existence of the thermodynamic limit for one statistical model, namely the canonical ensemble, goes back to a work by Van Hove in 1949, the true beginning of some continuous mathematical interest in the subject dates from the 1960s, with the works of Fisher, Ruelle, Lieb, and Lebowitz, amongst others.

Among the simplest systems that have been studied were the so-called lattice systems (Ruelle [48], see references therein). The Hamiltonian contains no kinetic energy term, is reduced to some potential term, and models, for instance, a quantum spin system that interacts with an external magnetic field; it acts on some Hilbert space associated to the lattice $\mathbf{Z}^d$ (typically $d = 3$). The potential is assumed to have finite range (see the details in [48]). In this setting, the thermodynamic limit exists, in the sense that the following limit exists

$$\lim_{\Lambda \to \infty} \frac{1}{N(\Lambda)} \, \mathrm{Log} \, Z_\Lambda,$$

where $N(\Lambda)$ denotes the number of sites of the subset $\Lambda \subset \mathbf{Z}^3$, and Z_Λ the corresponding partition function. With this simple example arises a first important feature which we shall often encounter in the sequel: the range of the interaction potential plays a significant role.

The existence of the thermodynamic limit for the (classical or quantum) microcanonical, canonical, and grand canonical ensembles for a system of particles in $\mathbf{R}^d$ is due to Fisher and Ruelle (see [48], and references therein). In order to exhibit a thermodynamic limit, a given system essentially must satisfy two conditions. The first condition is, as written by Ruelle, that *the interaction between distant particles must be negligible*, which is achieved in the work of Fisher and Ruelle through the assumption that the interaction between the particles should be tempered; that is to say, this interaction vanishes at infinity at least like $\frac{1}{|x|^\alpha}$, with $\alpha > d$, the dimension of the ambient space. Note in fact that this condition is not fulfilled by the Coulomb potential. The forces being not too repulsive at infinity, it therefore prevents the system from exploding. The second condition is, again quoting Ruelle, that *the interaction must not cause the collapse of infinitely many particles into a bounded region*, which is ensured by the so-called stability condition, which, generally speaking, says that the average energy $\frac{1}{N}U$ of N particles is bounded from below. From the quantum point of view, it has to be connected in some way with Heisenberg uncertainty principle: if particles get closer and closer, having their interaction potential go to $-\infty$, their kinetic energy increases, so their total energy is not too small. Note, considering a system of particles with equal charge, that this second condition is clearly not sufficient to ensure the existence of the thermodynamic limit. With the two assumptions of *tempered interaction* and *stability*, the thermodynamic limit exists in both classical and quantum settings. It is worth noticing that the proof is based upon the choice of a particular sequence of domains going to infinity, namely a sequence Λ_n of cubes of size 2^n. Next, using some monotonicity of the partition function Z_Λ with respect to the domain Λ, and the fact that the interaction is tempered, one breaks the cube Λ_{n+1} into smaller cubes of the type Λ_n, and obtains some essential decreasing of the sequence of the energies. The existence of the thermodynamic limit in the case of cubes follows, and then for general domains, packing these domains with cubes, and using the same kind of arguments as above (see the details in Ruelle [48]). This proof unfortunately does not cover the physically relevant case, firstly because, as pointed out above, the Coulomb interaction is not tempered, and secondly because, at the time this proof was issued, the stability was not known for a quantum system of point charges interacting with the Coulomb potential (although this stability with the Coulomb potential was known for hard core particles, and known to be false for a classical system of point particles of different sign charges; see Ruelle [48]).

In the late 1960s, using such a stability result by Dyson and Lenard (improved later by Lieb and Thirring [41]) and that holds in $\mathbf{R}^3$, if all the particles of one sign are fermions—no matter what the particles of the other sign are—which is the case in nature, the electrons being fermions, (see references in [36]), Lieb and Lebowitz proved the existence of the thermodynamic limit for real matter (see their proof in [36], its outlines in [32], [37], or the more concise version in [29]). In order to overcome the difficulty arising from the long-range nature of the Coulomb interaction, they argue as follows. Returning to the scheme of proof of Fisher and Ruelle, Lieb and Lebowitz pack in turn the cubes with balls, each ball being globally electrically neutral. Next, they notice it is sufficient to bound the average interaction between these balls, and this can be done using Newton's theorem: two such balls do not interact in average with each other, because of their neutrality and because of the spherical symmetry of the charge distribution in each ball. In other words, thanks to the spherical symmetry and the overall neutrality, they can circumvent the difficulty due to the fact that the Coulomb interaction is not tempered, and may proceed as Fisher and Ruelle. As pointed out by Lieb, this phenomenon, called screening, and which is a consequence of the global neutrality, is precisely what makes the Coulomb force behave as if it were short-range.

The proof of Lieb and Lebowitz has been extended in a rather straightforward way by Lieb and Narnhofer in 1974 to deal with the case of Jellium. The model of Jellium is a model used commonly in solid state physics: the electrons move in a uniform constant charge (and thus spherically symmetric!) background created by the nuclei in the domain Λ. The Hamiltonian is therefore

$$H = -\sum_i \Delta_i + \sum_{i<j} \frac{e^2}{|x_i - x_j|} - \rho \sum_i \int_\Lambda \frac{dy}{|x_i - y|} + \tfrac{1}{2}\rho^2 \iint_{\Lambda^2} \frac{dx\,dy}{|x - y|},$$

where ρ denotes the uniform background.

While the Jellium model still allows one to use the spherical symmetry tool of Lieb and Lebowitz, it is no longer possible for a crystal model; that is, in a model where the nuclei stand in fixed positions x_i in a lattice $\mathbf{Z}^3 \subset \mathbf{R}^3$. The strategy of Lieb and Lebowitz thus fails, and this difficulty is likely to explain why one had to wait until 1985 for a proof of the existence of the thermodynamic limit for a crystal, in the statistical setting. This proof is due to Fefferman [21]. It makes use of an auxiliary system, called an 'exploded system', where the interactions between the balls are turned off. Fefferman shows some equivalence between this system and the original one, and passes to the limit. With slight modifications (in particular, one must do without Newton's theorem), Fefferman's proof has been extended by Gregg in 1989 [25] to treat Coulomb-like interactions (i.e. $\frac{1}{|x|^\alpha}$ with $\alpha \neq 1$).

As far as we know, this is the state of the art concerning the existence of the thermodynamic limit in the statistical framework. At this point, it is worth emphasizing the main features of all the above situations. Notice first of all that the above works deal with the existence of a limit for the 'energy' (we refer by analogy to the quantum mechanics case) but, except in fact for the work by Ruelle (see [48]), not with the 'density' (i.e. the correlation functions or the density matrix), which suggests that the latter might be far more difficult. In order to prove the 'energy' limit, some conditions are required. Firstly, the two 'physical' conditions mentioned above must be recalled briefly:

(a) The interaction (or energy) is stable.

(b) The interaction is tempered, in the sense that either it is truly tempered or tempered in the average; that is to say, the spherical symmetry and neutrality tools may be used.

Secondly, there is also the following technical condition that certainly must not be considered in the same light. It may not be necessary, but we notice it is used in all the above situations, in one way or another (in fact, we do not know whether or not this condition is always satisfied):

(c) The energy (or free energy, or partition function, or ...) satisfies some monotonicity property with respect to the domain.

1.4.2 The Thomas–Fermi theory

These three features may also be found in the model we are now turning to, the Thomas–Fermi model of molecules (TF model for short). From the physical point of view, we now consider the zero-temperature limit. The quantum statistical model turns into a quantum mechanical model for matter. More precisely, for instance, the free energy has to be replaced by the ground state energy E_0, and the existence of the thermodynamic limit translates into the existence of the following limit:

$$\lim_{\Lambda \to \infty} \frac{1}{|\Lambda|} E_0(\Lambda). \tag{1.4}$$

In addition, if ρ_Λ denotes the TF ground state density, another question is the behaviour of ρ_Λ as Λ goes to infinity, and we are back to the formal setting we have defined in Section 1.2. Let us detail it now for the TF model. We set the nuclei at the points of $\Lambda \subset \mathbf{Z}^3$. The Coulomb potential they create becomes $\sum_{y \in \Lambda} \frac{1}{|x-y|}$. The molecular energy is therefore

$$E_\Lambda^{\mathrm{TF}}(\rho) + \frac{1}{2} \sum_{y \neq z \in \Lambda} \frac{1}{|y-z|} = \int_{\mathbf{R}^3} \rho^{5/3} - \int_{\mathbf{R}^3} \left(\sum_{k \in \Lambda} \frac{1}{|x-k|} \right) \rho(x) \mathrm{d}x$$

$$+ \frac{1}{2} \int \int_{\mathbf{R}^3 \times \mathbf{R}^3} \frac{\rho(x)\rho(y)}{|x-y|} \mathrm{d}x \, \mathrm{d}y$$

$$+ \frac{1}{2} \sum_{y \neq z \in \Lambda} \frac{1}{|y-z|}, \tag{1.5}$$

where we recall that the term in $\rho^{5/3}$ is a model for the kinetic energy of the electrons, while the third term on the right-hand side is an approximation (actually from above) of the coulombic repulsion between the electrons.

The molecular TF problem then reads

$$I_\Lambda^{\mathrm{TF}} = \inf \left\{ E_\Lambda^{\mathrm{TF}}(\rho) + \frac{1}{2} \sum_{y \neq z \in \Lambda} \frac{1}{|y-z|} \; ; \right.$$

$$\left. \rho \geq 0, \rho \in L^1 \cap L^{5/3}, \int_{\mathbf{R}^3} \rho = |\Lambda| \right\}. \tag{1.6}$$

We recall that the existence and the uniqueness of the minimizing density ρ_Λ of I_Λ^{TF} defined by (1.6) is known (see Lieb and Simon [40]). We may therefore ask questions (i), (ii), and (iii) of Section 1.2 in this setting.

In 1977, Lieb and Simon proved in [40] that for a large convenient family of sequences (Λ) (namely a modified notion of Van Hove sequence; see details in [40] and below), the three questions (i), (ii) and (iii) may be answered positively, all limits being independent of the sequence (Λ) considered.

More precisely,

$$\lim_{\Lambda \to \infty} \frac{1}{|\Lambda|} I_\Lambda^{\mathrm{TF}} = I_{\mathrm{per}}^{\mathrm{TF}} + \frac{M}{2}, \tag{1.7}$$

and

$$\lim_{\Lambda \to \infty} \rho_\Lambda = \rho_{\mathrm{per}} \qquad \text{in } L^1 \cap L^{5/3}(\Gamma_0),$$

and uniformly on the compact subsets of $\mathbf{R}^3$, where the minimization problem $I_{\mathrm{per}}^{\mathrm{TF}}$ is defined on the unit cubic cell $\Gamma_0 = \{(x_1, x_2, x_3) \in \mathbf{R}^3; -\frac{1}{2} < x_i \le \frac{1}{2}, \ i = 1, 2, 3\}$ by

$$I_{\mathrm{per}}^{\mathrm{TF}} = \inf \left\{ E_{\mathrm{per}}^{\mathrm{TF}}(\rho); \rho \ge 0, \rho \in L^1(\Gamma_0) \cap L^{5/3}(\Gamma_0), \int_{\Gamma_0} \rho = 1 \right\}, \tag{1.8}$$

$$E_{\mathrm{per}}^{\mathrm{TF}}(\rho) = \int_{\Gamma_0} \rho^{5/3} - \int_{\Gamma_0} \rho G + \frac{1}{2} \int \int_{\Gamma_0 \times \Gamma_0} \rho(x)\rho(y)G(x - y)\mathrm{d}x\,\mathrm{d}y, \tag{1.9}$$

where G is the periodic solution to

$$-\triangle G = 4\pi \Big(-1 + \sum_{y \in \mathbf{Z}^3} \delta(\cdot - y)\Big), \tag{1.10}$$

that satisfies

$$\int_{\Gamma_0} G = 0, \tag{1.11}$$

and with M in (1.7) defined by

$$M = \lim_{x \to 0} G(x) - \frac{1}{|x|},$$

and where ρ_{per} is the unique minimum of the strictly convex problem $I_{\mathrm{per}}^{\mathrm{TF}}$. In addition, ρ_{per} shares the periodicity of the nuclei.

Lieb and Simon's proof is based upon the following observation. There exists a pointwise relationship between the density ρ and the potential $\phi(x) = \sum_k \frac{1}{|x-k|} - \rho \star \frac{1}{|x|}$ the electrons are subjected to. This relationship is given by the Euler–Lagrange equation of the variational problem (1.6), and reads, in the neutral case,

$$\frac{5}{3}\rho_\Lambda^{2/3} = \phi_\Lambda. \tag{1.12}$$

As a consequence of this relationship, Teller's no-binding theorem holds. We recall that this theorem states that, in the TF setting, no molecular system is stable (see Lieb [33]). The proof of the existence of the thermodynamic limit in the TF theory makes use of this theorem, in particular in order to obtain the monotonicity of the sequence ϕ_Λ, and thus of the sequence ρ_Λ. (It is of course somewhat disturbing that the proof is based upon a highly non-physical property of Thomas–Fermi models.) Therefore we recover in this model the monotonicity (c) that we have identified above as an important factor for existing proofs. In this particular case, the monotonicity is so strong (it holds for the densities) that it implies in this setting that both conditions (a) and (b) are fulfilled.

It is, however, to be noticed that in the course of our work (see Chapter 6), we shall give another proof of the existence of the thermodynamic limit in the TF setting which does not make use of the monotonicity, and therefore that applies to more general sequences of domains than the ones considered in [40]. Moreover, this new proof also improves the convergence result for the sequence of densities.

Before putting an end to our brief overview of the state of the art in this field, let us finally mention, in order to be as comprehensive as possible, a recent work by Nakano [44] on the thermodynamic limit of the magnetic Thomas–Fermi model.

We now set out the problem we shall be considering in this work. It is an improvement of the TF model, namely the so-called Thomas–Fermi–von Weizsäcker model. We shall see below that the situation as far as the thermodynamic limit is concerned is much more complicated in this new setting. There are various reasons to consider such an extension. First of all, the Thomas–Fermi–von Weizsäcker model is a good example of what are called density dependent models, and it turns out that our proofs cover in fact many of these models. In particular, we shall mention in Chapter 3 what can be done on the so-called Thomas–Fermi–Dirac–von Weizsäcker model. Next, it has been shown in [15] that the Thomas–Fermi–von Weizsäcker model allows the binding of all neutral molecular systems. Therefore, we get closer to 'physical reality' and we cannot expect the proofs of Lieb and Simon [40] to carry through to this case—and, indeed, they do not. In other words, considering this particular test problem forces us to introduce completely new methods. And, in fact, these methods allow us to consider and treat much more realistic models (Hartree–Fock type models) involving reduced density matrices (and Bloch waves at the limit). This part of our work—which is our third motivation—will be detailed in a subsequent publication [17].

1.5   The thermodynamic limit in the Thomas–Fermi–von Weizsäcker setting

In this section, we define the mathematical problems we shall deal with in this book.

The thermodynamic limit problem that we have so far considered from a very general viewpoint (both physically and mathematically) is now restricted to the setting of the Thomas–Fermi–von Weizsäcker theory for a crystal. The questions we intend to answer in this setting are the three questions (i)–(iii) stated in Section 1.2. We use the same notations for the geometric quantities (Λ, $|\Lambda|$, Γ_k), and we aim at extending in this work the results obtained by Lieb and Simon for the Thomas–Fermi theory to the Thomas–Fermi–von Weizsäcker model.

1.5.1 *Description of the models and statement of the main results*

Before we define the models and state the main results, we need three definitions.

1.5.1.1 *Preliminaries*

Definition 1 We shall say here that a sequence Λ_i of finite subsets of $\mathbf{Z}^3$ goes to infinity if both of the following conditions hold:
  (a) For any finite subset $A \subset \mathbf{Z}^3$, there exists $i \in \mathbf{N}$ such that

$$\forall j \geq i, \qquad A \subset \Lambda_j.$$

  (b) If Λ^h is the set of points in $\mathbf{R}^3$ whose distance to $\partial\Gamma(\Lambda)$ is less than h, then

$$\lim_{i \to \infty} \frac{|\Lambda_i^h|}{|\Lambda_i|} = 0, \qquad \forall h > 0.$$

Condition (b) will be hereafter referred to as the Van Hove condition.

We find it useful to make a few comments on the above definition and illustrate what is and what is not a sequence satisfying the Van Hove condition (b). Roughly speaking, a sequence satisfying the Van Hove condition is a sequence for which the 'boundary' is negligible in front of the 'interior'.

For instance, a sequence of cubes in $\mathbf{R}^3$

$$\Lambda_i = \left\{ x \in \mathbf{Z}^3, -\frac{i}{2} < x_1 \leq \frac{i}{2}, -\frac{i}{2} < x_2 \leq \frac{i}{2}, -\frac{i}{2} < x_3 \leq \frac{i}{2} \right\} \qquad (1.13)$$

clearly satisfies the Van Hove condition. Indeed, the volume of Λ_i^h is of order $i^2 \times h$, while the volume of Λ_i is i^3.

On the contrary, we give below an example of a sequence of sets in $\mathbf{R}^2$, that does not satisfy the Van Hove condition; since, for instance, $\frac{|\Lambda_i^h|}{|\Lambda_i|} \longrightarrow \frac{2}{3}$, for $h = 1$ (see Figure 1.2 below).

Following the notation of [40], we shall write henceforth $\lim_{\Lambda \to \infty} f(\Lambda)$ instead of $\lim_{i \to \infty} f(\Lambda_i)$.

Definition 2 Λ being a sequence going to infinity according to Definition 1, we call $\Gamma'(\Lambda)$ a sequence of interior domains of $\Gamma(\Lambda)$, and we denote by $\Gamma'(\Lambda) \subset\subset \Gamma(\Lambda)$ any sequence of domains $\Gamma'(\Lambda) \subset \mathbf{R}^3$ also going to infinity in the sense that, for any compact subset $K \subset \mathbf{R}^3$, we have, for Λ large enough, $K \subset \Gamma'(\Lambda)$, and satisfying the following three conditions:

$$\Gamma'(\Lambda) \subset \Gamma(\Lambda);$$

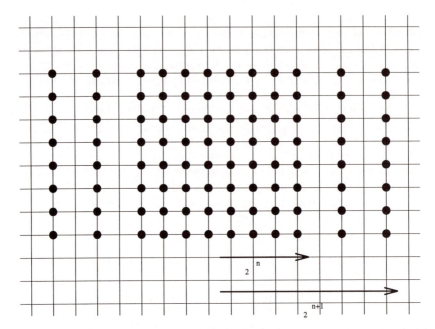

FIG. 1.2. **An example in two dimensions of a sequence that satisfies condition (a), that satisfies in addition $\Lambda_n \subset \Lambda_{n+1}$ but does not satisfy the Van Hove condition (b).** For every n, Λ_n is the union of a square of size 2^{n+1}, and of 2^n bars that are 2^{n+1} long.

$$\frac{1}{|\Lambda|} \text{ volume } (\Gamma'(\Lambda)) \longrightarrow 1 \qquad \text{as } \Lambda \text{ goes to infinity;}$$

$$\mathrm{d}(\Gamma'(\Lambda), \partial\Gamma(\Lambda)) \longrightarrow \infty \qquad \text{as } \Lambda \text{ goes to infinity.}$$

It will be proved in our work (see Chapter 5, Section 5.5) that, for any sequence Λ satisfying the conditions of Definition 1, there always exists at least one sequence of interior domains of $\Gamma(\Lambda)$. However, at this stage, the reader can already see, for instance, that, for the sequence of cubes (1.13), any sequence of the type

$$\Gamma'(\Lambda_i) = \left\{ x \in \mathbf{R}^3, -\frac{i - i^\alpha}{2} < x_j \le \frac{i - i^\alpha}{2}, \quad j = 1, 2, 3 \right\}, \tag{1.14}$$

with $0 < \alpha < 1$, is a sequence of interior domains.

Definition 3 For a given sequence Λ and a sequence ρ_Λ of densities, we call the following sequence of functions

$$\tilde{\rho}_\Lambda = \frac{1}{|\Lambda|} \sum_{k \in \Lambda} \rho_\Lambda(\cdot + k). \tag{1.15}$$

denoted by $\tilde{\rho}_\Lambda$, the $\sim$-transform of ρ_Λ.

1.5.1.2   *The models*   We consider an improved form of the Thomas–Fermi energy, the Thomas–Fermi–von Weizsäcker energy (TFW energy for short),

$$E_\Lambda^{\mathrm{TFW}}(\rho) = \int_{\mathbf{R}^3} |\nabla\sqrt{\rho}|^2 + \int_{\mathbf{R}^3} \rho^{5/3} - \int_{\mathbf{R}^3} \left(\sum_{k\in\Lambda} \frac{1}{|x-k|}\right)\rho(x)\mathrm{d}x$$
$$+ \frac{1}{2}\int\int_{\mathbf{R}^3\times\mathbf{R}^3} \frac{\rho(x)\rho(y)}{|x-y|}\mathrm{d}x\,\mathrm{d}y, \tag{1.16}$$

and the corresponding minimization problem

$$I_\Lambda^{\mathrm{TFW}} = \inf\left\{E_\Lambda^{\mathrm{TFW}}(\rho) + \frac{1}{2}\sum_{y\neq z\in\Lambda} \frac{1}{|y-z|};\right.$$
$$\left.\rho\geq 0,\ \sqrt{\rho}\in H^1(\mathbf{R}^3),\ \int_{\mathbf{R}^3}\rho = |\Lambda|\right\}. \tag{1.17}$$

It is a well known fact that the problem (1.16)–(1.17) has a unique minimizing density, denoted by ρ_Λ (see Lieb [33], Benguria, Brézis and Lieb [6], or Lions [42]), and that, denoting by $u_\Lambda = \sqrt{\rho_\Lambda}$, u_Λ is a solution to

$$-\Delta u_\Lambda + \left[\tfrac{5}{3}\rho_\Lambda^{2/3} - \Phi_\Lambda\right]u_\Lambda = -\theta_\Lambda u_\Lambda, \tag{1.18}$$

where we denote by

$$\Phi_\Lambda = \sum_{k\in\Lambda} \frac{1}{|x-k|} - \rho_\Lambda \star \frac{1}{|x|}$$

the effective potential the electrons experience, and where $\theta_\Lambda > 0$ is the Lagrange multiplier associated with the constraint in (1.17).

We may therefore ask the three questions mentioned in Section 1.2 about this model; namely:

(i) Does there exist a limit for the energy per unit volume $\frac{1}{|\Lambda|}I_\Lambda^{\mathrm{TFW}}$ when Λ goes to infinity?

(ii) Does the minimizing density ρ_Λ approach a limit ρ_∞ (in a sense to be made precise later) when Λ goes to infinity?

(iii) Does the limit density ρ_∞ have the same periodicity as the assumed periodicity of the nuclei?

It turns out that, for mathematical purposes (and in fact it is also used for numerical purposes, as the reader may have seen in Section 1.3 where we recalled some elements of solid state physics), it is convenient to consider first the above problem with a modified interaction potential. Indeed, as we shall soon see, and as we explained above, the slow decay at infinity of the Coulomb potential makes the problem mathematically more difficult. We therefore replace, in a first step towards the complete resolution of the Coulomb case, the Coulomb potential $\frac{1}{|x|}$ by a Yukawa potential

$$V(x) = \frac{1}{|x|} \exp(-a|x|), \tag{1.19}$$

for some $a > 0$ to be made precise later, and formulate the analogous energy

$$E_\Lambda(\rho) = \int_{\mathbf{R}^3} |\nabla\sqrt{\rho}|^2 + \int_{\mathbf{R}^3} \rho^{5/3} - \int_{\mathbf{R}^3} \rho V_\Lambda$$
$$+ \tfrac{1}{2} \int\int_{\mathbf{R}^3\times\mathbf{R}^3} \rho(x)\rho(y)V(x-y)\mathrm{d}x\,\mathrm{d}y, \tag{1.20}$$

with

$$V_\Lambda(x) = \sum_{k\in\Lambda} V(x-k), \tag{1.21}$$

and the corresponding minimization problem

$$I_\Lambda = \inf\left\{ E_\Lambda(\rho) + \tfrac{1}{2}\sum_{y\neq z\in\Lambda} V(y-z); \right.$$
$$\left. \rho \geq 0, \ \sqrt{\rho} \in H^1(\mathbf{R}^3), \int_{\mathbf{R}^3} \rho = |\Lambda| \right\}. \tag{1.22}$$

(Note that, for the sake of consistency, the Yukawa potential is used in (1.20), (1.21), and (1.22) for every interaction: nucleus–electron attraction, electron–electron repulsion, and nucleus–nucleus repulsion).

By considering this model, we choose to focus on the mathematical difficulties generated by the presence of the gradient term $\int |\nabla\sqrt{\rho}|^2$ (i.e. by the term $-\triangle$ in the Euler–Lagrange equation), while getting rid of the difficulties created by the non-integrability of the Coulomb potential. Despite the drastic simplification that comes from the fact that we consider a short-range potential, we wish to remark that, from the physical viewpoint, taking such a potential in this kind of study is not a surprising fact. Indeed, as mentioned in Lieb and Lebowitz [36], screening makes the Coulomb forces behave as if they were short-range.

This model turns out to be a 'good' model: the existence and uniqueness of the minimizing density are proved below. The questions (i)–(iii) on the problem (1.19)–(1.22) are hereafter referred to as the $(Y - \delta)$ program (Y for Yukawa, and δ to remind us of the fact that the nuclei are point nuclei, i.e. Dirac masses). On the other hand, the initial program, that is questions (i)–(iii) on the problem (1.16)–(1.17), is referred to as the $(Cb - \delta)$ program.

In an analogous way, we define the programs with smeared out nuclei, in which the point nuclei are replaced by nuclei described by a function $m \in \mathcal{D}(\mathbf{R}^3)$, supported in Γ_0, such that $\int_{\Gamma_0} m = 1$. Instead of V_Λ defined by $V_\Lambda = \sum_{k\in\Lambda} \frac{1}{|x-k|}$ and respectively by (1.19)–(1.21), V_Λ is defined in the Coulomb case by

$$V_\Lambda^m = \sum_{k\in\Lambda} m(x-k) \star \frac{1}{|x|}, \tag{1.23}$$

and, respectively, in the Yukawa case by

$$V_\Lambda^m = \sum_{k\in\Lambda} m(x-k) \star V. \tag{1.24}$$

The program (1.16)–(1.17) with nuclei (1.23) is called the $(Cb - m)$ program, while the program (1.19)–(1.22) with nuclei (1.24) is called the $(Y - m)$ program.

Let us point out that other cases of 'shape' of nuclei m will be treated in the course of our work. For instance, we shall consider less regular m such as non-negative measures (e.g. the sum of delta functions at different points of the unit cell). We refer the reader in particular to Subsection 1.5.2. However, when we mention the $(Cb - m)$ or the $(Y - m)$ programs in the sequel, we always mean, unless otherwise stated, that the shape m of the nuclei is a smooth function belonging to $\mathcal{D}(\mathbf{R}^3)$.

We shall also deal with slight modifications of these four problems we have just defined, but for the sake of simplicity we prefer to postpone their definitions until the next subsection.

In view of the results obtained by Lieb and Simon on the TF model, it is normal to guess (and hope) that if questions (i)–(iii) can be answered positively, then the limits should be related to the periodic problems we now describe.

First, for the $(Cb - \delta)$ program, we define the variational periodic problem on Γ_0 analogous to (1.8)–(1.9) :

$$I_{\text{per}}^{\text{TFW}} = \inf\left\{ E_{\text{per}}^{\text{TFW}}(\rho); \rho \geq 0, \sqrt{\rho} \in H_{\text{per}}^1(\mathbf{R}^3), \int_{\Gamma_0} \rho = 1 \right\}, \tag{1.25}$$

$$E_{\text{per}}^{\text{TFW}}(\rho) = \int_{\Gamma_0} |\nabla\sqrt{\rho}|^2 + \int_{\Gamma_0} \rho^{5/3} - \int_{\Gamma_0} \rho(x)G(x)\mathrm{d}x$$
$$+ \tfrac{1}{2} \int\int_{\Gamma_0\times\Gamma_0} \rho(x)\rho(y)G(x-y)\mathrm{d}x\,\mathrm{d}y, \tag{1.26}$$

where

$$H_{\text{per}}^1(\mathbf{R}^3) = \{u \in H_{\text{loc}}^1(\mathbf{R}^3), u \text{ periodic in } x_i, i = 1,2,3, \text{ of period } 1\}$$

and where G is the periodic potential defined by (1.10)–(1.11).

Next, for the $(Y - \delta)$ program, we define

$$I_{\mathrm{per}}(\mu) = \inf \left\{ E_{\mathrm{per}}(\rho); \rho \geq 0, \sqrt{\rho} \in H^1_{\mathrm{per}}(\mathbf{R}^3), \int_{\Gamma_0} \rho = \mu \right\}, \qquad (1.27)$$

for some $\mu > 0$, where the periodic energy is

$$\begin{aligned} E_{\mathrm{per}}(\rho) = &\int_{\Gamma_0} |\nabla \sqrt{\rho}|^2 + \int_{\Gamma_0} \rho^{5/3} - \int_{\Gamma_0} \rho V_\infty \\ &+ \tfrac{1}{2} \int\int_{\Gamma_0 \times \Gamma_0} \rho(x)\rho(y)V_\infty(x-y)\mathrm{d}x\,\mathrm{d}y \\ &+ \tfrac{1}{2} \sum_{y \neq 0 \in \mathbf{Z}^3} V(y), \end{aligned} \qquad (1.28)$$

with

$$V_\infty(x) = \sum_{k \in \mathbf{Z}^3} V(x - k). \qquad (1.29)$$

Recall that V is given by (1.19).

For reasons that will become clear in the sequel, we define in (1.27) the 'charge' per cell μ by

$$\mu = \min(\mu_0, 1) \; > 0, \qquad (1.30)$$

with $\mu_0 = \int_{\Gamma_0} \rho_0$, where ρ_0 denotes the unique minimizing density that satisfies

$$E_{\mathrm{per}}(\rho_0) = \inf\{E_{\mathrm{per}}(\rho); \rho \geq 0, \sqrt{\rho} \in H^1_{\mathrm{per}}(\mathbf{R}^3)\}. \qquad (1.31)$$

(We shall of course justify the existence and uniqueness of μ_0 (thus of μ) in the following: see Chapter 2.)

Let us remark here that, for all the above minimization problems, the set $H^1_{per}(\mathbf{R}^3)$ may be changed into $H^1(\Gamma_0)$; that is,

$$I^{\mathrm{TFW}}_{\mathrm{per}} = \inf \left\{ E^{\mathrm{TFW}}_{\mathrm{per}}(\rho); \rho \geq 0, \sqrt{\rho} \in H^1(\Gamma_0), \int_{\Gamma_0} \rho = 1 \right\}, \qquad (1.32)$$

$$I_{\mathrm{per}}(\mu) = \inf \left\{ E_{\mathrm{per}}(\rho); \rho \geq 0, \sqrt{\rho} \in H^1(\Gamma_0), \int_{\Gamma_0} \rho = \mu \right\}, \qquad (1.33)$$

$$E_{\mathrm{per}}(\rho_0) = \inf\{E_{\mathrm{per}}(\rho); \rho \geq 0, \sqrt{\rho} \in H^1(\Gamma_0)\}. \qquad (1.34)$$

Indeed, as we shall see in Chapter 3 using an argument based on the symmetry of the point nuclei (and therefore that will not necessarily hold in the $(Cb - m)$ and $(Y - m)$ programs if m does not have *ad hoc* properties), the minimizing densities of (1.32), (1.33), and (1.34) are periodic. We shall come back to this point below.

Replacing the point nuclei by the smeared out nuclei, we obtain the analogous periodic problems. First of all, for the $(Cb - m)$ program, we obtain problem (1.25) (denoted this time by $I_{\mathrm{per},m}^{\mathrm{TFW}}$) with the following energy

$$E_{\mathrm{per},m}^{\mathrm{TFW}}(\rho) = \int_{\Gamma_0} |\nabla\sqrt{\rho}|^2 + \int_{\Gamma_0} \rho^{5/3} - \int_{\Gamma_0} \rho(x)G_m(x)\mathrm{d}x$$
$$+ \frac{1}{2} \int\int_{\Gamma_0\times\Gamma_0} \rho(x)\rho(y)G(x-y)\mathrm{d}x\,\mathrm{d}y, \qquad (1.35)$$

where G_m is the periodic potential solution to

$$-\triangle G_m = 4\pi\Big(-1 + \sum_{y\in\mathbf{Z}^3} m(\cdot - y)\Big), \qquad (1.36)$$

that satisfies

$$\int_{\Gamma_0} G_m = 0. \qquad (1.37)$$

Note that G_m is also

$$G_m = G \star m,$$

with G defined by (1.10)–(1.11).

Similarly, for the $(Y - m)$ program, the periodic problem we get (denoted by I_{per}^m) is the analogue of problem (1.27) when the energy to consider (denoted by E_{per}^m) is obtained from (1.28) by replacing $\frac{1}{2}\sum_{y\neq 0\in\mathbf{Z}^3} V(y)$ by $\frac{1}{2}\sum_{k\neq 0\in\mathbf{Z}^3} \iint m(x)$ $m(y)V(x-y+k)\mathrm{d}x\,\mathrm{d}y)$ and V_∞ by V_∞^m defined by

$$V_\infty^m(x) = \sum_{k\in\mathbf{Z}^3} m \star V(x-k). \qquad (1.38)$$

In view of the remark we made above, the set $H_{\mathrm{per}}^1(\mathbf{R}^3)$ may be changed into $H^1(\Gamma_0)$ in the definitions of $I_{\mathrm{per},m}^{\mathrm{TFW}}$ and I_{per}^m, if m shares the same symmetries as the unit cell Γ_0. This will be detailed in Chapters 2 and 3.

1.5.1.3   *Main results*   The main results we have obtained on the above models are summarized in the following theorems.

Theorem 1.1 (Yukawa case) *Consider the $(Y - \delta)$ (respectively the $(Y - m)$) model. Take $a > 0$ small enough in (1.19). Then I_Λ, defined by (1.19)–(1.22) (respectively by (1.19),(1.20),(1.22), and (1.24)) has a unique minimum ρ_Λ. And we have the following:*

(1) *Convergence of the energy per unit volume*

$$\lim_{\Lambda\to\infty} \frac{1}{|\Lambda|}I_\Lambda = I_{\mathrm{per}}(\mu),$$

where $I_{\mathrm{per}}(\mu)$ is defined by (1.27)–(1.31) (respectively $= I_{\mathrm{per}}^m(\mu)$ defined by (1.27), (1.28), (1.30), (1.31), and (1.38)).

(2) Exact convergence of the density

As Λ goes to infinity, $\sqrt{\rho_\Lambda}$ converges to $\sqrt{\rho_{\mathrm{per}}}$, the unique (periodic) minimizing density of $I_{\mathrm{per}}(\mu)$ (respectively $I_{\mathrm{per}}^m(\mu)$), for the strong topology of $H_{\mathrm{loc}}^1(\mathbf{R}^3)$ and the strong topology of $L_{\mathrm{loc}}^p(\mathbf{R}^3)$, $1 \le p \le \infty$.

(3) Uniform convergence of the density on the interior domains

As Λ goes to infinity, ρ_Λ converges uniformly to ρ_{per} on any sequence of interior domains $\Gamma'(\Lambda)$ of $\Gamma(\Lambda)$ (see Definition 2); that is,

$$\lim_{\Lambda \to \infty} \sup_{x \in \Gamma'(\Lambda)} |\rho_\Lambda(x) - \rho_{\mathrm{per}}(x)| = 0.$$

Theorem 1.2 (Coulomb case) *Consider the $(Cb - \delta)$ (respectively $(Cb - m)$) model. Then we have the following:*

(1) Convergence of the energy per unit volume

$$\lim_{\Lambda \to \infty} \frac{1}{|\Lambda|} I_\Lambda^{\mathrm{TFW}} = I_{\mathrm{per}}^{\mathrm{TFW}} + \frac{M}{2},$$

where $I_{\mathrm{per}}^{\mathrm{TFW}}$ is defined by (1.25)–(1.26), and $M = \lim_{x \to 0} G(x) - \frac{1}{|x|}$ (respectively $= I_{\mathrm{per},m}^{\mathrm{TFW}} + \frac{M}{2}$, where $I_{\mathrm{per},m}^{\mathrm{TFW}}$ is defined by (1.25),(1.35),(1.36), and (1.37), and $M = \int \int_{\Gamma_0 \times \Gamma_0} m(x)m(y) \left[G(x - y) - \frac{1}{|x-y|} \right] \mathrm{d}x\,\mathrm{d}y).

(2) Exact convergence of the density

As Λ goes to infinity, $\sqrt{\rho_\Lambda}$ converges to $\sqrt{\rho_{\mathrm{per}}}$, the unique (periodic) minimizing density of $I_{\mathrm{per}}^{\mathrm{TFW}}$ (respectively, $I_{\mathrm{per},m}^{\mathrm{TFW}}$) for the strong topology of $H_{\mathrm{loc}}^1(\mathbf{R}^3)$ and the strong topology of $L_{\mathrm{loc}}^p(\mathbf{R}^3)$, $1 \le p \le \infty$.

(3) Uniform convergence of the density on the interior domains

As Λ goes to infinity, ρ_Λ converges uniformly to ρ_{per} on any sequence of interior domains $\Gamma'(\Lambda)$ of $\Gamma(\Lambda)$ (see Definition 2); that is,

$$\lim_{\Lambda \to \infty} \sup_{x \in \Gamma'(\Lambda)} |\rho_\Lambda(x) - \rho_{\mathrm{per}}(x)| = 0.$$

We now wish to make a few comments.

Let us begin with a remark that holds for both theorems.

As already mentioned and as will be shown in Chapter 6, there always exists a sequence of interior domains for any sequence satisfying Definition 1. Therefore, it is straightforward to see that (iii) implies (ii) in Theorem 1.1, and, respectively, that (iii) implies (ii) in Theorem 1.2. Moreover, we shall prove in Chapter 6 that (iii) implies (i) in Theorem 1.1, and, respectively, that (iii) implies (i) in Theorem 1.2. We shall return to this latter point below.

Let us now examine the Yukawa case.

In Theorem 1.1, we do not know whether $\mu = 1$ (this issue is in fact a compactness issue, since $\mu = 1$ means that no part of the electronic density goes to infinity or, in other words, that the whole density is bound by the crystal). We shall see in Chapter 2 that many situations may occur. Apart from this point, the answers to questions (i), (ii), and (iii) are affirmative for both settings, $(Y - \delta)$ and $(Y - m)$. The proof of the convergence per unit volume (see Chapter 2) is based upon the convexity of the energy and the integrability of the Yukawa potential that gives sense to (1.29). Paradoxically, this integrability is also the reason why we are not able to prove the compactness $\mu = 1$. This is related to the fact that the Yukawa potential is in some way not attractive enough at infinity. The convergence of the density (ii) comes from the fact that, due to this integrability of the Yukawa potential, uniform bounds on the density, in particular a L^∞ bound, and consequently a bound on the effective potential Φ_Λ, are available. One may thus pass to the limit locally in the Euler–Lagrange equation and prove, thanks to the short-range nature of the potential again (which is not in fact necessary, since we shall be able to circumvent this difficulty in the Coulomb case: see below), the uniqueness of the solution to the limit equation (see Chapter 4). Furthermore, the uniform convergence (iii) is obtained in Chapter 5. It is worth noticing that, without going into the deep nature of the limit equation, one can still prove some partial convergence results, that are of course weaker than the results mentioned in Theorem 1.1, and that are consequences of these results, but that can be obtained with arguments based only upon energetic estimations. These 'weak' results are collected in the following.

Proposition 1.3 ('Weak' results of convergence in the Yukawa case)

(1) Average convergence of the density

As Λ goes to infinity, $\sqrt{\tilde{\rho}_\Lambda}$ converges to $\sqrt{\rho_{\text{per}}}$, the unique (periodic) minimizing density of $I_{\text{per}}(\mu)$ (respectively, $I_{\text{per}}^m(\mu)$), for the strong topology of $L_{\text{loc}}^p(\mathbf{R}^3)$ $(1 \leq p < \infty)$ and $\nabla\sqrt{\tilde{\rho}_\Lambda}$ converges to $\nabla\sqrt{\rho_{\text{per}}}$, for the strong topology of $L_{\text{loc}}^2(\mathbf{R}^3)$

(2) Local convergence of the density up to a translation

There exists a sequence $y_\Lambda \in \mathbf{Z}^3$, possibly escaping at infinity, such that, as Λ goes to infinity, $\sqrt{\rho_\Lambda(\cdot + y_\Lambda)}$ converges to $\sqrt{\rho_{\text{per}}}$, the unique (periodic) minimizing density of $I_{\text{per}}(\mu)$ (respectively, $I_{\text{per}}^m(\mu)$), for the strong topology of $H^1(\Gamma_0)$.

On the other hand, the Coulomb case exhibits different difficulties. Compactness and the convergence of the energy hold, and thus the answer to question (i) is affirmative. Many proofs will be presented in Chapter 3, one of them being based upon a suitable limit of the Yukawa case when the coefficient a goes to 0. Let us only say that the compactness comes from the following formal argument: if some mass escapes at infinity, then the long-range nature of the Coulomb potential causes the energy per unit volume to explode, contradicting the fact that it is bounded from above. This may be seen either by taking the limit of the Yukawa problem to recover the Coulomb problem, or arguing on the Coulomb energy itself directly. It is noteworthy that for the proofs of the convergence of the energy per unit volume that we present in Chapter 3, we shall need some additional technical hypotheses. Indeed, we shall prove there a weaker result than (i), namely:

Proposition 1.4 (Convergence of the energy in the Coulomb case through a direct proof) *Consider the $(Cb - \delta)$ (respectively, $(Cb - m)$) model. Assume that the sequence (Λ) satisfies the condition*

$$(c) \qquad \qquad \frac{|\Lambda^h|}{|\Lambda|} Log(|\Lambda^h|) \overset{\Lambda \to \infty}{\longrightarrow} 0$$

in addition to the conditions (a) and (b) of Definition 1. Assume also in the $(Cb - m)$ case that the 'shape' of the nucleus m shares the symmetry of the unit cube. Then we have the convergence (i) of Theorem 1.2 of the energy per unit volume.

The additional assumptions stated in the above proposition are useful for the arguments we make in Chapter 3. However, we want to emphasize two points. First, we do not know if they are really necessary for these arguments. Secondly, we can get rid of these additional assumptions by using a quite different technique. This is the purpose of Chapter 6, where we shall prove the convergence of the energy per unit volume with arguments based upon the uniform convergence (iii) of the density (shown in Chapter 5). Neither the uniform convergence (iii) nor the arguments used in Chapter 6 need the hypothesis of the above proposition, and we therefore obtain (i) of Theorem 1.2 without any additional assumption.

For the density, the situation is much harder than in the Yukawa case. However, we have also succeeded in proving the exact local convergence of the density, by a rather different technique from the one used in Chapter 4 that will in fact also apply to the Yukawa case (see Chapter 5). Likewise, we shall prove in Chapter 5 the uniform convergence (iii).

Again, as in the Yukawa case, we may obtain 'weak' results of convergence using only energetic arguments. Therefore these arguments only hold under the additional assumptions, and we have the following:

Proposition 1.5 *Consider the $(Cb - \delta)$ (respectively, $(Cb - m)$) model. Assume that the sequence (Λ) satisfies the condition*

$$(c) \qquad \qquad \frac{|\Lambda^h|}{|\Lambda|} Log(|\Lambda^h|) \overset{\Lambda \to \infty}{\longrightarrow} 0$$

in addition to the conditions (a) and (b) of Definition 1. Assume also in the $(Cb - m)$ case that the 'shape' of the nucleus m shares the symmetry of the unit cube. Then we have the following:

(1) Average convergence of the density

As Λ goes to infinity, $\sqrt{\bar{\rho}_\Lambda}$ converges to $\sqrt{\rho_{\mathrm{per}}}$, the unique (periodic) minimizing density of $I_{\mathrm{per}}^{\mathrm{TFW}}$ (respectively, $I_{\mathrm{per},m}^{\mathrm{TFW}}$), for the strong topology of $L_{\mathrm{loc}}^p(\mathbf{R}^3)$ $(1 \leq p < \infty)$ and $\nabla\sqrt{\bar{\rho}_\Lambda}$ converges to $\nabla\sqrt{\rho_{\mathrm{per}}}$, for the strong topology of $L_{\mathrm{loc}}^2(\mathbf{R}^3)$.

(2) Local convergence of the density up to a translation

There exists a sequence $y_\Lambda \in \mathbf{Z}^3$, possibly escaping at infinity, such that, as Λ goes to infinity, $\sqrt{\rho_\Lambda(\cdot + y_\Lambda)}$ converges to $\sqrt{\rho_{\mathrm{per}}}$, the unique (periodic) minimizing density of $I_{\mathrm{per}}^{\mathrm{TFW}}$ (respectively, $I_{\mathrm{per},m}^{\mathrm{TFW}}$), for the strong topology of $H^1(\Gamma_0)$.

It is worth noticing the striking difference compared to the TF case that has been introduced in Section 1.4. We have seen there that in the TF case there exists a pointwise relationship (1.12) between the density ρ and the potential $\phi(x) = \sum_k \frac{1}{|x-k|} - \rho \star \frac{1}{|x|}$ the electrons are subjected to. The proof of the existence of the thermodynamic limit in the TF theory given in [40] is based upon this relationship and its consequences, in particular in order to obtain monotonicity on the sequence ϕ_Λ, and thus on the sequence ρ_Λ. (Using monotonicity forces us to restrict our attention to the case of a sequence Λ_i satisfying $\Lambda_i \subset \Lambda_{i+1}$ in addition to the properties (a) and (b); see Lieb and Simon [40] for details.) Unfortunately, in the TFW case, the analogous equation to (1.12) is the partial differential equation (1.18). Therefore many of the techniques used in [40] do not apply here. Our strategy of proof for the above theorem is radically different from the approach used in [40]. Rather, it draws its inspiration from the arguments used for the study of the TFW theory of molecules in Benguria, Brézis, and Lieb [6], Benguria and Lieb [5], and in Solovej [53], at least in their spirit if not in their detail. Roughly speaking, we shall use a maximum principle for the equation (1.18) in every situation where the pointwise relationship (1.12) would have been useful.

Let us finally mention that in the course of proving the convergence of the density for the TFW model with a Yukawa potential (see Chapter 4) first, and with a Coulomb potential (see Chapter 5), we shall prove successively the following results that may be useful in other fields of applications of the theory of elliptic PDE's.

Theorem 4.19 *Let c be a positive constant, let $\lambda \geq 0$, and let Γ be a periodic potential in $L_{\mathrm{loc}}^p(\mathbf{R}^3)$, for some $p > \frac{21}{8}$. Assume that on the periodic cell of Γ (for the sake of simplicity, we assume this cell is the unit cube Γ_0), the first eigenvalue of the operator $-\Delta - \Gamma$ with periodic boundary conditions is negative. Let $V \in L_{\mathrm{loc}}^1(\mathbf{R}^3)$ be a non-negative potential satisfying $V(x) = O\left(\frac{1}{|x|^{3+\alpha}}\right)$ at infinity for some $\alpha > 0$ (Other short-ranged potentials are tractable). Then, there exists a unique solution to*

$$\begin{cases} -\Delta u - \Gamma u + cu^{7/3} + \lambda(u^2 \star V)u = 0, \\ u \geq 0, \quad u \not\equiv 0. \end{cases}$$

This solution satisfies $u > 0$ on $\mathbf{R}^3$ and is periodic.

In addition, the assumption on $\lambda_1(-\Delta - \Gamma,\mathrm{per})$ *is satisfied as soon as a solution to the above equation exists.*

Theorem 6.5 *Let $c > 0$, and let m be a non-negative measure on $\mathbf{R}^3$ satisfying*

$$(H1) \qquad \sup_{x \in \mathbf{R}^3} m(x + B_1) < \infty,$$

$$(H2) \qquad \lim_{R \longrightarrow +\infty} \inf_{x \in \mathbf{R}^3} \frac{1}{R} \, m(x + B_R) = +\infty,$$

where B_R denotes the ball of radius R centered at 0.

Then, there is a unique solution $(u; \Phi)$ on $\mathbf{R}^3$ of the system

$$\begin{cases} -\Delta u + cu^{7/3} - \Phi u = 0, \\ u \geq 0, \\ -\Delta \Phi = 4\pi \, [m - u^2], \end{cases}$$

with $u \in L^{7/3}_{\mathrm{loc}} \cap L^2_{\mathrm{unif}}(\mathbf{R}^3)$ and $\Phi \in L^1_{\mathrm{unif}}(\mathbf{R}^3)$.

In addition, $\Phi \in L^{3,\infty}_{\mathrm{unif}}(\mathbf{R}^3)$, $\inf_{\mathbf{R}^3} u > 0$, and $u \in L^\infty(\mathbf{R}^3) \cap C^{0,\alpha}(\mathbf{R}^3) \cap W^{2,p}_{\mathrm{unif}}(\mathbf{R}^3)$ for all $0 < \alpha < 1$, $1 \leq p < 3$.

Remark 1.6 Note that in the above theorem, the uniqueness implies the fact that, if m is periodic, u and Φ are also periodic, of the same period as m.

We shall make comments in Chapter 6 on the assumptions (H1) and (H2), give some insight into their physical meaning, and show how they allow us to deal with very general situations.

1.5.2 *Corollaries*

In the above definition of the thermodynamic limit problem for a crystal, many simplifications have been made, namely:

(a) Each nucleus is of charge 1.

(b) There is one nucleus per cell.

(c) The nuclei are either point nuclei in the centre of the cell (($Y - \delta$) and ($Cb - \delta$) programs) or a smooth smearing (($Y - m$) and ($Cb - m$) programs).

(d) Each unit cell is a cube.

(e) The cube has size 1.

(f) The total charge of the nuclei is equal to the charge of the electrons.

(g) The exponent p appearing in the kinetic energy term $\int \rho^p$ of the Thomas–Fermi theory is $\frac{5}{3}$.

(h) The TFW model does not contain the Dirac exchange term $-\int \rho^{4/3}$.

All these points may be modified, leading to connected results and questions that will be treated in the course of our work. We give here a short overview of these results, and we refer the reader to each chapter for more details.

Keeping a cubic cell, one may first change the charge of the nucleus, or take more than one point nucleus per cell—say, a finite fixed number of point nuclei— or reduce the regularity of the smeared nucleus by taking the distribution m defining the shape of the nucleus in a larger space than $\mathcal{D}(\mathbf{R}^3)$.

For a finite number of point nuclei of any charge, or a measure m that is regular enough—these two cases essentially covering those of physical interest— the modifications of our arguments of Chapters 2–5 are straightforward.

When m is less regular, some of our arguments still remain valid and allow us to conclude (see Chapter 6).

In all these cases (and up to the modifications that will be mentioned below on the number of electrons), charge neutrality must be satisfied: when Λ is fixed, the number of electrons is the same as the total nuclear charge.

Besides, in addition or not to the above modifications of the nuclei, one may change the size and/or the shape of the unit cell. This brings us to the following important comment: there is more than one unit cell for a periodic potential. For instance, Figure 1.3 below shows three different unit cells for a periodic two-dimensional lattice of point nuclei. In our whole work, when we deal with 'the' unit cell of the potential or of the lattice, this means any of the convenient unit cells. All the results we state with one of them will hold also, *mutatis mutandis*, for any other one. (Note in particular that the potential G appearing in (1.26) does not depend on the choice of the unit cell.) In addition, if we say 'when the cell is cubic', or 'when the cell has got this type of symmetry', we mean if one of the periodic cells satisfies such a condition. Likewise, by 'when the unit cell is not cubic', we mean when no unit cell is cubic, since we may always replace a non-cubic unit cell by a cubic one if it is possible.

If the cell is kept cubic, changing its size is a straightforward modification. When the cell is a general parallelepiped, the situation is a little less simple. The arguments and the results of the Yukawa case still hold, while, in the Coulomb case, Theorem 1.2 holds, but only by the proof of Chapter 6. The direct strategy of Chapter 3, leading to Proposition 1.4, fails (as far as we know).

By the way, let us observe that it is an interesting question (in some sense, a first step towards the question of the optimized geometry that we shall raise in the next section) to determine among all possible cells in the set of parallelepipeda which one gives the lowest energy (see Chapter 3).

Regarding these questions of the shape of the nuclei and of the unit cell, it is to be emphasized that when the 'nucleus' (we put quotes here to indicate that this word must be understood as a single nucleus, or the set of nuclei contained in the unit cell) does not share the same symmetries as the unit cell Γ_0, which may of course not be a cube, then the periodic problem obtained in the thermodynamic limit must be defined on the space

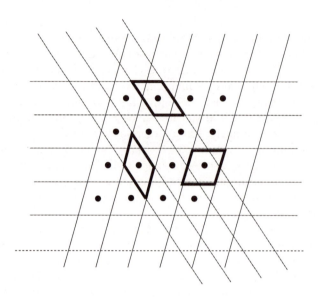

FIG. 1.3. **Some unit cells of a periodic lattice in two dimensions.**

$$H^1_{\text{per}}(\mathbf{R}^3) = \{u \in H^1_{\text{loc}}(\mathbf{R}^3); u \text{ periodic with periodic cell } \Gamma_0\},$$

and that this problem is not necessarily the same as the one posed on the function space $H^1(\Gamma_0)$. In other words, the equality between problems of the type (1.25) and (1.32) is not ensured. In all these questions where there is a symmetry breaking between the shape of the cell and the shape of the nucleus, the proofs and the results of the Yukawa case remain valid. On the other hand, the results of the Coulomb case remain valid, but the direct proofs of Chapter 3 fail (so far as we know), and the arguments of Chapter 6 are needed.

Now consider changes in the energy functional. From the mathematical viewpoint, considering the case of another exponent p instead of $p = \frac{5}{3}$, or even of a more general convex function $j(\rho)$ of ρ, in the kinetic energy term is of interest.

Adding the Dirac exchange term $-c \int \rho^{4/3}$ (with $c > 0$) to the TFW energy, thus obtaining the Thomas–Fermi–Dirac–von Weizsäcker model, *a priori* breaks the convexity of the model. We shall see in Chapter 3 that our arguments and results still hold when c is small enough.

For all the problems and modifications mentioned above, we must now make the following observation. The minimization problems of the type $\mathcal{I}_\Lambda$ that we have considered so far are defined with Dirichlet homogeneous conditions at infinity; that is to say that the functional space is $H^1(\mathbf{R}^3)$. It is possible to replace these Dirichlet boundary conditions in all our arguments by Neumann boundary conditions, or Dirichlet boundary conditions, or periodic boundary conditions on $\partial \Gamma(\Lambda)$ for instance. The reason, which will be explained in further detail in Chapters 2 and 3, and which is essentially based upon the fact that the sequence Λ satisfies the conditions of Definition 1, is that what happens on the boundary of $\Gamma(\Lambda)$ does not really matter. This observation brings us to a related one, namely the question of charge neutrality.

We have dealt so far with neutral molecular systems whose number of electrons (equal to the total nuclear charge) goes to infinity. Can we deal with ions? In the Yukawa case (with a small exponent a), any ionic system exists: a given nuclear charge $|\Lambda|$ can bound any number of electrons (see Chapter 2). In this case, the thermodynamic limit still exists and all our results still hold as soon as we assume that the default or excess of charge is a $o(|\Lambda|)$. On the other hand, in the Coulomb case, we cannot bound any arbitrary large number of electrons. We shall see in Chapter 6 that for negative ions we may, however, reach the same conclusion as in the neutral case. By the way, we will give some asymptotic upper bound on the excess of negative charge. The case of positive ions turns out to be more difficult. Indeed, in this case it is not clear that all the bounds we need are available. Some questions remain open (see Chapter 6). But it must already be emphasized that, if the thermodynamic limit exists, then necessarily the excess of positive charge must be negligible in front of $|\Lambda|$.

Let us also mention here that in Chapter 6 we shall also treat some cases when the nuclei are not on a periodic lattice but present some different kinds of geometries (related to questions of physical interest, such as the theory of quasicrystals).

However, even more important than all the above questions, extending our work to the Hartree–Fock setting is, we believe, a fundamental question. From the applications viewpoint, it is clearly crucial, and this is why we prefer to devote the next section to the presentation of this setting.

1.6 The thermodynamic limit in the Hartree–Fock setting: open problems

The Hartree–Fock model (HF model for short) is one of the most commonly used models in quantum molecular chemistry, both for the theory and in the numerical computations. It is also a common model in solid state theory. This is why it is quite natural to ask questions about the thermodynamic limit in this setting.

We shall study HF type models in a forthcoming publication [17], which will, of course, contain a more detailed presentation. Here, we restrict ourselves to giving a short introduction to the HF program.

Take, when Λ is fixed, with $|\Lambda|$ an integer, the molecule we have introduced in Section 1.2. In the HF setting, its ground state energy reads

$$
I_\Lambda^{\text{HF}} = \inf \left\{ E_\Lambda^{\text{HF}}(\varphi_1, ..., \varphi_{|\Lambda|}) + \frac{1}{2} \sum_{a \neq b \in \Lambda} \frac{1}{|a - b|}; \right.
$$

$$
\left. \varphi_i \in H^1(\mathbf{R}^3), \int \varphi_i \varphi_j^* = \delta_{i,j}, \{i, j\} \subset \{1, ..., |\Lambda|\} \right\}, \quad (1.39)
$$

where

$$E_\Lambda^{\mathrm{HF}}(\varphi_1, ..., \varphi_{|\Lambda|}) = \sum_{i=1}^{|\Lambda|} \int |\nabla \varphi_i|^2 - \sum_{i=1}^{|\Lambda|} \sum_{k \in \Lambda} \int \frac{1}{|x-k|} |\varphi_i|^2$$

$$+ \tfrac{1}{2} \iint \frac{\rho(x)\rho(y)}{|x-y|} \, dx \, dy - \tfrac{1}{2} \iint \frac{|\rho(x,y)|^2}{|x-y|} \, dx \, dy, \quad (1.40)$$

$$\rho(x,y) = \sum_{i=1}^{|\Lambda|} \varphi_i(x)\varphi_i(y)^*, \quad (1.41)$$

$$\rho(x) = \rho(x,x) = \sum_{i=1}^{|\Lambda|} |\varphi_i(x)|^2. \quad (1.42)$$

In the above formula, the φ_i's are the one-electron wave functions (φ_i^* is the conjugate complex of φ_i), $\rho(x,y)$ is the density matrix, and $\rho(x)$ is the electronic density.

The existence of a minimizing $(\varphi_1, ..., \varphi_{|\Lambda|})$ is due to Lieb and Simon [40] and Lions [42], the uniqueness of $(\varphi_1, ..., \varphi_{|\Lambda|})$ (up to an orthogonal transform) still being an open problem.

In our forthcoming work, we shall study the existence of the thermodynamic limit for the HF energy per unit volume $\frac{I^{\mathrm{HF}}}{|\Lambda|}$, and for various related models (restricted Hartree, Hartree, and reduced Hartree–Fock models).

By the way, we shall connect this model with the common tools of solid state physics like Bloch functions (see, for instance, Callaway [14], Kittel [27], or Madelung [43]). Our study extends to a model rather close from the mathematical point of view to the HF model, namely the Kohn–Sham model (KS model for short), however with some slight modifications of this latter model (see the details in [17]).

Apart from the interaction term, that we forget in this brief formal argument, the HF energy (or the KS energy) can be thought to be close to the sum of the first $|\Lambda|$ smallest eigenvalues of the Hamiltonian $-\Delta - \sum_{k \in \Lambda} \frac{1}{|x-k|} + V$, where $V = \frac{1}{|x|} \star \rho$ and ρ is a periodic density. Therefore, the study of the thermodynamic limit for the HF energy per unit volume is somewhat related to the question of the asymptotic behaviour of $\frac{1}{|\Lambda|} \sum_{k=1}^{|\Lambda|} \lambda_i$, where the λ_i are the smallest eigenvalues, for Dirichlet boundary conditions on a large box of volume $|\Lambda|$, of a Schrödinger operator with periodic potential $-\Delta + V_{\mathrm{per}}$. This question is related to questions addressed by Gérard [23].

Note also that, on our way towards the Hartree–Fock model, we may also replace the somewhat complicated exchange term of the Hartree–Fock energy by some approximation of it, such as the Dirac exchange term $- \int \rho^{4/3}$. This motivates our particular interest in models involving this kind of non-convex term such as the Thomas–Fermi–Dirac–von Weizsäcker model (see Chapter 3).

The question of the convergence of the density matrix seems to be much more complicated and we shall not address it here (see [17]).

The last question we want to raise concerns the question of the geometry of the nuclei. Let us first recall that, in the molecular case (Λ fixed), the proof of the existence of the electronic ground state may be seen as a first step in the proof of the existence of a global minimizer, i.e. of both a minimal geometry and a minimal electronic density, for the global energy. In the TFW theory (and also in the TFDW theory and in the Hartree theory), this existence question has been solved by two of us (see Catto and Lions [15]). This allows us to ask the following questions. Take, for each Λ, the optimized nuclei configuration, the optimized electronic density with respect to this configuration, and the ground state energy thus obtained. Do the energy per unit volume and the density approach a limit when Λ goes to infinity? Does the geometry of the nuclei converge to a regular lattice? These questions are, of course, of great mathematical and physical interest. We hope our work will be a preliminary step towards their solution.

2

CONVERGENCE OF THE ENERGY FOR THE THOMAS–FERMI–VON WEIZSÄCKER MODEL WITH YUKAWA POTENTIAL

2.1 Introduction

This chapter is devoted to the proof of the existence of a limit for the energy per unit volume in the setting of the TFW model with Yukawa potential, and in the setting of some related models.

Let us first recall in this introduction the setting of the TFW problem with Yukawa potential, that we have introduced above in Section 1.5 of Chapter 1. We denote by

$$V(x) = \frac{1}{|x|} \exp(-a|x|), \tag{2.1}$$

the Yukawa potential, for some $a > 0$ to be determined later on (Proposition 2.2). The energy reads

$$E_\Lambda(\rho) = \int_{\mathbf{R}^3} |\nabla \sqrt{\rho}|^2 + \int_{\mathbf{R}^3} \rho^{5/3} - \int_{\mathbf{R}^3} \rho V_\Lambda$$

$$+ \tfrac{1}{2} \int\int_{\mathbf{R}^3 \times \mathbf{R}^3} \rho(x)\rho(y)V(x-y)\mathrm{d}x\,\mathrm{d}y, \tag{2.2}$$

with

$$V_\Lambda(x) = \sum_{k \in \Lambda} V(x-k).$$

We shall denote in this part

$$D(f,g) = \int\int_{\mathbf{R}^3 \times \mathbf{R}^3} f(x)g(y)V(x-y)\mathrm{d}x\,\mathrm{d}y,$$

whenever it makes sense.

The minimization problem at Λ fixed is

$$I_\Lambda = \inf \left\{ E_\Lambda(\rho) + \tfrac{1}{2} \sum_{y \neq z \in \Lambda} V(y-z) \; ; \right.$$

$$\rho \geq 0, \ \sqrt{\rho} \in H^1(\mathbf{R}^3), \int_{\mathbf{R}^3} \rho = |\Lambda| \Big\}. \tag{2.3}$$

Questions (i), (ii), and (iii) of Section 1.2 of Chapter 1 for (2.1)–(2.3) are the so-called $(Y - \delta)$ problem. The periodic problem that we intend to obtain in the thermodynamic limit is the following:

$$I_{\mathrm{per}}(\mu) = \inf\{E_{\mathrm{per}}(\rho); \rho \geq 0, \sqrt{\rho} \in H^1_{\mathrm{per}}(\mathbf{R}^3), \int_{\Gamma_0} \rho = \mu\}, \tag{2.4}$$

for some $\mu > 0$, where the periodic energy is

$$\begin{aligned}
E_{\mathrm{per}}(\rho) = {} & \int_{\Gamma_0} |\nabla \sqrt{\rho}|^2 + \int_{\Gamma_0} \rho^{5/3} - \int_{\Gamma_0} \rho V_\infty \\
& + \frac{1}{2} \iint_{\Gamma_0 \times \Gamma_0} \rho(x)\rho(y)V_\infty(x - y)\mathrm{d}x\,\mathrm{d}y \\
& + \frac{1}{2} \sum_{y \neq 0 \in \mathbf{Z}^3} V(y),
\end{aligned} \tag{2.5}$$

with

$$V_\infty(x) = \sum_{k \in \mathbf{Z}^3} V(x - k). \tag{2.6}$$

Recall that the function space $H^1_{\mathrm{per}}(\mathbf{R}^3)$ is defined by

$$H^1_{\mathrm{per}}(\mathbf{R}^3) = \{u \in H^1_{\mathrm{loc}}(\mathbf{R}^3), u \text{ periodic in } x_i, \ i = 1, 2, 3, \text{ of period } 1\}.$$

We define, in (2.4), the 'charge' μ by

$$\mu = \min(\mu_0, 1) > 0, \tag{2.7}$$

with $\mu_0 = \int_{\Gamma_0} \rho_0$, where ρ_0 denotes the unique minimizing density that satisfies

$$E_{\mathrm{per}}(\rho_0) = \inf\{E_{\mathrm{per}}(\rho); \rho \geq 0, \sqrt{\rho} \in H^1_{\mathrm{per}}(\mathbf{R}^3)\}. \tag{2.8}$$

Let us already remark that we also have

$$I_{\mathrm{per}}(\mu) = \inf\left\{E_{\mathrm{per}}(\rho); \rho \geq 0, \sqrt{\rho} \in H^1(\Gamma_0), \int_{\Gamma_0} \rho = \mu\right\} \tag{2.9}$$

and

$$E_{\mathrm{per}}(\rho_0) = \inf\{E_{\mathrm{per}}(\rho); \rho \geq 0, \sqrt{\rho} \in H^1(\Gamma_0)\}, \tag{2.10}$$

because the minimizing densities of the right-hand sides of (2.9) and (2.10) are in fact in $H^1_{\mathrm{per}}(\mathbf{R}^3)$. Indeed, let us argue on (2.9) for instance. The minimizing density ρ exists (because the minimization problem is compact; see the proof of

Proposition 2.2 below) and is unique (because (2.5) is a strictly convex energy functional). Therefore, we have

$$\rho(-x_1, x_2, x_3) = \rho(x_1, x_2, x_3),$$

because the function $(x_1, x_2, x_3) \longrightarrow \rho(-x_1, x_2, x_3)$ clearly has the same energy as ρ, the point being to check that the terms involving both ρ and V_∞ remain unchanged, since $V_\infty(-x_1, x_2, x_3) = V_\infty(x_1, x_2, x_3)$. This latter fact is due to the symmetry of the point nucleus. Likewise, $\rho(x_1, x_2, -x_3) = \rho(x_1, -x_2, x_3) = \rho(x_1, x_2, x_3)$; thus ρ satisfies the periodic boundary conditions on Γ_0. Therefore, in the point nuclei case, it is equivalent to deal with (2.4) and (2.8) or with (2.9) and (2.10). However, it is wise to keep in mind that we have in fact used some symmetry of the nucleus.

Alternatively, we shall also consider in this part the $(Y - m)$ program that deals with smeared nuclei; that is,

$$V_\Lambda^m = \sum_{k \in \Lambda} m(x - k) \star V, \tag{2.11}$$

together with (2.1)–(2.2). The corresponding minimization problem is

$$I_\Lambda^m = \inf \left\{ E_\Lambda^m(\rho) + \tfrac{1}{2} \sum_{y \neq z \in \Lambda} \iint m(x + y)m(t + z)V(x - t + y - z)\mathrm{d}x\,\mathrm{d}t; \right.$$

$$\left. \rho \geq 0, \sqrt{\rho} \in H^1(\mathbf{R}^3), \int_{\mathbf{R}^3} \rho = |\Lambda| \right\}, \tag{2.12}$$

and the periodic problem that is likely to be the thermodynamic limit is

$$E_{\mathrm{per}}^m(\rho) = \int_{\Gamma_0} |\nabla \sqrt{\rho}|^2 + \int_{\Gamma_0} \rho^{5/3} - \int_{\Gamma_0} \rho V_\infty^m$$

$$+ \tfrac{1}{2} \iint_{\Gamma_0 \times \Gamma_0} \rho(x)\rho(y)V_\infty^m(x - y)\mathrm{d}x\,\mathrm{d}y$$

$$+ \tfrac{1}{2} \sum_{k \neq 0 \in \mathbf{Z}^3} \iint m(x)m(y)V(x - y + k), \tag{2.13}$$

together with (2.4) (in which E_{per} is replaced by E_{per}^m), (2.7), (2.8) (in which E_{per} is replaced by E_{per}^m), and

$$V_\infty^m(x) = \sum_{k \in \mathbf{Z}^3} (m \star V)(x - k). \tag{2.14}$$

It is important to note here again that, if the function m shares the same symmetries as Γ_0, then, for the same reasons as above, (2.4)–(2.13) is also (2.9)–(2.13) and (2.8)–(2.13) is also (2.10)–(2.13).

We now state the main result of this chapter.

Theorem 2.1 (Yukawa case, convergence of the energy per unit volume) *Consider the* $(Y - \delta)$ *(respectively, the* $(Y - m)$*) model. Take* $a > 0$ *small enough in (2.1). Then* I_Λ *defined by (2.1)–(2.3) (respectively, by (2.1), (2.2), (2.11), and (2.12)) has a unique minimum* ρ_Λ. *Furthermore, we have*

$$\lim_{\Lambda \to \infty} \frac{1}{|\Lambda|} I_\Lambda = I_{\mathrm{per}}(\mu), \tag{2.15}$$

where $I_{\mathrm{per}}(\mu)$ *is defined by (2.4)–(2.8) (respectively,* $= I_{\mathrm{per}}^m(\mu)$ *defined by (2.7)–(2.14)).*

The following five sections, 2.2–2.6, are devoted to the proof of the above theorem in the $(Y - \delta)$ model, and to various comments about it, while Section 2.7 deals with the $(Y - m)$ program and some extensions.

In Section 2.2, we state the properties of the minimization problem I_Λ (existence and uniqueness of a minimum and so on) we need in the sequel. We give the first *a priori* estimates on V_Λ and ρ_Λ and we motivate the use of the $\sim$-transform to obtain a lower bound on $\frac{I_\Lambda}{|\Lambda|}$ as Λ goes to infinity. As we shall see by the end of this section, it mainly remains to study the term

$$\frac{1}{|\Lambda|} \int\!\!\!\int_{\mathbf{R}^3 \times \mathbf{R}^3} \rho_\Lambda(x)\rho_\Lambda(y)V(x - y)\,\mathrm{d}x\,\mathrm{d}y$$

in order to obtain a lower bound on $\frac{I_\Lambda}{|\Lambda|}$.

Section 2.3 is precisely devoted to the study of the term

$$\frac{1}{|\Lambda|} \int\!\!\!\int_{\mathbf{R}^3 \times \mathbf{R}^3} \rho_\Lambda(x)\rho_\Lambda(y)V(x - y)\mathrm{d}x\,\mathrm{d}y$$

as Λ goes to infinity, which is the most delicate term to handle. Once we have determined how this term behaves, we are able to obtain a lower bound of the energy per unit volume. This bound involves the periodic minimization problem defined above.

In Section 2.4, we prove that the upper limit of $\frac{I_\Lambda}{|\Lambda|}$ as Λ goes to infinity is indeed bounded from above by the expected periodic minimization problem $I_{per}(\mu)$. In fact, this section is completely independent of the preceding ones, and the (rather technical) proofs may be skipped in a first reading.

In Section 2.5, we put together the whole of the information obtained so far concerning $\frac{I_\Lambda}{|\Lambda|}$ and ρ_Λ as Λ goes to infinity, to complete the proof of the main theorem in the point nuclei case.

In Section 2.6, we show first that, as expected and announced in Chapter 1, the periodic Yukawa problem (both energy and density) converges to the coulombic one as the parameter a, entering the definition of the Yukawa potential, goes to zero.

Next, we develop some arguments to convince ourselves there is no way to prove that the absolute charge μ_0 is greater than 1, since we exhibit various situations where either it holds or the contrary does.

Finally, we briefly explain in Section 2.7 where and how we have to modify our arguments in the case of smeared out nuclei, and in a few other situations.

Before beginning our argument, let us give some notation that we shall use hereafter.

Notation We shall denote

$$D_{\Gamma_0}(\rho,\rho) = \int\int_{\Gamma_0 \times \Gamma_0} \rho(x)\rho(y)V_\infty(x-y)\mathrm{d}x\,\mathrm{d}y.$$

The nuclear repulsion term will be denoted by

$$U_\Lambda = \sum_{\substack{y,z\in\Lambda \\ y\neq z}} V(y-z)$$

and

$$U_\infty = \sum_{y\in\mathbf{Z}^3\setminus\{0\}} V(y).$$

2.2   Preliminaries: a priori estimates

We begin this study with some elementary properties of the minimization problems introduced above.

We first check that the minimization problems admit minima (Proposition 2.2). Next, we investigate the properties of these minima, which will be used extensively in the following sections.

Bearing in mind that our objective is to pass to the limit when Λ goes to infinity, it is clear that we are primarily interested in finding uniform bounds (with respect to Λ) on the various quantities involved in the problem: the potential V_Λ (Lemma 2.4), the density ρ_Λ, and the Lagrange multiplier θ_Λ (Proposition 2.5). Next, from these bounds, we deduce in Corollary 2.7 and Proposition 2.9 preliminary convergence results on the densities and estimates from below for all the terms of the energy except the convolution term (Corollary 2.11).

Proposition 2.2 *There exists a constant $0 < a_c < \infty$ such that for any constant a satisfying $0 \leq a \leq a_c$, the infimum I_Λ defined by (2.1)–(2.3) is achieved, for all $\Lambda \subset \mathbf{Z}^3$, by a unique positive density ρ_Λ satisfying $\int_{\mathbf{R}^3} \rho_\Lambda = |\Lambda|$ and $\sqrt{\rho_\Lambda} = u_\Lambda \in H^1(\mathbf{R}^3)$. In addition, u_Λ is a solution to the following Euler–Lagrange equation:*

$$-\Delta u_\Lambda - V_\Lambda u_\Lambda + \tfrac{5}{3}u_\Lambda^{7/3} + (\rho_\Lambda \star V)u_\Lambda = -\theta_\Lambda u_\Lambda, \tag{2.16}$$

with a positive Lagrange multiplier θ_Λ.

Proof of Proposition 2.2 For the case $a = 0$, we refer to [6].

We begin this proof with a basic remark: the energy functional E_Λ is strictly convex with respect to ρ. Indeed, the only non-standard term is the term $D(\rho, \rho)$, since all the other terms may be treated as in [6] for instance. Note that one of them at least, $\int_{\mathbf{R}^3} \rho^{5/3}$, is strictly convex.

In order to show that $D(\rho, \rho)$ is convex, it suffices to show, since it is quadratic, that the function $f \longrightarrow D(f, f)$ is non-negative. For this purpose, we remark that the potential V satisfies $-\triangle V + a^2 V = 4\pi \delta_0$ on $\mathbf{R}^3$; thus its Fourier transform is given by $\hat{V}(\zeta) = \frac{1}{a^2 + |\zeta|^2}$ up to some irrelevant positive constants. Hence,

$$D(f, f) = \int_{\mathbf{R}^3} |\hat{f}(\zeta)|^2 \hat{V}(\zeta) \, d\zeta$$

and is thus non-negative.

Let us first show that $E_\Lambda(\rho)$ is always bounded from below on the set $\int_{\mathbf{R}^3} \rho = |\Lambda|$. It is straightforward to see that, for all $a \geq 0$, $V_\Lambda \in L^{5/2} + L^\infty(\mathbf{R}^3)$. Thus, from Hölder's inequality,

$$\int_{\mathbf{R}^3} V_\Lambda \, \rho \leq \alpha \left(\int_{\mathbf{R}^3} \rho^{5/3} \right)^{3/5} + \beta |\Lambda|, \qquad (2.17)$$

for some constants α and β that depend only on V_Λ.

It follows that

$$E_\Lambda(\rho) \geq \int_{\mathbf{R}^3} \rho^{5/3} - \alpha \left(\int_{\mathbf{R}^3} \rho^{5/3} \right)^{3/5} - \beta |\Lambda|,$$

whence $I_\Lambda > -\infty$.

We now remark that we also have

$$I_\Lambda = \inf \left\{ \int_{\mathbf{R}^3} |\nabla u|^2 - \int_{\mathbf{R}^3} V_\Lambda u^2 + \int_{\mathbf{R}^3} |u|^{10/3} + \tfrac{1}{2} D(u^2, u^2); \right.$$

$$\left. u \in H^1(\mathbf{R}^3), \int_{\mathbf{R}^3} u^2 = |\Lambda| \right\}$$

$$+ \tfrac{1}{2} \sum_{y \neq z} V(y - z). \qquad (2.18)$$

Let u_n be a minimizing sequence of the problem (2.18) above. Using (2.17) and the Sobolev inequality, it is easy to see that, for n large enough,

$$I_\Lambda + 1 \geq \int_{\mathbf{R}^3} |\nabla u_n|^2 - C|\Lambda|^{2/5} \left(\int_{\mathbf{R}^3} |\nabla u_n|^2 \right)^{3/5} - \beta |\Lambda|. \qquad (2.19)$$

Therefore, u_n is bounded in $H^1(\mathbf{R}^3)$, and thus, extracting a subsequence if necessary, is weakly convergent in $H^1(\mathbf{R}^3)$ and in $L^p(\mathbf{R}^3)$ for $2 \leq p \leq 6$ to some $u \in H^1(\mathbf{R}^3)$, which satisfies $\int_{\mathbf{R}^3} u^2 \leq |\Lambda|$.

Since the potential V_Λ vanishes at infinity, we know by standard arguments (see, for instance, Lions [42]) that I_Λ is also given by

$$I_\Lambda = \inf\left\{\int_{\mathbf{R}^3} |\nabla u|^2 - \int_{\mathbf{R}^3} V_\Lambda u^2 + \int_{\mathbf{R}^3} |u|^{10/3} + \tfrac{1}{2}D(u^2, u^2);\right.$$
$$\left. u \in H^1(\mathbf{R}^3), \int_{\mathbf{R}^3} u^2 \le |\Lambda|\right\}$$
$$+ \tfrac{1}{2}\sum_{y \ne z} V(y - z). \qquad (2.20)$$

Then, we have

$$\int_{\mathbf{R}^3} |\nabla u|^2 + \int_{\mathbf{R}^3} |u|^{10/3} + \tfrac{1}{2}D(u^2, u^2)$$
$$\le \liminf_{n \to \infty} \int_{\mathbf{R}^3} |\nabla u_n|^2 + \int_{\mathbf{R}^3} |u_n|^{10/3} + \tfrac{1}{2}D(u_n^2, u_n^2),$$

while, since $V_\Lambda \in L^{3+\varepsilon} + L^{3-\varepsilon}$,

$$\int_{\mathbf{R}^3} V_\Lambda u_n^2 \overset{n \to \infty}{\longrightarrow} \int_{\mathbf{R}^3} V_\Lambda u^2.$$

Therefore u is a minimum of (2.20). We always may assume it is non-negative. In addition, u is a solution to the Euler–Lagrange equation of problem (2.20); that is,

$$-\triangle u - V_\Lambda u + \tfrac{5}{3}u^{7/3} + (u^2 \star V)u = -\theta u, \qquad (2.21)$$

for some Lagrange multiplier θ.

By standard elliptic regularity arguments, we then deduce that u is C^∞ away from Λ, and, by the strong maximum principle, that either $u > 0$ or $u \equiv 0$.

We claim that, for a small enough, we cannot have $u \equiv 0$ (when Λ is non-empty of course).

For this purpose, we embed I_Λ in the family of minimization problems defined by

$$I_\Lambda(\lambda) = \inf\left\{E_\Lambda(u) + \tfrac{1}{2}U_\Lambda; u \in H^1(\mathbf{R}^3), \int_{\mathbf{R}^3} u^2 = \lambda\right\},$$

for any real $\lambda \ge 0$. We thus have $I_\Lambda(|\Lambda|) = I_\Lambda$.

As noticed for I_Λ, $I_\Lambda(\lambda)$ is also given by

$$\inf\left\{E_\Lambda(u) + \tfrac{1}{2}U_\Lambda; u \in H^1(\mathbf{R}^3), \int_{\mathbf{R}^3} u^2 \le \lambda\right\},$$

and thus $I_\Lambda(\lambda)$ is a non-increasing function of λ.

We shall now prove that we may choose a small enough such that $I_\Lambda(\lambda) < \tfrac{1}{2}U_\Lambda$, for all $\Lambda \subset \mathbf{Z}^3$ and $\lambda \in \mathbf{R}_+$.

Let $\varepsilon > 0$ be fixed, $\varphi \in \mathcal{D}(\mathbf{R}^3)$, with $\int_{\mathbf{R}^3} \varphi^2 = 1$. Then

$$I_\Lambda(\varepsilon) - \tfrac{1}{2}U_\Lambda \le E_\Lambda\left(\sqrt{\varepsilon}\varphi\right) - \tfrac{1}{2}U_\Lambda$$
$$= \varepsilon\left(\int_{\mathbf{R}^3}|\nabla\varphi|^2 - \int_{\mathbf{R}^3}V_\Lambda\varphi^2\right) + \frac{\varepsilon^2}{2}D(\varphi^2, \varphi^2) + \varepsilon^{5/3}\int_{\mathbf{R}^3}\varphi^{10/3}.$$

The right-hand side is negative for ε small enough provided there exists some $\varphi \in \mathcal{D}(\mathbf{R}^3)$ with $\int_{\mathbf{R}^3}\varphi^2 = 1$ such that

$$\int_{\mathbf{R}^3}|\nabla\varphi|^2 - \int_{\mathbf{R}^3}V_\Lambda\varphi^2 < 0.$$

In order to prove our claim, it is thus sufficient to ensure the existence of $\varphi \in \mathcal{D}(\mathbf{R}^3)$ with $\int_{\mathbf{R}^3}\varphi^2 = 1$ such that

$$\int_{\mathbf{R}^3}|\nabla\varphi|^2 - \int_{\mathbf{R}^3}\frac{e^{-a|x|}}{|x|}\varphi^2 < 0,$$

for small a.

This claim follows from a simple scaling argument. Choosing $\varphi_0 \in \mathcal{D}(\mathbf{R}^3)$ with $\int_{\mathbf{R}^3}\varphi_0^2 = 1$, then $\varphi_a = a^{3/2}\varphi_0(ax)$ satisfies

$$\int_{\mathbf{R}^3}|\nabla\varphi_a|^2 - \int_{\mathbf{R}^3}\frac{e^{-a|x|}}{|x|}\varphi_a^2 = a^2\int_{\mathbf{R}^3}|\nabla\varphi_0|^2 - a\int_{\mathbf{R}^3}\frac{e^{-|x|}}{|x|}\varphi_0^2 < 0,$$

for a small enough.

Now suppose that $\theta \le 0$ in (2.21). It follows that $u > 0$ satisfies

$$-\Delta u + Wu \ge 0, \qquad (2.22)$$

where the potential

$$W = \tfrac{5}{3}u^{4/3} + u^2 \star V$$

is such that the positive part of its spherical average $[W]_+$ belongs to $L^{3/2}(\mathbf{R}^3)$.

We may apply, for instance, Theorem 7.18 of [33], and we reach the contradiction $u \notin L^2(\mathbf{R}^3)$. Therefore $\theta > 0$, which in particular implies that $\int_{\mathbf{R}^3}u^2 = |\Lambda|$.

We define $\rho_\Lambda = u^2$, $u_\Lambda = \sqrt{\rho_\Lambda}$, and $\theta_\Lambda = \theta$. The uniqueness of ρ_Λ follows from the strict convexity of the energy with respect to ρ. $\diamond$

Remark 2.3 (On the electronic charge $\int_{\mathbf{R}^3}\rho$ and the exponent p)

(1) It is clear that the above argument also shows that, when $0 < a < a_c$, then, for all $\lambda > 0$ and for any Λ non-empty, the problem

$$I_\Lambda(\lambda) = \inf\left\{E_\Lambda(u^2); u \in H^1(\mathbf{R}^3), \int_{\mathbf{R}^3}u^2 = \lambda\right\}$$

has a unique (up to a change of sign) minimum, with a positive Lagrange multiplier. On the other hand, in the 'true' TFW case (that is, with a Coulomb potential), the situation is completely different, since $I_\Lambda(\lambda)$ has a minimum if and only if λ is less than or equal to a critical charge $\lambda_c(\Lambda)$ (depending on Λ) which is always larger than $|\Lambda|$.

(2) The case when the power $\frac{5}{3}$ is replaced by p with $\frac{5}{3} < p \le 3$ can be treated exactly as the case $p = \frac{5}{3}$. Indeed, we only use $p > 1$ in order to obtain (2.16), $1 \le p \le 3$ for (2.19), and $p \ge \frac{5}{3}$ in order to obtain a potential W in (2.22) which belongs to $L^{3/2}$ at infinity. For other cases of exponent p, we refer the reader to Section 2.7.2.

In the following, we shall always assume that the constant a appearing in the definition (2.1) of the potential V belongs to the interval $]0, a_c]$.

In the following lemma, we regroup some technical results on the sequence V_Λ that we shall need in the sequel.

Lemma 2.4 *Let $a > 0$ be fixed in (2.1). There exists a constant C independent of $\Lambda \subset \mathbf{Z}^3$, such that:*

(i) $\|V_\Lambda\|_{L^p} \le C|\Lambda|^{\frac{1}{p}}$, *for all* $1 \le p < 3$.

In addition, for any Van Hove sequence Λ, we have:

(ii) $\dfrac{1}{|\Lambda|} \displaystyle\int_{\Gamma(\Lambda)^c} V_\Lambda^p \overset{\Lambda\to\infty}{\longrightarrow} 0$, *for all* $1 \le p < +\infty$;

(iii) $\dfrac{1}{|\Lambda|} \displaystyle\int_{\Gamma(\Lambda)} (V_\infty - V_\Lambda)^p \overset{\Lambda\to\infty}{\longrightarrow} 0$, *for all* $1 \le p < +\infty$; *and*

(iv) $\dfrac{1}{|\Lambda|} \displaystyle\int_{\mathbf{R}^3} |V_\Lambda(\cdot + k) - V_\Lambda|^p \overset{\Lambda\to\infty}{\longrightarrow} 0$, *for all* $1 \le p < 3$, *and for all* $k \in \mathbf{Z}^3$.

Proof of Lemma 2.4 (i) Let $x \notin \Lambda$. By definition,

$$V_\Lambda(x) = \sum_{y \in \Lambda} \frac{\exp(-a|x-y|)}{|x-y|}.$$

We may split this sum into two terms. Indeed, denoting by $[x]$ the point x of Λ such that $x \in \Gamma_y$, we have, for $x \in \Gamma(\Lambda)$,

$$V_\Lambda(x) \le \frac{\exp(-a|x-[x]|)}{|x-[x]|} + 2 \sum_{y \in \Lambda, |y-x| \ge 1/2} \exp(-a|x-y|).$$

On the other hand, we have, for $x \notin \Gamma(\Lambda)$,

$$V_\Lambda(x) \le 2 \sum_{y \in \Lambda} \exp(-a|x-y|).$$

For $p \ge 1$, we deduce in all cases

$$(V_\Lambda(x))^p \leq 2^{p-1}\left(C(x)\frac{\exp(-ap|x-[x]|)}{|x-[x]|^p} + \left(2\sum_{y\in\Lambda}\exp(-a|x-y|)\right)^p\right),$$

with $C(x) = 1$ if $x \in \Gamma(\Lambda)$, $= 0$ otherwise.

It is worth noticing at this stage that the periodic function defined by

$$\sum_{y\in\mathbf{Z}^3}\exp(-a|x-y|)$$

is continuous on $\mathbf{R}^3$, and hence bounded on $\mathbf{R}^3$. So, V_Λ is uniformly bounded on $\Gamma(\Lambda)^c$.

Thus,

$$\left(\sum_{y\in\Lambda}\exp(-a|x-y|)\right)^p \leq C\sum_{y\in\Lambda}\exp(-a|x-y|).$$

Hence

$$(V_\Lambda(x))^p \leq C'(x)\frac{\exp(-ap|x-[x]|)}{|x-[x]|^p} + C\sum_{y\in\Lambda}\exp(-a|x-y|), \qquad (2.23)$$

where $C'(x) = 0$ outside $\Gamma(\Lambda)$ and is a constant on $\Gamma(\Lambda)$.

For $p < 3$, we now integrate over $\mathbf{R}^3$, and obtain

$$\int_{\mathbf{R}^3}(V_\Lambda(x))^p \leq \int_{\Gamma(\Lambda)}\frac{\exp(-ap|x-[x]|)}{|x-[x]|^p} + C\int_{\mathbf{R}^3}\sum_{y\in\Lambda}\exp(-a|x-y|)$$

$$\leq 2^p \exp\left(-\frac{ap}{2}\right)|\Lambda| + C|\Lambda|\int_{\mathbf{R}^3}\exp(-a|x|)$$

$$\leq C|\Lambda|.$$

Thus, (i) holds.

(ii) In addition, we can use the same inequality (2.23), and integrate it over the set $\Gamma(\Lambda)^c$ to obtain:

$$\int_{\Gamma(\Lambda)^c}V_\Lambda^p \leq C\int_{\Gamma(\Lambda)^c}\sum_{y\in\Lambda}\exp(-a|x-y|).$$

For $h > 0$ fixed, we have, using the notations of Definition 1,

$$\int_{\Gamma(\Lambda)^c}\sum_{y\in\Lambda\cap\Lambda^h}\exp(-a|x-y|) \leq C|\Lambda^h|,$$

while

$$\int_{\Gamma(\Lambda)^c}\sum_{y\in\Lambda, y\notin\Lambda^h}\exp(-a|x-y|) \leq \exp\left(-\frac{a}{2}h\right)\sum_{y\in\Lambda}\int_{\mathbf{R}^3}\exp\left(-\frac{a}{2}|x-y|\right)$$

$$\leq C|\Lambda|\exp\left(-\frac{a}{2}h\right),$$

for some constant C that is independent of Λ.

If Λ goes to infinity (as prescribed in Definition 1), we then reach the desired conclusion, letting first Λ, then h, go to infinity.

(iii) We introduce the notation

$$V_{\Lambda^c}(x) = V_\infty(x) - V_\Lambda(x) = \sum_{y \in \mathbf{Z}^3 \setminus \Lambda} \frac{\exp(-a|x - y|)}{|x - y|}.$$

Then, for all x in $\Gamma(\Lambda)$,

$$V_{\Lambda^c}(x) \leq 2 \sum_{y \in \mathbf{Z}^3 \setminus \Lambda} \exp(-a|x - y|) \leq 2 \sum_{y \in \mathbf{Z}^3} \exp(-a|x - y|).$$

Therefore, V_{Λ^c} is uniformly bounded on $\Gamma(\Lambda)$ by some positive constant, say C_∞, and we just have to prove (iii) for $p = 1$.

As in the proof of (ii), we fix $h > 0$. Thus, we obtain on the one hand,

$$\int_{\Gamma(\Lambda) \cap \Lambda^h} V_{\Lambda^c}(x) \, \mathrm{d}x \leq C_\infty |\Lambda^h|,$$

while, on the other hand,

$$\int_{\Gamma(\Lambda) \setminus \Lambda^h} V_{\Lambda^c}(x) \, \mathrm{d}x \leq \frac{C_\infty}{h + \frac{1}{2}} |\Lambda|,$$

since $|x - y| \geq h + \frac{1}{2}$, for all x, y such that $\mathrm{dist}(x; \partial\Gamma(\Lambda)) \geq h$ and $y \in \Gamma(\Lambda)^c$.

We complete the proof of (iii) as before, letting first Λ, then h, go to infinity.

(iv) The proof of (iv) follows the same lines as the proof of (i). Let $k \in \mathbf{Z}^3$ be fixed. We have

$$V_\Lambda(x + k) - V_\Lambda(x) = \sum_{y \in \Lambda} V(x + k - y) - \sum_{y \in \Lambda} V(x - y)$$

$$= \sum_{y \in (\Lambda - k) \setminus \Lambda} V(x - y) - \sum_{y \in \Lambda \setminus (\Lambda - k)} V(x - y),$$

where we denote $A \setminus B = \{x \in A, x \notin B\}$ and $A + k = \{x + k / x \in A\}$.

We now remark that there are at most $o(|\Lambda|)$ terms in each sum. Indeed, if $y \in (\Lambda - k) \setminus \Lambda$, we necessarily have $\mathrm{dist}(y, \partial\Gamma(\Lambda)) \leq |k|$, and thus

$$\left((\Lambda - k) \setminus \Lambda \right) \cup \left(\Lambda \setminus (\Lambda - k) \right) \subset \Lambda^{|k|}.$$

Next, we argue as we did for the proof of (i). We have

$$|V_\Lambda(x + k) - V_\Lambda(x)| \leq \sum_{y \in \Lambda^{|k|}} V(x - y)$$

$$\leq C(x) V(x - [x]) + c \sum_{y \in \Lambda^{|k|}} \exp(-a|x - y|),$$

where $C(x)$ vanishes outside the set $\{x / \exists y \in \Lambda^{|k|}, x \in \Gamma_y\}$.

Therefore, as before,

$$|V_\Lambda(x+k) - V_\Lambda(x)|^p \leq C(x)V(x-[x])^p + c \sum_{y \in \Lambda^{|k|}} \exp(-a|x-y|).$$

Integrating over $\mathbf{R}^3$ when $p < 3$, we obtain

$$\int_{\mathbf{R}^3} |V_\Lambda(x+k) - V_\Lambda(x)|^p \leq C|\Lambda^{|k|}|,$$

and thus (iv) holds. ◇

Proposition 2.5 *Let $a > 0$ be fixed in (2.1). There exist various constants C independent of $\Lambda \subset \mathbf{Z}^3$, such that, for all $\Lambda \subset \mathbf{Z}^3$, the following holds:*

(i) $0 < \theta_\Lambda \leq C$;

(ii) $\|\rho_\Lambda\|_{L^p} \leq C|\Lambda|^{\frac{1}{p}}$, *for all* $1 \leq p < \infty$;

(iii) $\|\rho_\Lambda\|_{L^\infty} \leq C$;

(iv) $\dfrac{1}{|\Lambda|} \displaystyle\int_{\mathbf{R}^3} |\nabla \sqrt{\rho_\Lambda}|^2 \leq C$;

(v) $\dfrac{1}{|\Lambda|} \displaystyle\int_{\mathbf{R}^3} V_\Lambda \rho_\Lambda \leq C$;

(vi) $\dfrac{1}{|\Lambda|} \displaystyle\iint_{\mathbf{R}^3 \times \mathbf{R}^3} \rho_\Lambda(x)\rho_\Lambda(y)V(x-y)\,\mathrm{d}x\,\mathrm{d}y \leq C$;

(vii) $\left|\dfrac{I_\Lambda}{|\Lambda|}\right| \leq C$.

Proof of Proposition 2.5 Let ρ_Λ be the minimum of I_Λ and $u_\Lambda = \sqrt{\rho_\Lambda}$. Multiplying the Euler–Lagrange equation (2.16) by u_Λ itself, and next integrating over $\mathbf{R}^3$, we obtain

$$\int_{\mathbf{R}^3} |\nabla \sqrt{\rho_\Lambda}|^2 + \tfrac{5}{3}\int_{\mathbf{R}^3} \rho_\Lambda^{5/3} + D(\rho_\Lambda, \rho_\Lambda) + \theta_\Lambda|\Lambda| \leq \int_{\mathbf{R}^3} V_\Lambda \rho_\Lambda. \qquad (2.24)$$

In particular, this implies

$$\frac{1}{|\Lambda|}\int_{\mathbf{R}^3} \rho_\Lambda^{5/3} \leq C\frac{1}{|\Lambda|}\int_{\mathbf{R}^3} V_\Lambda \rho_\Lambda$$

$$\leq C\frac{1}{|\Lambda|}\|V_\Lambda\|_{L^{5/2}}\|\rho_\Lambda\|_{L^{5/3}}, \qquad (2.25)$$

thus, using Lemma 2.4(i),

$$\frac{1}{|\Lambda|}\int_{\mathbf{R}^3} \rho_\Lambda^{5/3} \leq C\left(\frac{1}{|\Lambda|}\int_{\mathbf{R}^3} \rho_\Lambda^{5/3}\right)^{3/5},$$

and (ii) holds for $p = \tfrac{5}{3}$.

From Hölder's inequality and the constraint $\|\rho_\Lambda\|_{L^1} = |\Lambda|$, (ii) immediately follows for all $1 \le p \le \frac{5}{3}$. Because of (2.25), we also deduce (v) and we bound the right-hand side of (2.24) by some $C|\Lambda|$ in order to obtain (i), (iv), and (vi).

We now turn to the proof of (iii). Once (iii) is proved, (ii) follows for $p > \frac{5}{3}$, using Hölder's inequality together with (ii) for $p = \frac{5}{3}$.

We use a general argument which applies if we replace the exponent 5/3 in the definition of the Thomas–Fermi term $\int_{\mathbf{R}^3} \rho^{5/3}$ by any exponent p, with $\frac{3}{2} < p < 3$. (Actually, as shown in Remark 2.6 below, the result also holds for any $1 < p < \infty$.)

From the Euler–Lagrange equation (2.16) satisfied by u_Λ, we deduce

$$-\triangle u_\Lambda + p\, u_\Lambda^{2p-1} \le V_\Lambda\, u_\Lambda \le V_\infty\, u_\Lambda,$$

almost everywhere on $\mathbf{R}^3$, for $\frac{3}{2} < p$. We now choose $r_0 \in \,]\frac{\sqrt{3}}{2}, 1)$ and observe that, for all $y \in \mathbf{Z}^3$,

(P1) $\Gamma_y \subset B(y; r_0)$,

and

(P2) $B(y; r_0) \cap \mathbf{Z}^3 = \{y\}$.

We first claim that there exists a positive constant $C_0 = C_0(r_0)$ such that, for all y in $\mathbf{Z}^3$,

$$V_\infty(x) \le \frac{C_0}{|x - y|} \qquad \text{for all } x \text{ in } B(y; r_0). \tag{2.26}$$

We shall check this inequality for $y = 0$, the general case following from the periodicity of V_∞.

Indeed, for all x in $B(0; r_0)$, we may write

$$V_\infty(x) = \sum_{y \in \mathbf{Z}^3} \frac{\exp(-a|x - y|)}{|x - y|}$$

$$= \frac{\exp(-a|x|)}{|x|} + \sum_{y \ne 0 \in \mathbf{Z}^3} \frac{\exp(-a|x - y|)}{|x - y|}$$

$$\le \frac{1}{|x|} + \frac{1}{1 - r_0} \sum_{y \in \mathbf{Z}^3} \exp(-a|x - y|),$$

$$\text{since } 1 - r_0 = \min\{|x - y|/x \in B(0; r_0), y \in \mathbf{Z}^3\},$$

$$\le \frac{1}{|x|} + \frac{1}{1 - r_0} \exp(a|x|) \sum_{y \in \mathbf{Z}^3} \exp(-a|y|),$$

$$\text{since } |x - y| \ge |y| - |x|, \text{ and, finally,}$$

$$\le \frac{1}{|x|} + \frac{\exp(ar_0)}{1 - r_0} \sum_{y \in \mathbf{Z}^3} \exp(-a|y|),$$

$$\le \frac{1}{|x|} + C_1(r_0),$$

where $C_1(r_0)$ is a positive constant depending only on r_0. In order to conclude, we take, for example, $C_0 = C_1(r_0) \cdot r_0 + 1$.

Then, from (2.26) and Jensen's inequality, we find some positive constant $C = C(p; r_0)$ such that

$$V_\infty u_\Lambda \leq \frac{p}{2} u_\Lambda^{2p-1} + \frac{C}{|x - y|^{p'}} \, ,$$

for all x in $B(y; r_0)$, and for all y in $\mathbf{Z}^3$, with $p' = \frac{2p-1}{2p-2}$, the conjugate exponent of $2p - 1$.

Finally, u_Λ satisfies

$$-\Delta u_\Lambda + \frac{p}{2} u_\Lambda^{2p-1} \leq \frac{C}{|x - y|^{p'}}, \tag{2.27}$$

for all y in $\mathbf{Z}^3$, for all x in $B(y; r_0)$, for some positive constant C, independent of y and Λ.

Our strategy of proof is now the following.

We translate y to 0 and use a comparison argument on $B(0; r_0)$ to find a uniform (that means independent of Λ) L^∞-bound for u_Λ on $B(0; r_0')$, where $r_0' < r_0$ satisfies (P1)–(P2) ($r_0 > r_0' > \frac{\sqrt{3}}{2}$, so that $\Gamma_y \subset B(y; r_0')$). Then, we know that the same bound holds on $B(y; r_0')$, for all y in $\mathbf{Z}^3$, since $u_\Lambda (\cdot + y)$ satisfies (2.27) in $B(0; r_0)$. We then recover a uniform L^∞ bound on $\mathbf{R}^3$.

For this purpose, we build two positive radial functions u_1 and u_2 in L^∞_{loc} $(B(0; r_0))$ satisfying, respectively,

$$\begin{cases} -\Delta u_1 = \frac{C}{|x|^{p'}} & \text{in } B(0; r_0), \\ u_1 = 0 & \text{on } \partial B(0; r_0), \end{cases}$$

and

$$\begin{cases} -\Delta u_2 + \frac{p}{2} u_2^{2p-1} \geq 0 & \text{in } B(0; r_0), \\ u_2 = +\infty & \text{on } \partial B(0; r_0). \end{cases} \tag{2.28}$$

Therefore, $\bar{u}$ defined by $\bar{u} = u_1 + u_2$ satisfies

$$-\Delta \bar{u} + \frac{p}{2} \bar{u}^{2p-1} = -\Delta u_1 - \Delta u_2 + \frac{p}{2}(u_1 + u_2)^{2p-1}$$

$$\geq \frac{C}{|x|^{p'}} - \Delta u_2 + \frac{p}{2} u_2^{2p-1}$$

$$\geq \frac{C}{|x|^{p'}} \quad \text{in } B(0; r_0),$$

together with

$$\bar{u} = +\infty \quad \text{on } \partial B(0; r_0).$$

In view of the maximum principle, we then have

$$u_\Lambda \leq \bar{u} \quad \text{on } B(0; r_0).$$

It now just remains to check that $\bar{u}$ is bounded on $B(0; r_0')$.

Standard calculations show that

$$u_1(x) = \frac{C}{(2-p')(3-p')}(r_0^{2-p'} - |x|^{2-p'})$$

and

$$u_2(x) = C\,r_0^{\frac{1}{p-1}}(r_0^2 - |x|^2)^{-\frac{1}{p-1}},$$

work, where C denotes various positive constants depending only on p and r_0 (the general form expected for a solution to (2.28) is taken from [57]). We may notice at this stage that u_1 is bounded at 0, since $p > \frac{3}{2}$ implies $p' < 2$. Then, u_2 and $\bar{u}$ are uniformly bounded on $\bar{B}(0; r_0')$.

This concludes the proof of (iii).

It only remains to prove (vii). In view of (ii), (iii), (v), and (vi), proving (vii) amounts to proving that $\dfrac{1}{2}\dfrac{U_\Lambda}{|\Lambda|}$ is bounded independently of Λ.

This point is clear, since we have

$$\frac{1}{2|\Lambda|}U_\Lambda \le \frac{1}{|\Lambda|}\sum_{y\in\Lambda}\sum_{z\in\Lambda}e^{-a|y-z|}$$

$$\le \frac{1}{|\Lambda|}\sum_{y\in\Lambda}\sum_{z\in\mathbf{Z}^3}e^{-a|y-z|}$$

$$= \sum_{z\in\mathbf{Z}^3}e^{-a|z|}.$$

In fact, as we shall see later on, $\dfrac{1}{2}\dfrac{U_\Lambda}{|\Lambda|}\xrightarrow{\Lambda\to\infty}\dfrac{1}{2}\displaystyle\sum_{y\neq 0\in\mathbf{Z}^3}\dfrac{\exp(-a|y|)}{|y|}.$

$\diamondsuit$

Remark 2.6 In the proof of (iii), we have shown that, for $p > \frac{3}{2}$, any $u \in L^\infty$, $u \ge 0$, that satisfies an inequality of the type

$$-\Delta u + u^{2p-1} \le \frac{C_0}{|x|}u$$

on the ball $B(0, r_0)$ may be bounded on any ball $B(0, r_0')$ (with $r_0' < r_0$) by a constant that only depends on p, r_0', and C_0, but not on u itself. For this purpose, we have in fact only used the fact that the singularity $\frac{1}{|x|}$ belongs to $L^{\frac{3}{2}\frac{2p-1}{2p-2}}$. The argument therefore also holds for other convenient singularities (for related considerations, see Chapter 4). Actually, using the particular form of the singularity $\frac{1}{|x|}$, we may extend this result for any $1 < p < \infty$.

Indeed, if $u \in L^\infty$, $u \ge 0$ satisfies

$$-\Delta u + u^{2p-1} \le \frac{C_0}{|x|}u,$$

the usual argument that we have used above first shows that, for $0 < \varepsilon \le |x| \le r_0 - \varepsilon$, $0 \le u \le C_\varepsilon$, where the constant C_ε only depends on C_0 and ε.

Let $0 < r_0' < r_0$ be fixed. Using the above bound, we have $u \le C_1$ on $\partial B(0, r_0')$. By a straightforward calculation, we build a supersolution to

$$-\Delta u + u^{2p-1} = \frac{C_0}{|x|} u,$$

on $B(0, r_0')$ by defining $w = Ke^{-\mu r}$ with $\mu = \frac{C_0}{2}$, $K \ge max\left(C_1, \mu^{\frac{1}{p-1}}\right)e^{\mu r_0'}$. Next, we show that $u \le w$ by a comparison argument. The fact that $w \ge 0$ on $\partial B(0, r_0')$ and $w > 0$ in $B(0, r_0')$ implies that $\lambda_1(-\Delta + w^{2(p-1)} - \frac{C_0}{|x|}, B(0, r_0')) > 0$. From

$$-\Delta(w - u) + (w^{2p-1} - u^{2p-1}) - \frac{C_0}{|x|}(w - u) \ge 0,$$

and $w^{2p-1} - u^{2p-1} \le (2p-1)w^{2(p-1)}(w - u)$, we then deduce

$$[-\Delta + (2p-1)w^{2(p-1)} - \frac{C_0}{|x|}](w - u) \ge 0.$$

Since $w - u \ge 0$ on $\partial B(0, r_0')$, $2p - 1 > 1$ and $\lambda_1 > 0$, it follows that $w \ge u$ on $B(0, r_0')$.

Corollary 2.7 (u_Λ) *is bounded in* $H^1_{loc}(\mathbf{R}^3) \cap L^\infty(\mathbf{R}^3)$. *Then, up to a subsequence,* u_Λ *converges to* $u_\infty \in H^1_{loc}(\mathbf{R}^3)$, *strongly in* $L^p_{loc}(\mathbf{R}^3)$, *for all* $1 \le p < +\infty$, *almost everywhere on* $\mathbf{R}^3$, *and* ∇u_Λ *converges to* ∇u_∞ *weakly in* $\left(L^2_{loc}(\mathbf{R}^3)\right)^3$.

Remark 2.8 Note that the limit may depend on the sequence Λ. We shall prove below that u_∞ does not depend on the sequence (Λ).

We define $\rho_\infty = u_\infty^2$.

Proof of Corollary 2.7 Since u_Λ is uniformly bounded on $\mathbf{R}^3$ (see Proposition 2.5), u_Λ is clearly bounded in $L^p_{loc}(\mathbf{R}^3)$, for all $1 \le p \le \infty$, independently of Λ.

It now remains to prove that, independently of Λ, (∇u_Λ) is bounded in $\left(L^2_{loc}(\mathbf{R}^3)\right)^3$. Then, Corollary 2.7 will follow immediately from Rellich's theorem and Hölder's inequality.

Let $R > 0$ be fixed and take $\varphi \in \mathcal{D}(B_{2R})$ such that $\varphi \equiv 1$ on B_R and $|\varphi|$, $|\nabla \varphi| \le C$, almost everywhere on $\mathbf{R}^3$, where C depends only on φ.

We multiply the Euler–Lagrange equation (2.16) by $u_\Lambda \varphi^2$ and integrate over $\mathbf{R}^3$. Next, we use the uniform L^∞ bound on u_Λ and the uniform bound on θ_Λ, together with $\|\rho_\Lambda \star V\|_{L^\infty(\mathbf{R}^3)} \le C\|V\|_{L^1(\mathbf{R}^3)}$ and $\|V_\Lambda\|_{L^1(B_{2R})} \le \|V_\infty\|_{L^1(B_{2R})}$, to deduce

$$\left|\int_{\mathbf{R}^3}(-\Delta u_\Lambda)u_\Lambda \varphi^2\right| \le C,$$

where C denotes various positive constants independent of Λ.

Then, we write that

$$\int_{\mathbf{R}^3} (-\Delta u_\Lambda) u_\Lambda \varphi^2 = \int_{\mathbf{R}^3} |\nabla u_\Lambda|^2 \varphi^2 + 2 \int_{\mathbf{R}^3} \varphi \nabla \varphi \cdot u_\Lambda \nabla u_\Lambda.$$

Next, we apply the Cauchy–Schwarz inequality and obtain

$$\left| \int_{\mathbf{R}^3} \varphi \nabla \varphi \cdot u_\Lambda \nabla u_\Lambda \right| \leq \left(\int_{\mathbf{R}^3} \varphi^2 |\nabla u_\Lambda|^2 \right)^{1/2} \left(\int_{\mathbf{R}^3} u_\Lambda^2 |\nabla \varphi|^2 \right)^{1/2}$$
$$\leq C \left(\int_{\mathbf{R}^3} \varphi^2 |\nabla u_\Lambda|^2 \right)^{1/2}.$$

We deduce from the above equalities that $\int_{\mathbf{R}^3} \varphi^2 |\nabla u_\Lambda|^2 \leq C$, which implies in particular that $\int_{B_R} |\nabla u_\Lambda|^2 \leq C$, for any $R > 0$. The proof is complete. $\diamond$

From the estimates we have obtained above, we may deduce the strong convergence (up to the extraction of subsequences) of the sequence $(\tilde{\rho}_\Lambda)$ defined in Chapter 1, which will be our main idea for obtaining a lower bound on $\frac{I_\Lambda}{|\Lambda|}$ as $\Lambda \longrightarrow \infty$. More precisely, we have the following:

Proposition 2.9 *The sequence $\sqrt{\tilde{\rho}_\Lambda}$ is bounded in $H^1(\Gamma_0)$ and in $L^p(\Gamma_0)$, for all $1 \leq p \leq +\infty$.*

In particular, up to a subsequence, we may assume that $\sqrt{\tilde{\rho}_\Lambda}$ converges to some non-negative function $\sqrt{\tilde{\rho}}$ strongly in $L^p(\Gamma_0)$, for all $1 \leq p < +\infty$, almost everywhere on $\mathbf{R}^3$, and that $\nabla \sqrt{\tilde{\rho}_\Lambda}$ converges to $\nabla \sqrt{\tilde{\rho}}$ weakly in $(L^2(\Gamma_0))^3$.

Remark 2.10 We shall prove below that the limit function $\tilde{\rho}$ is uniquely defined as the minimizer of $I_{per}(\mu)$, for some $0 < \mu \leq 1$, independent of the sequence Λ. Hence, the whole sequence $(\tilde{\rho}_\Lambda)$ is in fact convergent to a unique limit that does not depend on the particular sequence (Λ) we consider.

As a consequence of Proposition 2.9, we also have the following:

Corollary 2.11 *For any Van Hove sequence (Λ), we have*

$$\liminf_{\Lambda \to \infty} \frac{1}{|\Lambda|} \int_{\mathbf{R}^3} |\nabla \sqrt{\rho_\Lambda}|^2 \geq \int_{\Gamma_0} |\nabla \sqrt{\tilde{\rho}}|^2, \tag{2.29}$$

$$\liminf_{\Lambda \to \infty} \frac{1}{|\Lambda|} \int_{\mathbf{R}^3} \rho_\Lambda^p \geq \int_{\Gamma_0} \tilde{\rho}^p \quad for \ all \ \ 1 \leq p < +\infty, \tag{2.30}$$

and

$$\lim_{\Lambda \to \infty} \frac{1}{|\Lambda|} \int_{\mathbf{R}^3} V_\Lambda \rho_\Lambda = \int_{\Gamma_0} V_\infty \tilde{\rho}. \tag{2.31}$$

Proof of Proposition 2.9 and Corollary 2.11 Recall that

$$\tilde{\rho}_\Lambda = \frac{1}{|\Lambda|} \sum_{y \in \Lambda} \rho_\Lambda(\cdot + y).$$

Then, we use the convexity of the function $t \to t^p$, for $1 \le p < +\infty$, to obtain the following inequality:

$$\frac{1}{|\Lambda|} \int_{\mathbf{R}^3} \rho_\Lambda^p \ge \frac{1}{|\Lambda|} \int_{\Gamma(\Lambda)} \rho_\Lambda^p(x)\, dx = \frac{1}{|\Lambda|} \sum_{y \in \Lambda} \int_{\Gamma_0} \rho_\Lambda(x+y)^p\, dx$$

$$\ge \int_{\Gamma_0} \left(\frac{1}{|\Lambda|} \sum_{y \in \Lambda} \rho_\Lambda(x+y) \right)^p dx.$$

Thus,

$$\frac{1}{|\Lambda|} \int_{\mathbf{R}^3} \rho_\Lambda^p \ge \int_{\Gamma_0} \tilde{\rho}_\Lambda^p. \tag{2.32}$$

For the same reason, we see that

$$\frac{1}{|\Lambda|} \int_{\mathbf{R}^3} |\nabla \sqrt{\rho_\Lambda}|^2 \ge \frac{1}{|\Lambda|} \int_{\Gamma(\Lambda)} |\nabla \sqrt{\rho_\Lambda}|^2 \ge \int_{\Gamma_0} |\nabla \sqrt{\tilde{\rho}_\Lambda}|^2. \tag{2.33}$$

Moreover, we obviously have

$$\|\tilde{\rho}_\Lambda\|_{L^\infty(\mathbf{R}^3)} \le \|\rho_\Lambda\|_{L^\infty(\mathbf{R}^3)} \le C.$$

We then complete the proof of Proposition 2.9, with Proposition 2.5 and Rellich's theorem. Passing to the limit in (2.32) and (2.33), we deduce (2.29) and (2.30).

It just remains to check (2.31).

Since

$$\frac{1}{|\Lambda|} \int_{\Gamma(\Lambda)^c} V_\Lambda(x)\, dx \to 0 \qquad \text{when } \Lambda \to \infty$$

(see Lemma 2.4) and $\|\rho_\Lambda\|_{L^\infty(\mathbf{R}^3)} \le C$, we deduce

$$\frac{1}{|\Lambda|} \int_{\Gamma(\Lambda)^c} V_\Lambda(x)\rho_\Lambda(x)\, dx \to 0 \qquad \text{as } \Lambda \to \infty.$$

Using Lemma 2.4 once again, we recall that we have

$$\frac{1}{|\Lambda|} \int_{\Gamma(\Lambda)} |V_\Lambda(x) - V_\infty(x)|\, dx \to 0 \qquad \text{as } \Lambda \to \infty,$$

and thus, for the same reason as above,

$$\frac{1}{|\Lambda|} \int_{\Gamma(\Lambda)} |V_\Lambda(x) - V_\infty(x)|\rho_\Lambda(x)\, dx \to 0 \qquad \text{as } \Lambda \to \infty.$$

Now, the periodicity of V_∞ yields

$$\frac{1}{|\Lambda|} \int_{\Gamma(\Lambda)} V_\infty(x)\rho_\Lambda(x)\, dx = \int_{\Gamma_0} V_\infty(x)\tilde{\rho}_\Lambda(x)\, dx$$

which converges to $\int_{\Gamma_0} V_\infty(x)\tilde{\rho}(x)\, dx$, since, for example, $V_\infty \in L^2(\Gamma_0)$ and $\tilde{\rho}_\Lambda$ converges to $\tilde{\rho}$ in $L^2(\Gamma_0)$-weak.   ◇

2.3   Lower bound for the energy

In this section, we obtain a bound from below for the energy per unit volume $\frac{I_\Lambda}{|\Lambda|}$, the next section being devoted to finding a bound from above.

In order to prove that we can asymptotically bound $\frac{I_\Lambda}{|\Lambda|}$ from below by the periodic problem I_{per}, we are going to use extensively the $\sim$-transform of ρ_Λ. Thanks to the convexity of each term of the energy, we are able to compare $\liminf_{\Lambda \to \infty} E_{\text{per}}(\tilde{\rho}_\Lambda)$ and $\liminf_{\Lambda \to \infty} \frac{1}{|\Lambda|} E_\Lambda(\rho_\Lambda)$. In this comparison, the main difficulty is the behaviour of the convolution term. We manage to analyze it by using some 'periodicity' estimate in some average sense (obtained in Proposition 2.15 below). This 'periodicity' is a consequence of the fact that the energy $E_\Lambda(\rho)$ is strictly convex in the neighbourhood of its minimizing density ρ_Λ. Once this periodicity is known, we are able to find a sharp bound from below for the convolution term, and thus for the whole energy (Proposition 2.17 below).

Lemma 2.12 *Let $C > 0$ be fixed. There exists some constant $\alpha > 0$ such that we have, for all $0 \leq u \leq v \leq C$:*

(i) $v^{5/3} - u^{5/3} - \frac{5}{3}u^{2/3}(v - u) \geq \alpha(v - u)^2$;

(ii) $v^{5/3} - u^{5/3} - \frac{5}{3}v^{2/3}(v - u) \leq -\alpha(v - u)^2$.

Remark 2.13 (1) This lemma is based only upon the fact that $1 < \frac{5}{3} \leq 2$, and thus also holds mutatis mutandis for any exponent $1 < p \leq 2$. For $p > 2$, the same result holds true provided that we assume that $0 < \mu \leq u \leq v \leq C$, the constant α then depending on μ and C.

(2) The assumption $u, v \in [0, C]$ is necessary in order to obtain the existence of some constant $\alpha > 0$.

Proof of Lemma 2.12   We define the function

$$f_\alpha(u, v) = v^{5/3} - u^{5/3} - \frac{5}{3}u^{2/3}(v - u) - \alpha(v - u)^2.$$

When $u = 0$, $f_\alpha(0, v) \geq 0$ on $[0, C]$ provided that $\alpha \leq \frac{1}{C^{1/3}}$.

When $u > 0$, we set $x = \frac{v}{u}$ and divide (i) by $u^{5/3}$. We thus want to show that

$$g_\alpha(x) = x^{5/3} - \frac{5}{3}x + \frac{2}{3} - \alpha(x - 1)^2 u^{1/3} \geq 0,$$

for all $u \in [0, C]$ and $x \in [1, \frac{C}{u}]$.

It is now easy to check that, as soon as the condition

$$\alpha \leq \frac{5}{9C^{1/3}}$$

holds, g_α is a convex function on $[1, \frac{C}{u}]$ satisfying $g_\alpha(1) = g_\alpha'(1) = 0$. It is then non-negative on $[1, \frac{C}{u}]$, and the proof of (i) is complete.

The proof of (ii) is very similar and gives the same condition on α, so we skip it. $\diamondsuit$

Lemma 2.14 *Let $C > 0$ be fixed. There exists some constant $\alpha > 0$ such that, for all u and v belonging to $[0, C]$, and for all $t \in [0, 1]$, we have*

$$(tv + (1 - t)u)^{5/3} \leq tv^{5/3} + (1 - t)u^{5/3} - \alpha t(1 - t)(v - u)^2.$$

Proof of Lemma 2.14 Let u and v be fixed in $[0, C]$ such that $u \leq v$. Using Lemma 2.12(i) for $tv + (1 - t)u$ and v, we have

$$v^{5/3} - (tv + (1 - t)u)^{5/3}$$
$$- \tfrac{5}{3}(tv + (1 - t)u)^{2/3}(1 - t)(v - u)$$
$$- \alpha(1 - t)^2(v - u)^2 \geq 0,$$

while using Lemma 2.12(ii) for u and $tv + (1 - t)u$, we obtain

$$(tv + (1 - t)u)^{5/3} - u^{5/3} - \tfrac{5}{3}(tv + (1 - t)u)^{2/3}t(v - u) + \alpha t^2(v - u)^2 \leq 0.$$

Combining the above two inequalities, we obtain

$$tv^{5/3} + (1 - t)u^{5/3} - (tv + (1 - t)u)^{5/3} - \alpha t(1 - t)(v - u)^2 \geq 0,$$

and Lemma 2.14 holds. When $u \geq v$, we use Lemma 2.12(i) for $tv + (1 - t)u$ and u on the one hand, and Lemma 2.12(ii) for v and $tv + (1 - t)u$ on the other hand—and the same result holds. $\diamondsuit$

Proposition 2.15 *For all $k \in \mathbf{Z}^3$, and for all $1 < q < \infty$, we have*

$$\lim_{\Lambda \to \infty} \frac{1}{|\Lambda|} \int_{\mathbf{R}^3} |\rho_\Lambda(\cdot + k) - \rho_\Lambda|^q = 0. \tag{2.34}$$

Proof of Proposition 2.15 We begin by proving that

$$\lim_{\Lambda \to \infty} \frac{1}{|\Lambda|} \left[E_\Lambda(\rho_\Lambda) - E_\Lambda(\rho_\Lambda(\cdot + k)) \right] = 0. \tag{2.35}$$

It is clear from the expression of E_Λ that (2.35) reduces to

$$\lim_{\Lambda \to \infty} \frac{1}{|\Lambda|} \int_{\mathbf{R}^3} V_\Lambda(\rho_\Lambda - \rho_\Lambda(\cdot + k)) = 0. \tag{2.36}$$

Now (2.36) is a straightforward consequence of the Hölder inequality and of the assertions in Lemma 2.4(iv) and Proposition 2.5(ii).

We now turn to the proof of (2.34).

We use the convexity with respect to the density ρ of the terms $\int_{\mathbf{R}^3} |\nabla\sqrt{\rho}|^2$, $-\int_{\mathbf{R}^3} V_\Lambda \rho$, and $D(\rho, \rho)$ of the energy, and Lemma 2.14 for the term $\int_{\mathbf{R}^3} \rho^{5/3}$. We thus have, for any $t \in [0, 1]$, for some $\alpha > 0$ (independent of Λ, k, t)

$$E_\Lambda(t\rho_\Lambda(\cdot + k) + (1 - t)\rho_\Lambda) \leq tE_\Lambda(\rho_\Lambda(\cdot + k)) + (1 - t)E_\Lambda(\rho_\Lambda)$$
$$- \alpha t(1 - t) \int_{\mathbf{R}^3} |\rho_\Lambda(\cdot + k) - \rho_\Lambda|^2.$$

For this inequality, we have used the fact that we know that the sequence ρ_Λ is bounded uniformly in L^∞, which allows us to apply Lemmas 2.12 and 2.14.

Next, we remark that, since ρ_Λ is the minimum of the energy on the set $\int_{\mathbf{R}^3} \rho = |\Lambda|$, and $\int_{\mathbf{R}^3} t\rho_\Lambda(\cdot + k) + (1-t)\rho_\Lambda = |\Lambda|$, we have

$$E_\Lambda(t\rho_\Lambda(\cdot + k) + (1-t)\rho_\Lambda) \geq E_\Lambda(\rho_\Lambda),$$

and hence

$$\alpha \int_{\mathbf{R}^3} |\rho_\Lambda(\cdot + k) - \rho_\Lambda|^2 - \alpha t \int_{\mathbf{R}^3} |\rho_\Lambda(\cdot + k) - \rho_\Lambda|^2$$
$$\leq E_\Lambda(\rho_\Lambda(\cdot + k)) - E_\Lambda(\rho_\Lambda),$$

provided that $t \neq 0$. Taking $t = \frac{1}{2}$ for instance, and using (2.35), we obtain

$$\lim_{\Lambda \to \infty} \frac{1}{|\Lambda|} \int_{\mathbf{R}^3} |\rho_\Lambda(\cdot + k) - \rho_\Lambda|^2 = 0.$$

Using the uniform L^∞ bound, it immediately follows that (2.34) holds for all $q \geq 2$. In addition, for $1 < q < 2$, we have, using Hölder's inequality,

$$\int_{\mathbf{R}^3} |\rho_\Lambda(\cdot + k) - \rho_\Lambda|^q$$
$$\leq \left(\int_{\mathbf{R}^3} |\rho_\Lambda(\cdot + k) - \rho_\Lambda| \right)^{2-q} \left(\int_{\mathbf{R}^3} |\rho_\Lambda(\cdot + k) - \rho_\Lambda|^2 \right)^{q-1}$$
$$\leq \left(\int_{\mathbf{R}^3} \rho_\Lambda(\cdot + k) + \rho_\Lambda \right)^{2-q} \left(\int_{\mathbf{R}^3} |\rho_\Lambda(\cdot + k) - \rho_\Lambda|^2 \right)^{q-1}$$
$$= O(|\Lambda|)^{2-q} o(|\Lambda|)^{q-1} = o(|\Lambda|).$$

$\diamondsuit$

Remark 2.16 In the proof of (2.35), it is not necessary to know that $\|\rho_\Lambda\|_{L^\infty}$ is uniformly bounded, since the use of (ii) (for $p > \frac{3}{2}$) of Proposition 2.5 and Lemma 2.4 allows to conclude. On the other hand, the L^∞ bound is necessary for the rest of the argument.

The main result of this section is the following:

Proposition 2.17 *For any Van Hove sequence* (Λ), *we have*

$$E_{per}(\tilde{\rho}) \leq \liminf_{\Lambda \to \infty} E_{per}(\tilde{\rho}_\Lambda) \leq \liminf_{\Lambda \to \infty} \frac{1}{|\Lambda|} E_\Lambda(\rho_\Lambda).$$

Proof of Proposition 2.17 In order to prove this claim, it only remains to show that

$$\liminf_{\Lambda \to \infty} D_{\Gamma_0}(\tilde{\rho}_\Lambda, \tilde{\rho}_\Lambda) \leq \liminf_{\Lambda \to \infty} \frac{1}{|\Lambda|} D(\rho_\Lambda, \rho_\Lambda), \qquad (2.37)$$

since, in view of Proposition 2.5 above and its proof, we have

$$\liminf_{\Lambda \to \infty} \left[\int_{\Gamma_0} |\nabla \sqrt{\tilde{\rho}_\Lambda}|^2 + \int_{\Gamma_0} \tilde{\rho}_\Lambda^{5/3} - \int_{\Gamma_0} V_\infty \tilde{\rho}_\Lambda \right]$$

$$\leq \liminf_{\Lambda \to \infty} \left[\frac{1}{|\Lambda|} \int_{\mathbf{R}^3} |\nabla \sqrt{\rho_\Lambda}|^2 + \frac{1}{|\Lambda|} \int_{\mathbf{R}^3} \rho_\Lambda^{5/3} - \frac{1}{|\Lambda|} \int_{\mathbf{R}^3} V_\Lambda \rho_\Lambda \right].$$

For the proof of (2.37), it clearly suffices to show that, for any $\varepsilon > 0$, we have

$$\liminf_{\Lambda \to \infty} D_{\Gamma_0}(\tilde{\rho}_\Lambda, \tilde{\rho}_\Lambda) \leq \liminf_{\Lambda \to \infty} \frac{1}{|\Lambda|} D(\rho_\Lambda, \rho_\Lambda) + \varepsilon. \qquad (2.38)$$

The proof of (2.38) is divided into three steps, and is only based upon Propositions 2.5 and 2.15. Let $\varepsilon > 0$ be fixed.

Step 1. There exists some finite subset $A \subset \mathbf{Z}^3$, depending only on ε and not on Λ, such that

$$\frac{1}{|\Lambda|} \sum_{z \in \Lambda} \sum_{k \in A^c} \iint_{\Gamma_0 \times \Gamma_0} \rho_\Lambda(x+z)\rho_\Lambda(y+z)V(x-y+k) \, dx \, dy \leq \varepsilon.$$

Indeed, the left-hand side is bounded from above by

$$\frac{1}{|\Lambda|} \|\rho_\Lambda\|_{L^\infty}^2 \sum_{z \in \Lambda} \sum_{k \in A^c} \iint_{\Gamma_0 \times \Gamma_0} V(x-y+k) \, dx \, dy$$

$$= \|\rho_\Lambda\|_{L^\infty}^2 \sum_{k \in A^c} \iint_{\Gamma_0 \times \Gamma_0} V(x-y+k) \, dx \, dy$$

$$\leq \varepsilon \|\rho_\Lambda\|_{L^\infty}^2,$$

since the series $\sum_{k \in \mathbf{Z}^3} \iint_{\Gamma_0 \times \Gamma_0} V(x-y+k) \, dx \, dy$ converges (to $\iint_{\Gamma_0 \times \Gamma_0} V_\infty(x-y) \, dx \, dy = \int_{\Gamma_0} V_\infty(x) \, dx$).

Step 2. If $k \neq 0 \in \mathbf{Z}^3$ is fixed (when $k = 0$, the result below is obvious),

$$\frac{1}{|\Lambda|} \sum_{z \in \Lambda} \iint_{\Gamma_0 \times \Gamma_0} \rho_\Lambda(y+z)\rho_\Lambda(x+z+k)V(x-y+k) \, dx \, dy$$

$$= \frac{1}{|\Lambda|} \sum_{z \in \Lambda} \iint_{\Gamma_0 \times \Gamma_0} \rho_\Lambda(y+z)\rho_\Lambda(x+z)V(x-y+k) \, dx \, dy$$

$$+ o(1),$$

where the remainder term $o(1)$ depends only on k, and not on Λ.

Indeed, the difference between the above two terms may be bounded as follows:

$$\left| \sum_{z \in \Lambda} \iint_{\Gamma_0 \times \Gamma_0} \rho_\Lambda(y+z) \left(\rho_\Lambda(x+z) - \rho_\Lambda(x+z+k) \right) V(x-y+k) \, dx \, dy \right|$$

$$\leq \|\rho_\Lambda\|_{L^\infty} \|V\|_{L^1(\mathbf{R}^3)} \sum_{z \in \Lambda} \int_{\Gamma_0} |\rho_\Lambda(x+z+k) - \rho_\Lambda(x+z)| \, dx$$

$$= \|\rho_\Lambda\|_{L^\infty} \|V\|_{L^1(\mathbf{R}^3)} \int_{\Gamma(\Lambda)} |\rho_\Lambda(x+k) - \rho_\Lambda(x)| \, dx$$

$$\le \|\rho_\Lambda\|_{L^\infty} \|V\|_{L^1(\mathbf{R}^3)} |\Lambda|^{1/2} \left(\int_{\mathbf{R}^3} |\rho_\Lambda(x+k) - \rho_\Lambda(x)|^2 \right)^{1/2},$$

and we conclude using Proposition 2.15.

Step 3. The term $D_{\Gamma_0}(\rho, \rho)$ is convex, for the same reason as for the term $D(\rho, \rho)$, and thus we may write

$$D_{\Gamma_0}\left(\tilde\rho_\Lambda, \tilde\rho_\Lambda\right)$$
$$\le \frac{1}{|\Lambda|} \sum_{z\in\Lambda} D_{\Gamma_0}(\rho_\Lambda(\cdot + z), \rho_\Lambda(\cdot + z))$$
$$= \frac{1}{|\Lambda|} \sum_{z\in\Lambda} \sum_{k\in\mathbf{Z}^3} \iint_{\Gamma_0\times\Gamma_0} \rho_\Lambda(x+z)\rho_\Lambda(y+z)V(x-y+k)\,\mathrm{d}x\,\mathrm{d}y.$$

In this latter sum, we distinguish three kinds of terms: the term $k = 0$, the terms $(k \ne 0, k \in A)$, and the terms $k \in A^c$. We leave the first one unchanged. We remark that the second ones are in finite number and thus may be bounded from above using Lemma 2.4, by

$$\frac{1}{|\Lambda|} \sum_{z\in\Lambda} \sum_{k\ne 0\in A} \iint_{\Gamma_0\times\Gamma_0} \rho_\Lambda(x+z)\rho_\Lambda(y+z)V(x-y+k)\,\mathrm{d}x\,\mathrm{d}y + o(1).$$

where $o(1)$ depends only on ε. Finally, we bound the third ones from above by ε using step 1. Gathering these estimates, we obtain

$$D_{\Gamma_0}(\tilde\rho_\Lambda, \tilde\rho_\Lambda) \le \frac{1}{|\Lambda|} \sum_{z\in\Lambda} \sum_{k\in A} \iint_{\Gamma_0\times\Gamma_0} \rho_\Lambda(x+z+k)\rho_\Lambda(y+z)V(x-y+k)\,\mathrm{d}x\,\mathrm{d}y$$
$$+ o(1) + \varepsilon. \tag{2.39}$$

Next, we develop $D(\rho_\Lambda, \rho_\Lambda)$ as follows

$$\frac{1}{|\Lambda|} D\left(\rho_\Lambda, \rho_\Lambda\right)$$
$$= \frac{1}{|\Lambda|} \sum_{z\in\mathbf{Z}^3} \sum_{t\in\mathbf{Z}^3} \iint_{\Gamma_0\times\Gamma_0} \rho_\Lambda(y+z)\rho_\Lambda(x+t)V(x-y-z+t)\,\mathrm{d}x\,\mathrm{d}y$$
$$= \frac{1}{|\Lambda|} \sum_{z\in\mathbf{Z}^3} \sum_{k\in\mathbf{Z}^3} \iint_{\Gamma_0\times\Gamma_0} \rho_\Lambda(y+z)\rho_\Lambda(x+z+k)V(x-y+k)\,\mathrm{d}x\,\mathrm{d}y$$
$$= \frac{1}{|\Lambda|} \sum_{z\in\Lambda} \sum_{k\in A} \iint_{\Gamma_0\times\Gamma_0} \rho_\Lambda(y+z)\rho_\Lambda(x+z+k)V(x-y+k)\,\mathrm{d}x\,\mathrm{d}y$$
$$+ \frac{1}{|\Lambda|} \sum_{z\notin\Lambda\cup k\notin A} \iint_{\Gamma_0\times\Gamma_0} \rho_\Lambda(y+z)\rho_\Lambda(x+z+k)V(x-y+k)\,\mathrm{d}x\,\mathrm{d}y.$$

Leaving apart the second sum that is non-negative, we obtain, recognizing in the first sum the right-hand side of (2.39),

$$D_{\Gamma_0}(\tilde{\rho}_\Lambda, \tilde{\rho}_\Lambda) \leq \frac{1}{|\Lambda|} D(\rho_\Lambda, \rho_\Lambda) + o(1) + \varepsilon.$$

Hence, passing to the lower limit, we deduce (2.37). $\diamond$

At this stage, all the information we have obtained on $\liminf_{\Lambda \to \infty} \frac{I_\Lambda}{|\Lambda|}$ is contained in the following:

Corollary 2.18 *Let* $\mu = \int_{\Gamma_0} \tilde{\rho}(x)\,\mathrm{d}x$. *Then,* $0 \leq \mu \leq 1$ *and*

$$I_{\mathrm{per}}(\mu) \leq E_{\mathrm{per}}(\tilde{\rho}) \leq \liminf_{\Lambda \to \infty} \frac{I_\Lambda}{|\Lambda|},$$

where

$$I_{\mathrm{per}}(\mu) = \inf \left\{ E_{\mathrm{per}}(\rho); \rho \geq 0, \sqrt{\rho} \in H^1(\Gamma_0), \int_{\Gamma_0} \rho(x)\,\mathrm{d}x = \mu \right\}$$

$$= \inf \left\{ E_{\mathrm{per}}(\rho); \rho \geq 0, \sqrt{\rho} \in H^1_{\mathrm{per}}(\mathbf{R}^3), \int_{\Gamma_0} \rho(x)\,\mathrm{d}x = \mu \right\}. \quad (2.40)$$

Remark 2.19 In order to prove

$$I_{\mathrm{per}}(\mu) \leq E_{\mathrm{per}}(\tilde{\rho}) \quad (2.41)$$

in the above corollary, we have used the fact that $H^1_{\mathrm{per}}(\mathbf{R}^3)$ may be replaced by $H^1(\Gamma_0)$ in the definition (2.40) of $I_{per}(\mu)$, because of the symmetry of the point nucleus. However, without using this latter fact (we recall that this symmetry will not necessarily hold when we deal with smeared nuclei; see Section 2.7 below), we can directly prove (2.41) with $I_{\mathrm{per}}(\mu)$ defined by (2.40). Indeed, it is possible to show that $\tilde{\rho}$ is necessarily periodic (see Section 2.7).

2.4 Upper bound for the energy

We are now going to address the problem of finding an upper bound on $\frac{1}{|\Lambda|} I_\Lambda$. In fact, we are going to prove that this limit is less than the periodic energy $I_{\mathrm{per}}(\mu)$ defined in (2.4)–(2.8). Our strategy of proof is to build a convenient sequence of functions whose energy approaches the ground state energy when Λ is fixed, and that converges to the periodic minimizing density when Λ goes to infinity. Let us emphasize that, in this part of our argument, we extensively use the fact that the sequence Λ satisfies the Van Hove condition given in Definition 1.

We begin this section with Proposition 2.20, which states that, in the case of some unit charge per cell, the upper bound may be obtained with the help of the periodic problem. For the sake of simplicity, we have chosen to state such a result in the case of the unit charge. However, the result is more general and actually we shall not directly use Proposition 2.20 but, rather, Corollary 2.21 below and the main result of this section which is contained in Corollary 2.26 below.

Proposition 2.20 *For any Van Hove sequence* (Λ), *we have*

$$\limsup_{\Lambda \to \infty} \frac{1}{|\Lambda|} I_\Lambda \leq I_{\text{per}}(1),$$

where we recall that $I_{\text{per}}(1)$ *is defined by (2.4)-(2.6).*

Corollary 2.21 *Let* $\lambda > 0$. *We consider the problems*

$$I_\Lambda(\lambda|\Lambda|) = \inf \left\{ E_\Lambda(\rho) + \tfrac{1}{2} \sum_{y \neq z \in \Lambda} V(y - z); \right.$$

$$\left. \rho \geq 0, \sqrt{\rho} \in H^1(\mathbf{R}^3), \int_{\mathbf{R}^3} \rho = \lambda|\Lambda| \right\},$$

and

$$I_{\text{per}}(\lambda) = \inf \left\{ E_{\text{per}}(\rho); \rho \geq 0, \sqrt{\rho} \in H^1(\Gamma_0), \int_{\Gamma_0} \rho = \lambda \right\}. \qquad (2.42)$$

Then, we have

$$\limsup_{\Lambda \to \infty} \frac{1}{|\Lambda|} I_\Lambda(\lambda|\Lambda|) \leq I_{per}(\lambda).$$

Remark 2.22 It has been shown that the periodic problem $I_{\text{per}}(\lambda)$ also reads

$$I_{\text{per}}(\lambda) = \inf \left\{ E_{\text{per}}(\rho); \rho \geq 0, \sqrt{\rho} \in H^1_{\text{per}}(\mathbf{R}^3), \int_{\Gamma_0} \rho = \lambda \right\}, \qquad (2.43)$$

by symmetry considerations.

The proof of this corollary is in fact exactly the same as the proof of Proposition 2.20. Therefore we shall skip it and, for the sake of simplicity, rather give the proof of Proposition 2.20.

Before turning to the proof of Proposition 2.20, we summarize in the following proposition the properties of the family of minimization problems we have introduced above in (2.42), and that contains the particular problem $I_{\text{per}}(\mu)$ defined in Section 2.1 by (2.4)-(2.6).

Proposition 2.23 *Consider for* $\lambda \geq 0$ *the above problems (2.42). Then we have the following:*

(i) For all $\lambda \geq 0$, $I_{\text{per}}(\lambda)$ *admits a unique minimum, which we denote by* ρ_λ.

(ii) This minimum shares the same symmetries as the unit cube Γ_0, *and satisfies* $\sqrt{\rho_\lambda} \in H^1_{\text{per}}(\Gamma_0)$, *the subspace of* $H^1(\Gamma_0)$ *consisting of periodic functions.*

(iii) The function $\lambda \to I_{\text{per}}(\lambda)$ *is strictly convex with respect to* λ, *and it satisfies*

$$\begin{cases} I_{\text{per}}(\lambda) < \tfrac{1}{2} \sum_{y \neq 0 \in \mathbf{Z}^3} V(y), & \text{for } \lambda > 0 \text{ small enough,} \\ I_{\text{per}}(1) < \tfrac{1}{2} \sum_{y \neq 0 \in \mathbf{Z}^3} V(y), & \text{for a small enough,} \\ I_{\text{per}}(\lambda) \to +\infty & \text{as } \lambda \to +\infty. \end{cases}$$

(iv) There exists a unique function $\bar{\rho}$ that minimizes $E_{\mathrm{per}}(\rho)$ on the set $\{\rho \geq 0, \sqrt{\rho} \in H^1(\Gamma_0)\}$. We denote by $\mu_0 = \int_{\Gamma_0} \bar{\rho}$. We have $\mu_0 > 0$ and $I_{\mathrm{per}}(\mu_0) < \frac{1}{2} \sum_{y \neq 0 \in \mathbf{Z}^3} V(y)$.
(v) ρ_λ belongs to $L^\infty(\Gamma_0)$.

Proof of Proposition 2.23 (i) It is straightforward to see that, for any $\lambda \geq 0$, $I_{\mathrm{per}}(\lambda)$ has a unique minimum. Recall that the problem is set on the bounded set Γ_0, and compactness follows from Rellich's theorem and Sobolev's theorem.

Since $E_{\mathrm{per}}(\rho)$ is a strictly convex function of ρ, it follows that the minimizer ρ_λ is unique.

(ii) Since ρ_λ is unique, we know that it shares the same symmetries as the unit cube, and also that $\sqrt{\rho_\lambda} \in H^1_{\mathrm{per}}(\Gamma_0)$.

(iii) From (i) and the strict convexity of the energy, we easily deduce that $I_{\mathrm{per}}(\lambda)$ is strictly convex with respect to λ.

Taking $\rho = \lambda . 1_{\Gamma_0}$, we now remark that

$$E_{\mathrm{per}}(\rho) = \lambda^{5/3} - \lambda \int_{\Gamma_0} V_\infty + \tfrac{1}{2}\lambda^2 \int_{\Gamma_0} V_\infty + \tfrac{1}{2} \sum_{y \neq 0 \in \mathbf{Z}^3} V(y),$$

and thus

$$I_{\mathrm{per}}(\lambda) < \tfrac{1}{2} \sum_{y \neq 0 \in \mathbf{Z}^3} V(y),$$

if λ is small enough.

In addition, taking $\lambda = 1$ we see that

$$E_{\mathrm{per}}(\rho) = 1 - \tfrac{1}{2} \int_{\Gamma_0} V_\infty + \tfrac{1}{2} \sum_{y \neq 0 \in \mathbf{Z}^3} V(y)$$

$$< \tfrac{1}{2} \sum_{y \neq 0 \in \mathbf{Z}^3} V(y),$$

for a small enough, since $\int_{\Gamma_0} V_\infty = \frac{4\pi}{a^2}$. Besides, for any ρ, we have

$$E_{\mathrm{per}}(\rho) \geq \int_{\Gamma_0} \rho^{5/3} - \|V_\infty\|_{L^{5/2}(\Gamma_0)} \left(\int_{\Gamma_0} \rho^{5/3} \right)^{3/5}.$$

Since $t \longrightarrow t - Ct^{3/5}$ is an increasing function for t large enough, and since $\int_{\Gamma_0} \rho^{5/3} \geq \left(\int_{\Gamma_0} \rho \right)^{5/3}$, the right-hand side is bounded from below, for $\int_{\Gamma_0} \rho$ large enough, by

$$\left(\int_{\Gamma_0} \rho \right)^{5/3} - \|V_\infty\|_{L^{5/2}(\Gamma_0)} \left(\int_{\Gamma_0} \rho \right).$$

Therefore, for λ large enough,

$$I_{\mathrm{per}}(\lambda) \geq \lambda^{5/3} - \|V_\infty\|_{L^{5/2}(\Gamma_0)} \lambda,$$

and thus $I_{\mathrm{per}}(\lambda)$ goes to infinity as $\lambda \to +\infty$.

(iv) From (i) and (iii), we easily deduce the existence of a unique absolute minimizing density $\bar{\rho}$, satisfying $\int_{\Gamma_0} \bar{\rho} > 0$ and

$$E_{\text{per}}(\bar{\rho}) = \inf\{E_{\text{per}}(\rho); \rho \geq 0, \sqrt{\rho} \in H^1(\Gamma_0)\} < \frac{1}{2} \sum_{y \neq 0 \in \mathbf{Z}^3} V(y).$$

(v) The Euler–Lagrange equation satisfied by $u_\lambda = \sqrt{\rho_\lambda}$ reads

$$-\Delta u_\lambda - V_\infty u_\lambda + \tfrac{5}{3} u_\lambda^{7/3} + \left(\int_{\Gamma_0} V_\infty(x - y)\rho_\lambda(y)dy \right) u_\lambda + \theta_\lambda u_\lambda = 0$$

on Γ_0, where θ_λ is the Lagrange multiplier associated with $I_{\text{per}}(\lambda)$, together with the periodic boundary condition on $\partial\Gamma_0$. Therefore, it may be viewed as a partial differential equation on $\mathbf{R}^3$, provided that we extend u_λ by periodicity on the whole space. Then, it just remains to copy the proof of Proposition 2.5(iii) to obtain the L^∞ bound on u_λ (if $\theta_\lambda < 0$, replace V_∞ by $V_\infty - \theta_\lambda$ in that argument). $\diamondsuit$

Proof of Proposition 2.20   When Λ is fixed, we are going to build a test function that will approach the ground-state energy at Λ fixed in a convenient way, in order to obtain an upper bound on this energy.

Let $h > \frac{1}{2}$ be fixed independently of Λ; for instance, $h = 1$. We denote by $\overset{\circ}{\Lambda}$ the set of points in Λ that are at least at a distance h from $\partial\Gamma(\Lambda)$.

Next, we denote $\Gamma(\overset{\circ}{\Lambda}) = \bigcup_{y \in \overset{\circ}{\Lambda}} \Gamma_y$.

On $\Gamma(\overset{\circ}{\Lambda})$, we define the function $\bar{\rho}_\Lambda$ by

$$\bar{\rho}_\Lambda(x) = \rho_{\text{per}}(x - [x]),$$

where ρ_{per} is the minimum of the problem $I_{\text{per}}(1)$, and where $[x]$ denotes, as above, the point $y \in \Lambda$ for which $x \in \Gamma_y$.

We have

$$\int_{\Gamma(\overset{\circ}{\Lambda})} \bar{\rho}_\Lambda = |\Gamma(\overset{\circ}{\Lambda})|.$$

Next, we extend the function $\bar{\rho}_\Lambda$ over $\mathbf{R}^3$ in such a way that it satisfies the following properties (see also Figure 2.1 below):

$$\begin{cases} \int_{\mathbf{R}^3} \bar{\rho}_\Lambda = |\Lambda|, \\[1mm] \int_{\Gamma_y} \bar{\rho}_\Lambda = 1, \quad \text{for all } y \in \partial\Lambda \\[1mm] \sqrt{\bar{\rho}_\Lambda} \in H^1(\mathbf{R}^3), \\[1mm] \|\bar{\rho}_\Lambda\|_{L^\infty} \leq C \quad \text{independent of } \Lambda, \\[1mm] \bar{\rho}_\Lambda = 0 \quad \text{i.e. outside } \Gamma(\Lambda), \\[1mm] \int_{\Gamma(\partial\Lambda)} |\nabla\sqrt{\bar{\rho}_\Lambda}|^2 \leq C|\Lambda^h|, \quad C \text{ independent of } \Lambda, \end{cases} \qquad (2.44)$$

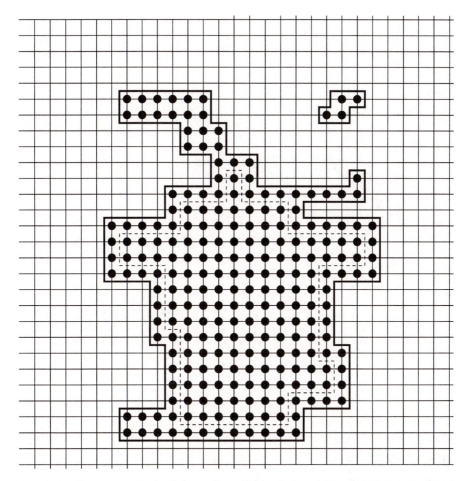

FIG. 2.1. **Construction of $\bar{\rho}_\Lambda$ in two dimensions.** We choose $h = 1$. Inside
the dashed line, $\bar{\rho}_\Lambda$ is the periodic density ρ_{per}. The solid curve represents
the boundary $\partial\Gamma(\Lambda)$. Outside this curve, $\bar{\rho}_\Lambda \equiv 0$.

where $\partial\Lambda = \Lambda \backslash \overset{\circ}{\Lambda}$.

We now claim that

$$\frac{1}{|\Lambda|}\left(E_\Lambda(\bar{\rho}_\Lambda) - E_{\overset{\circ}{\Lambda}}(\bar{\rho}_\Lambda)\right) \longrightarrow 0, \tag{2.45}$$

and

$$\frac{1}{|\Lambda|}E_{\overset{\circ}{\Lambda}}(\bar{\rho}_\Lambda) \longrightarrow I_{\text{per}}(1), \tag{2.46}$$

as Λ goes to infinity.

Both assertions are a consequence of the fact that, for a sequence satisfying
the criteria of Definition 1, what happens on the 'boundary' of $\Gamma(\Lambda)$ can be

neglected in front of what happens 'inside' $\Gamma(\Lambda)$. This property is contained in the following technical lemma (that will also be useful in the Coulomb case; see Chapter 3).

Lemma 2.24 *Let Λ be a sequence going to infinity as prescribed in Definition 1. Let $h > 0$ and denote $\partial\Lambda = \Lambda\backslash\overset{\circ}{\Lambda}$, where $\overset{\circ}{\Lambda}$ is the set of points of Λ that are at least at a distance h from $\partial\Gamma(\Lambda)$. Let f be a continuous (for example) function on $\mathbf{R}^3\backslash\mathbf{Z}^3$.*

When Λ goes to infinity, the following convergences hold, uniformly on compact subsets of $\mathbf{R}^3\backslash\mathbf{Z}^3$ and in $L^1_{\mathrm{loc}}(\mathbf{R}^3)$:

(i) $\dfrac{1}{|\Lambda|}\displaystyle\sum_{y\in\Lambda}\sum_{z\in\Lambda} f(x+y-z) \longrightarrow \sum_{k\in\mathbf{Z}^3} f(x+k),$

$$\text{as soon as } f = O\left(\tfrac{1}{|x|^4}\right) \text{ as } |x| \longrightarrow \infty;$$

(ii) $\dfrac{1}{|\Lambda|}\displaystyle\sum_{y\in\Lambda}\sum_{z\in\partial\Lambda} f(x+y-z) \longrightarrow 0,$

$$\text{as soon as } f = O\left(\tfrac{1}{|x|^4}\right) \text{ as } |x| \longrightarrow \infty.$$

Assume in addition that $\frac{|\Lambda^h|}{|\Lambda|}\mathrm{Log}(|\Lambda^h|)$ goes to 0 as Λ goes to infinity. Then:

(iii) $\dfrac{1}{|\Lambda|}\displaystyle\sum_{y\in\partial\Lambda}\sum_{z\in\partial\Lambda} f(x+y-z) \longrightarrow 0,$

$$\text{as soon as } f = O\left(\tfrac{1}{|x|^2}\right) \text{ as } |x| \longrightarrow \infty.$$

Remark 2.25 (1) In (i) and (ii) (respectively in (iii)), the condition $f = O\left(\tfrac{1}{|x|^p}\right)$ with $p > 3$ (respectively $f = O\left(\tfrac{1}{|x|^p}\right)$ with $p > 1$) as $|x|$ goes to infinity is enough to conclude, but the proof is somewhat more technical.

(2) In (ii), one can also consider $\frac{1}{|\overset{\circ}{\Lambda}|}\sum_{y\in\overset{\circ}{\Lambda}}\sum_{z\in\partial\Lambda}$. In addition, in (i), one can consider, for instance, $\frac{1}{|\overset{\circ}{\Lambda}|}\sum_{y\in\overset{\circ}{\Lambda}}\sum_{z\in\Lambda}$. Finally, Λ may be replaced by $\overset{\circ}{\Lambda}$.

(3) The additional hypothesis made on $|\Lambda^h|$ for (iii) might not be optimal. Nevertheless, the additional assumption is satisfied by a wide range of sequences Λ: balls, cubes, and so on.

(4) Of course, when $f = O\left(\tfrac{1}{|x|^4}\right)$ as $|x|$ goes to infinity, (iii) is a straightforward consequence of (ii), and one does not need the additional assumption on $|\Lambda^h|$.

Proof of Lemma 2.24 For (i)–(iii), since we are only dealing with absolutely convergent series, it is of course enough to prove the result for $f \geq 0$. We shall thus assume that this is the case.

(i) We may write

$$\frac{1}{|\Lambda|} \sum_{y \in \Lambda} \sum_{z \in \Lambda} f(x + y - z) = \sum_{k \in \Lambda - \Lambda} a(k, \Lambda) f(x + k),$$

with

$$|\Lambda| a(k, \Lambda) = \sharp \{ y \in \Lambda / y - k \in \Lambda \}$$
$$= \sharp (\Lambda \cap (k + \Lambda)),$$

where we denote by $\sharp B$ the number of elements of the set B.

Since $\Lambda \cap (k + \Lambda) \subset \Lambda$, we have

$$a(k, \Lambda) \le 1. \tag{2.47}$$

Besides,

$$\sharp (\Lambda \cap (k + \Lambda)) = vol(\Gamma(\Lambda) \cap (\Gamma(\Lambda) + k))$$
$$= vol \ \Gamma(\Lambda) - vol(\Gamma(\Lambda) \backslash \Gamma(\Lambda) + k)$$
$$\ge vol \ \Gamma(\Lambda) - vol \ \Lambda^{|k|}$$
$$= |\Lambda| - |\Lambda^{|k|}|,$$

and thus

$$a(k, \Lambda) \ge 1 - \frac{|\Lambda^{|k|}|}{|\Lambda|}. \tag{2.48}$$

Then, (2.47) and (2.48) together show that

$$a(k, \Lambda) \longrightarrow 1,$$

when k is fixed and Λ goes to infinity, the convergence being uniform in k if k belongs to a fixed bounded set.

We denote $S_\Lambda(x) = \frac{1}{|\Lambda|} \sum_{y \in \Lambda} \sum_{z \in \Lambda} f(x + y - z)$ and $S(x) = \sum_{k \in \mathbf{Z}^3} f(x + k)$.

The assumption made on f ensures that the series defining S is uniformly convergent on the compact subsets of $\mathbf{R}^3 \backslash \mathbf{Z}^3$.

Then, for every x in $\mathbf{R}^3$, we have

$$0 \le S(x) - S_\Lambda(x)$$
$$\le \sum_{k \in \Lambda - \Lambda} (1 - a(k, \Lambda)) f(x + k) + \sum_{k \in \mathbf{Z}^3 \backslash (\Lambda - \Lambda)} f(x + k).$$

We first choose a finite subset Λ_1 of $\mathbf{Z}^3$ in order to have, for all Λ containing Λ_1,

$$0 \le S(x) - S_\Lambda(x)$$

$$\le \sum_{k \in \Lambda_1 - \Lambda_1} (1 - a(k, \Lambda)) f(x + k) + \sum_{k \in \mathbf{Z}^3 \setminus (\Lambda_1 - \Lambda_1)} f(x + k)$$

$$\le \sup_{k \in \Lambda_1} \left(\frac{|\Lambda^{|k|}|}{|\Lambda|} \right) S(x) + \sum_{k \in \mathbf{Z}^3 \setminus (\Lambda_1 - \Lambda_1)} f(x + k).$$

The upper bound converges to 0 letting $|\Lambda|$, and then $|\Lambda_1|$, go to infinity, and using the definition of a Van Hove sequence. The convergence is clearly uniform on the compact subsets of $\mathbf{R}^3 \setminus \mathbf{Z}^3$. The proof of (i) is then complete.

(ii) The proof of (ii) mimics the proof of (i). We write

$$\frac{1}{|\Lambda|} \sum_{y \in \Lambda} \sum_{z \in \partial \Lambda} f(x + y - z) = \sum_{k \in \Lambda - \partial \Lambda} b(k, \Lambda) f(x + k),$$

where

$$|\Lambda| b(k, \Lambda) = \sharp(\partial \Lambda \cap (\Lambda + k)) \le \sharp \partial \Lambda \le |\Lambda^h|,$$

and thus $b(k, \Lambda)$ goes to 0 independently of k.

Next,

$$\frac{1}{|\Lambda|} \sum_{y \in \Lambda} \sum_{z \in \partial \Lambda} f(x + y - z) = \sum_{k \in \Lambda - \partial \Lambda} b(k, \Lambda) f(x + k)$$

$$\le \frac{|\Lambda^h|}{|\Lambda|} \sum_{k \in \mathbf{Z}^3} f(x + k).$$

We conclude once more letting Λ go to infinity.

(iii) As above, we write

$$\frac{1}{|\Lambda|} \sum_{y \in \partial \Lambda} \sum_{z \in \partial \Lambda} f(x + y - z) = \sum_{k \in \partial \Lambda - \partial \Lambda} c(k, \Lambda) f(x + k),$$

with

$$|\Lambda| c(k, \Lambda) = \sharp(\partial \Lambda \cap (\partial \Lambda + k)) \le |\partial \Lambda|.$$

Since $\partial \Lambda$ is essentially two-dimensional, the major contribution to the above sum comes from a subset A_Λ of $\partial \Lambda - \partial \Lambda$ which has $O(|\partial \Lambda|)$ elements, and which is in a one-to-one correspondence with a subset of $\mathbf{Z}^2$, of volume $O(|\partial \Lambda|)$.

Therefore,

$$\sum_{k \in \partial \Lambda - \partial \Lambda} c(k, \Lambda) f(x + k) \le C \frac{|\Lambda^h|}{|\Lambda|} \sum_{k \in \partial \Lambda - \partial \Lambda} \frac{1}{|x + k|^2}$$

$$\leq C\frac{|\Lambda^h|}{|\Lambda|}\sum_{k\in A_\Lambda}\frac{1}{|k|^2}.$$

Next, we bound $\displaystyle\sum_{k\in A_\Lambda}\frac{1}{|k|^2}$ as follows.

Since A_Λ is in a one-to-one correspondence with a subset of $\mathbf{Z}^2$, namely $\tilde{A}_\Lambda$, we may associate to any $k\in A_\Lambda$ a point j of $\mathbf{Z}^2$ of integer coordinates. In addition, we may bound $\frac{1}{|k|^2_{\mathbf{R}^3}}$ from above by $\frac{C}{|j|^2_{\mathbf{R}^2}}$, where C is a fixed constant that does not depend either on k or on Λ, and where $|\cdot|_{\mathbf{R}^n}$ denotes the Euclidean norm in $\mathbf{R}^n$, $n=2,3$.

Next,

$$\sum_{j\in\tilde{A}_\Lambda}\frac{1}{|j|^2}\leq\sum_{j\in B_\Lambda}\frac{1}{|j|^2},$$

where B_Λ denotes the ball centered at 0 in $\mathbf{R}^2$ that has the same volume as $\tilde{A}_\Lambda$ (The 'worst' situation one may have to face is the one where all points of $\tilde{A}_\Lambda$ are as close as possible to 0).

Finally,

$$\sum_{j\in B_\Lambda}\frac{1}{|j|^2}=O\big(Log|\partial\Lambda|\big).$$

Therefore,

$$\sum_{k\in A_\Lambda}\frac{1}{|k|^2_{\mathbf{R}^3}}\leq C\sum_{k\in B_\Lambda}\frac{1}{|k|^2_{\mathbf{R}^2}}=O(Log(|\partial\Lambda|)),$$

where B_Λ is the ball of $\mathbf{R}^2$ centered at 0, of volume $|\partial\Lambda|$. This is how we obtain an upper bound of the form

$$O\left(\frac{|\Lambda^h|}{|\Lambda|}Log(|\partial\Lambda|)\right),$$

that goes to 0 as Λ goes to infinity. $\diamond$

We now come back to the proof of Proposition 2.20.

In view of the 'periodicity' of $\bar{\rho}_\Lambda$ over $\Gamma(\overset{\circ}{\Lambda})$, we have

$$E_{\overset{\circ}{\Lambda}}(\bar{\rho}_\Lambda)=|\overset{\circ}{\Lambda}|\left(\int_{\Gamma_0}|\nabla\sqrt{\rho_{\text{per}}}|^2+\int_{\Gamma_0}\rho_{\text{per}}^{5/3}\right)$$

$$+\sum_{y\in\partial\Lambda}\left(\int_{\Gamma_y}|\nabla\sqrt{\bar{\rho}_\Lambda}|^2+\int_{\Gamma_y}\bar{\rho}_\Lambda^{5/3}\right)$$

$$-\int_{\Gamma_0}\sum_{y\in\overset{\circ}{\Lambda}}\sum_{z\in\overset{\circ}{\Lambda}}V(x+y-z)\rho_{\text{per}}(x)\,dx$$

$$-\int_{\Gamma_0}\sum_{y\in\overset{\circ}{\Lambda}}\sum_{z\in\partial\Lambda}V(x+y-z)\bar{\rho}_\Lambda(x+z)\,dx$$

$$+ \frac{1}{2} \iint_{\Gamma_0 \times \Gamma_0} \rho_{\mathrm{per}}(x) \rho_{\mathrm{per}}(y) \left[\sum_{z \in \overset{\circ}{\Lambda}} \sum_{t \in \overset{\circ}{\Lambda}} V(x - y + z - t) \right] dx\, dy$$

$$+ \iint_{\Gamma_0 \times \Gamma_0} \sum_{z \in \overset{\circ}{\Lambda}} \sum_{t \in \partial\Lambda} V(x - y + z - t) \bar\rho_\Lambda(x + z) \bar\rho_\Lambda(y + t) dx\, dy$$

$$+ \frac{1}{2} \iint_{\Gamma_0 \times \Gamma_0} \sum_{z \in \partial\Lambda} \sum_{t \in \partial\Lambda} V(x - y + z - t) \bar\rho_\Lambda(x + z) \bar\rho_\Lambda(y + t)\, dx\, dy$$

$$+ \frac{1}{2} U_{\overset{\circ}{\Lambda}}. \tag{2.49}$$

From assertion (i) of Lemma 2.24 (and Remark 2.25), and from (2.44), we deduce

$$\frac{1}{|\Lambda|} \int_{\Gamma_0} \sum_{y \in \overset{\circ}{\Lambda}} \sum_{z \in \overset{\circ}{\Lambda}} V(x + y - z) \rho_{\mathrm{per}}(x)\, dx \longrightarrow \int_{\Gamma_0} V_\infty \rho_{\mathrm{per}},$$

$$- \frac{1}{|\Lambda|} \int_{\Gamma_0} \sum_{y \in \overset{\circ}{\Lambda}} \sum_{z \in \partial\Lambda} V(x + y - z) \bar\rho_\Lambda(x + z)\, dx \longrightarrow 0,$$

$$\frac{1}{2} \frac{1}{|\Lambda|} \iint_{\Gamma_0 \times \Gamma_0} \rho_{\mathrm{per}}(x) \rho_{\mathrm{per}}(y) \left[\sum_{z \in \overset{\circ}{\Lambda}} \sum_{t \in \overset{\circ}{\Lambda}} V(x - y + z - t) \right] dx\, dy$$

$$\longrightarrow \frac{1}{2} \iint_{\Gamma_0 \times \Gamma_0} \rho_{\mathrm{per}}(x) \rho_{\mathrm{per}}(y) V_\infty(x - y) dx\, dy,$$

$$\frac{1}{|\Lambda|} \iint_{\Gamma_0 \times \Gamma_0} \sum_{z \in \overset{\circ}{\Lambda}} \sum_{t \in \partial\Lambda} V(x - y + z - t) \bar\rho_\Lambda(x + z) \bar\rho_\Lambda(y + t) dx\, dy$$

$$+ \frac{1}{2} \frac{1}{|\Lambda|} \iint_{\Gamma_0 \times \Gamma_0} \sum_{z \in \partial\Lambda} \sum_{t \in \partial\Lambda} V(x - y + z - t) \bar\rho_\Lambda(x + z) \bar\rho_\Lambda(y + t)\, dx\, dy$$

$$\longrightarrow 0,$$

while

$$\frac{1}{2} \frac{1}{|\Lambda|} U_{\overset{\circ}{\Lambda}} \longrightarrow \frac{1}{2} \sum_{k \neq 0 \in \mathbf{Z}^3} V(k).$$

In addition, the properties (2.44) alone allow us to show that

$$\frac{1}{|\Lambda|} \sum_{y \in \partial\Lambda} \left(\int_{\Gamma_y} |\nabla \sqrt{\bar\rho_\Lambda}|^2 + \int_{\Gamma_y} \bar\rho_\Lambda^{5/3} \right) \longrightarrow 0.$$

Therefore, we have

$$E_{\overset{\circ}{\Lambda}}(\bar\rho_\Lambda) = |\overset{\circ}{\Lambda}| \left(\int_{\Gamma_0} |\nabla \sqrt{\rho_{\mathrm{per}}}|^2 + \int_{\Gamma_0} \rho_{\mathrm{per}}^{5/3} \right)$$

$$- \int_{\Gamma_0} \sum_{y \in \overset{\circ}{\Lambda}} \sum_{z \in \overset{\circ}{\Lambda}} V(x + y - z) \rho_{\mathrm{per}}(x) \mathrm{d}x + \tfrac{1}{2} U_{\overset{\circ}{\Lambda}}$$

$$+ \tfrac{1}{2} \iint_{\Gamma_0 \times \Gamma_0} \rho_{\mathrm{per}}(x) \rho_{\mathrm{per}}(y) \left[\sum_{z \in \overset{\circ}{\Lambda}} \sum_{t \in \overset{\circ}{\Lambda}} V(x - y + z - t) \right] \mathrm{d}x \, \mathrm{d}y + o(|\Lambda|)$$

$$= |\Lambda| E_{\mathrm{per}}(\rho_{\mathrm{per}}) + o(|\Lambda|).$$

Of course, (2.45) follows from the same arguments, which show that

$$\frac{1}{|\Lambda|} \int_{\mathbf{R}^3} (V_\Lambda - V_{\overset{\circ}{\Lambda}}) \bar{\rho}_\Lambda \longrightarrow 0.$$

Now that we have established (2.45) and (2.46), Proposition 2.20 follows in a straightforward way:

$$\limsup_{\Lambda \to \infty} \frac{1}{|\Lambda|} I_\Lambda \leq \limsup_{\Lambda \to \infty} \frac{1}{|\Lambda|} E_\Lambda(\bar{\rho}_\Lambda) = I_{\mathrm{per}}(1),$$

since $\int_{\mathbf{R}^3} \bar{\rho}_\Lambda = |\Lambda|$.   ◇

As a conclusion of this section, we are able to prove the following:

Corollary 2.26 *(i) With the notation of Corollary 2.18, we have*

$$\lim_{\Lambda \to \infty} \frac{I_\Lambda}{|\Lambda|} = I_{\mathrm{per}}(\mu). \tag{2.50}$$

(ii) In addition, $\sqrt{\bar{\rho}_\Lambda}$ converges to $\sqrt{\tilde{\rho}}$ strongly in $L^p(\Gamma_0)$, for all $1 \leq p < \infty$, $\nabla \sqrt{\bar{\rho}_\Lambda}$ converges to $\nabla \sqrt{\tilde{\rho}}$ strongly in $L^2(\Gamma_0)^3$, and $\tilde{\rho}$ is the unique minimizer ρ_{per} of $I_{\mathrm{per}}(\mu)$.

(iii) Moreover, the sequence of Lagrange multipliers θ_Λ converges to θ_{per}, the Lagrange multiplier associated with $I_{\mathrm{per}}(\mu)$.

Remark 2.27 Observe that we have not yet proved that μ and $\tilde{\rho}$ do not depend on the sequence Λ.

Proof of Corollary 2.26 (i) Let ρ_{per} be the unique minimizer of $I_{\mathrm{per}}(\mu)$. We may apply, step by step, the argument of Proposition 2.20 to ensure that

$$\limsup_{\Lambda \to \infty} \frac{I_\Lambda(\mu|\Lambda|)}{|\Lambda|} \leq I_{\mathrm{per}}(\mu).$$

Besides, as the function $I_\Lambda(\lambda)$ is non-increasing with respect to λ, we know that

$$\frac{I_\Lambda}{|\Lambda|} = \frac{I_\Lambda(|\Lambda|)}{|\Lambda|} \leq \frac{I_\Lambda(\mu|\Lambda|)}{|\Lambda|}.$$

Therefore, we deduce, letting Λ go to infinity, that

$$I_{\mathrm{per}}(\mu) \leq \liminf_{\Lambda \to \infty} \frac{I_\Lambda}{|\Lambda|} \leq \liminf_{\Lambda \to \infty} \frac{I_\Lambda(\mu|\Lambda|)}{|\Lambda|}$$

and (2.50) follows easily.

(ii) is a direct consequence of (i), which yields

$$\lim_{\Lambda \to \infty} \frac{1}{|\Lambda|} \int_{\mathbf{R}^3} |\nabla u_\Lambda|^2 = \int_{\Gamma_0} |\nabla \sqrt{\rho_{\mathrm{per}}}|^2, \qquad (2.51)$$

$$\lim_{\Lambda \to \infty} \frac{1}{|\Lambda|} \int_{\mathbf{R}^3} \rho_\Lambda^p = \int_{\Gamma_0} \rho_{\mathrm{per}}^p, \qquad \text{for all } 1 < p < \infty,$$

$$\lim_{\Lambda \to \infty} \frac{1}{|\Lambda|} D(\rho_\Lambda, \rho_\Lambda) = D_{\Gamma_0}(\rho_{\mathrm{per}}, \rho_{\mathrm{per}}). \qquad (2.52)$$

To prove (iii), we first multiply the Euler–Lagrange equation (2.16) by u_Λ and integrate over $\mathbf{R}^3$, in order to obtain

$$\frac{1}{|\Lambda|} \int_{\mathbf{R}^3} |\nabla u_\Lambda|^2 + \frac{5}{3} \frac{1}{|\Lambda|} \int_{\mathbf{R}^3} \rho_\Lambda^{5/3} - \frac{1}{|\Lambda|} \int_{\mathbf{R}^3} V_\Lambda \rho_\Lambda + \frac{1}{|\Lambda|} D(\rho_\Lambda, \rho_\Lambda) + \theta_\Lambda = 0. \quad (2.53)$$

As θ_Λ is bounded, we may assume that it converges, up to a subsequence, to some $\theta_\infty \geq 0$.

Next, we let Λ go to infinity in (2.53) and we use (2.51) and (2.52), together with (2.31), to deduce that

$$\int_{\Gamma_0} |\nabla \sqrt{\rho_{\mathrm{per}}}|^2 + \frac{5}{3} \int_{\Gamma_0} \rho_{\mathrm{per}}^{5/3} - \int_{\Gamma_0} V_\infty \rho_{\mathrm{per}} + D_{\Gamma_0}(\rho_{\mathrm{per}}, \rho_{\mathrm{per}}) + \theta_\infty = 0. \quad (2.54)$$

On the other hand, the Euler–Lagrange equation satisfied by $u_{\mathrm{per}} = \sqrt{\rho_{\mathrm{per}}}$ is

$$-\Delta u_{\mathrm{per}} + \frac{5}{3} u_{\mathrm{per}}^{7/3} - V_\infty u_{\mathrm{per}} + \left(\int_{\Gamma_0} V_\infty(x - y) u_{\mathrm{per}}(y) dy \right) u_{\mathrm{per}}$$
$$+ \theta_{\mathrm{per}} u_{\mathrm{per}} = 0. \qquad (2.55)$$

If we integrate (2.55) against u_{per} on Γ_0, we have

$$\int_{\Gamma_0} |\nabla \sqrt{\rho_{\mathrm{per}}}|^2 + \frac{5}{3} \int_{\Gamma_0} \rho_{\mathrm{per}}^{5/3} - \int_{\Gamma_0} V_\infty \rho_{\mathrm{per}} + D_{\Gamma_0}(\rho_{\mathrm{per}}, \rho_{\mathrm{per}})$$
$$+ \theta_{\mathrm{per}} \mu = 0. \qquad (2.56)$$

Now comparing (2.54) and (2.56), we obtain

$$\theta_\infty = \mu \, \theta_{\mathrm{per}}.$$

If $\mu = 1$, it is clear that $\theta_\infty = \theta_{\mathrm{per}}$; while $0 < \mu < 1$ implies that θ_{per}, and thus θ_∞, vanish. (Note that $\mu > 0$ by Proposition 2.23(iii).) $\diamond$

2.5 Proof of the main theorem

We conclude in this section the proof of the main theorem given in the introduction.

For this purpose, we regroup in the following proposition results that are direct consequences of the preceding two sections.

Proposition 2.28 *Let Λ be a sequence going to infinity as prescribed in Definition 1. Then, there exists a function μ, which may a priori depend on the sequence Λ, such that:*

(i) For all $c > 0$,

$$\lim_{\Lambda \to \infty} \frac{I_\Lambda(c|\Lambda|)}{|\Lambda|} = I_{\mathrm{per}}(\mu(c)).$$

(ii) $c \longmapsto I_{\mathrm{per}}(\mu(c))$ is a non-increasing function.

(iii) For all $c > 0$,

$$I_{\mathrm{per}}(\mu(c)) \leq I_{\mathrm{per}}(c),$$

the equality being true if and only if $\mu(c) \leq \mu_0$.

(iv) $\mu(c) = \mu_0$, for all $c \geq \mu_0$.

Proof of Proposition 2.28 Fix $c > 0$. Let $\rho_{\Lambda,c}$ be the minimum of the problem $I_\Lambda(c|\Lambda|)$. Let $\mu(c) = \liminf_{\Lambda \to \infty} \int_{\Gamma_0} \tilde{\rho}_{\Lambda,c}$. Clearly, $\mu(c) \leq c$. By the same arguments as those used in the proof of Proposition 2.17, we have

$$I_{\mathrm{per}}(\mu(c)) \leq \liminf_{\Lambda \to \infty} \frac{I_\Lambda(c|\Lambda|)}{|\Lambda|}.$$

Since I_Λ is non-increasing and $\mu(c) \leq c$,

$$\liminf_{\Lambda \to \infty} \frac{I_\Lambda(c|\Lambda|)}{|\Lambda|} \leq \limsup_{\Lambda \to \infty} \frac{I_\Lambda(\mu(c)|\Lambda|)}{|\Lambda|}.$$

Next, by Corollary 2.21,

$$\limsup_{\Lambda \to \infty} \frac{I_\Lambda(\mu(c)|\Lambda|)}{|\Lambda|} \leq I_{\mathrm{per}}(\mu(c)).$$

We thus obtain (i).

(ii) is a straightforward consequence of (i) and of the monotonicity of I_Λ.

(iii) follows from (i) and the proof of Corollary 2.26.

Recall now that μ_0 is the charge of the absolute minimum of I_{per}. Take $c \geq \mu_0$. We have

$$I_{\mathrm{per}}(\mu_0) \leq I_{\mathrm{per}}(\mu(c)) \leq I_{\mathrm{per}}(\mu(\mu_0)),$$

the last inequality holding by (ii).

Now, with (iii),

$$I_{per}(\mu(\mu_0)) \leq I_{per}(\mu_0).$$

Therefore

$$I_{per}(\mu_0) = I_{per}(\mu(c)),$$

and (iv) follows.   ◇

We now turn to the proof of the main theorem.

Assume first that $\mu_0 \leq 1$: by (iv) of Proposition 2.28, we have $\mu(1) = \mu_0$.

On the contrary, assume $1 < \mu_0$. Then, we have $\mu(1) \leq 1 < \mu_0$, and thus $I_{per}(\mu(1)) \geq I_{per}(1)$, because I_{per} is decreasing on $[0, \mu_0]$.

Now, $I_{per}(\mu(1)) \leq I_{per}(1)$ by (iii) of Proposition 2.28, and thus $I_{per}(1) = I_{per}(\mu(1))$. Therefore $\mu(1) = 1$.

In both cases, we obtain

$$\mu(1) = min(\mu_0, 1).$$

It follows that the limit charge does not depend on the sequence Λ. Finally, all the convergences of the main theorem follow from Corollaries 2.4 and 4.5.

Remark 2.29 We deduce from (i)–(iv) in Proposition 2.28 that we have

$$\lim_{\Lambda \to \infty} \frac{I_\Lambda(c|\Lambda|)}{|\Lambda|} = I_{per}(\mu_0) < I_{per}(c),$$

as soon as $c > \mu_0$.

This means that we cannot expect to bind as many electrons as we want to the nuclei. There exists a maximum 'number' of electrons, μ_0, allowed per cell to form a stable crystal. We refer the reader to Section 2.7.3 for further comments on this kind of phenomenon.

In the following section, we address the problem of comparing μ_0 and 1, in order to decide whether or not $\mu(1)$ equals 1.

2.6   Recovering the Coulomb potential: compactness

When the parameter a arising in the Yukawa potential goes to zero, we may expect that the Yukawa periodic problem will converge to the Coulomb periodic problem. We show in this section that this is indeed the case. We shall also study here whether $\mu = 1$ or $\mu < 1$. From the physical viewpoint, we want to know if some charge escapes at infinity while taking the thermodynamic limit for a Yukawa interaction potential.

From now on, we recall the dependence on a by the superscript a, the superscript 0 referring to the Coulomb case.

We begin with the following:

Proposition 2.30 *For all $\lambda \in [0, +\infty)$,*

$$I^a_{per}(\lambda) - \frac{2\pi}{a^2}(\lambda - 1)^2 \longrightarrow I^0_{per}(\lambda) + \frac{M}{2} \qquad as \quad a \to 0^+. \qquad (2.57)$$

In particular,

$$I^a_{per}(1) \longrightarrow I^0_{per}(1) + \frac{M}{2} \qquad as \quad a \to 0^+.$$

Moreover,

$$I^a_{per}(\mu_a) \longrightarrow I^0_{per}(1) + \frac{M}{2} \qquad as \quad a \to 0^+, \qquad (2.58)$$

where $\mu_a \equiv \mu_0(a)$ is the charge of the absolute minimum of $I^a_{per}(\lambda)$, and

$$\mu_a \longrightarrow 1 \qquad as \quad a \to 0^+. \qquad (2.59)$$

Furthermore, if $\rho_a \equiv \rho^a_{per}$ is the corresponding absolute minimizer, ρ_a converges to ρ_0 the (unique) minimizer of $I^0_{per}(1)$, strongly in $L^p(\Gamma_0)$, for all $1 \le p < +\infty$ and $\nabla\sqrt{\rho_a}$ converges to $\nabla\sqrt{\rho_0}$ strongly in $L^2(\Gamma_0)$.

Remark 2.31 The convergence (2.58) which describes exactly how the Yukawa case approximates the Coulomb case will be very useful in the rest of this work. In Chapter 3, a proof of the existence of the thermodynamic limit for the Coulomb energy per unit volume will be based upon (2.58).

Proof of Proposition 2.30 Let

$$V^a_\infty(x) = \sum_{y \in \mathbf{Z}^3} \frac{\exp(-a|x - y|)}{|x - y|}$$

be the periodic Yukawa potential.
    Then,

$$\int_{\Gamma_0} V^a_\infty(x)\,dx = \frac{4\pi}{a^2} = \int_{\mathbf{R}^3} V^a(x)\,dx.$$

We set

$$\widetilde{V^a_\infty} = V^a_\infty - \frac{4\pi}{a^2},$$

so that $\widetilde{V^a_\infty}$ is the unique periodic solution to

$$-\Delta\widetilde{V^a_\infty} + a^2\widetilde{V^a_\infty} = 4\pi \left(\sum_{y \in \mathbf{Z}^3} \delta_y - 1 \right)$$

in $\mathcal{D}'(\mathbf{R}^3)$, and it has zero mean value on Γ_0.

Now recall that the periodic Coulomb potential G is the unique periodic solution to

$$-\triangle G = 4\pi \left(\sum_{y \in \mathbf{Z}^3} \delta_y - 1 \right) \quad \text{in } \mathcal{D}'(\mathbf{R}^3),$$

satisfying

$$\int_{\Gamma_0} G(x)\,\mathrm{d}x = 0.$$

We now prove that

$$\|\widetilde{V_\infty^a} - G\|_{L^\infty(\Gamma_0)} \longrightarrow 0 \qquad \text{as} \quad a \to 0^+. \tag{2.60}$$

Then, the same convergence holds in every $L^p_{\mathrm{loc}}(\mathbf{R}^3)$, for all $1 \le p \le +\infty$.

For this purpose, we rewrite $\widetilde{V_\infty^a}$ and G with the help of their Fourier coefficients as follows:

$$\widetilde{V_\infty^a}(x) = \frac{1}{\pi} \sum_{n \in \mathbf{Z}^3 \setminus \{0\}} \frac{1}{|n|^2 + a^2} \exp(2i\pi(n, x))$$

and

$$G(x) = \frac{1}{\pi} \sum_{n \in \mathbf{Z}^3 \setminus \{0\}} \frac{1}{|n|^2} \exp(2i\pi(n, x)).$$

Then, for $x \neq 0$,

$$
\begin{aligned}
|\widetilde{V_\infty^a}(x) - G(x)| &\le \frac{1}{\pi} \sum_{n \in \mathbf{Z}^3 \setminus \{0\}} \left| \frac{1}{a^2 + |n|^2} - \frac{1}{|n|^2} \right| \\
&\le \frac{1}{\pi} \sum_{n \in \mathbf{Z}^3 \setminus \{0\}} \frac{a^2}{|n|^2 \, (a^2 + |n|^2)} \\
&\le \frac{a^2}{\pi} \sum_{n \in \mathbf{Z}^3 \setminus \{0\}} \frac{1}{|n|^4},
\end{aligned}
\tag{2.61}
$$

which proves (2.60).

In order to prove (2.57), we first rewrite the expression of $E_{\mathrm{per}}^a(\rho)$ in terms of $\widetilde{V_\infty^a}$, for any ρ in $H^1(\Gamma_0)$, with $\int_{\Gamma_0} \rho(x)\,\mathrm{d}x = \lambda$.

Clearly, we have

$$
\begin{aligned}
E_{\mathrm{per}}^a(\rho) = {}& \int_{\Gamma_0} |\nabla\sqrt{\rho}|^2 + \int_{\Gamma_0} \rho^{5/3}(x)\,\mathrm{d}x - \int_{\Gamma_0} \widetilde{V_\infty^a}(x)\rho(x)\,\mathrm{d}x \\
& + \frac{1}{2} \iint_{\Gamma_0 \times \Gamma_0} \rho(x)\widetilde{V_\infty^a}(x - y)\rho(y)\,\mathrm{d}x\,\mathrm{d}y
\end{aligned}
$$

$$+ \tfrac{1}{2}\left[U_\infty^a - \frac{4\pi}{a^2}\right] + \frac{2\pi}{a^2}(\lambda - 1)^2.$$

Thus, (2.57) follows immediately from (2.60). By the way, let us note that, for $\lambda = 0$, (2.57) reads

$$U_\infty^a - \frac{4\pi}{a^2} \longrightarrow M \qquad \text{as } a \to 0^+.$$

Now let ρ_a be the absolute minimizer of $I_{\text{per}}^a(\lambda)$ with charge μ_a. On the one hand,

$$\limsup_{a\to 0^+} I_{\text{per}}^a(\mu_a) \le \limsup_{a\to 0^+} I_{\text{per}}^a(1) = I_{\text{per}}^0(1) + \frac{M}{2}. \tag{2.62}$$

In particular, this ensures that $I_{\text{per}}^a(\mu_a)$ is bounded from above independently of a, as $a \to 0^+$.

On the other hand,

$$\liminf_{a\to 0^+} I_{\text{per}}^a (\mu_a)$$

$$= \liminf_{a\to 0^+} E_{\text{per}}^a(\rho_a)$$

$$= \liminf_{a\to 0^+} \left[\int_{\Gamma_0} |\nabla\sqrt{\rho_a}|^2 + \tfrac{1}{2}\iint_{\Gamma_0\times\Gamma_0} \rho_a(x)\widetilde{V_\infty^a}(x-y)\rho_a(y)\,dx\,dy \right.$$

$$+ \int_{\Gamma_0} \rho_a^{5/3}(x)\,dx - \int_{\Gamma_0} \widetilde{V_\infty^a}(x)\rho_a(x)\,dx$$

$$\left. + \tfrac{1}{2}\left(U_\infty^a - \frac{4\pi}{a^2}\right) + \frac{2\pi}{a^2}(\mu_a - 1)^2 \right] \tag{2.63}$$

$$\ge M + \int_{\Gamma_0} \rho_a^{5/3} - \int_{\Gamma_0} \widetilde{V_\infty^a}(x)\rho_a(x)\,dx, \tag{2.64}$$

since the other terms in (2.63) are non-negative ($\widetilde{V_\infty^a}$ is the kernel of a non-negative, semi-definite operator).

However, it is easily deduced from (2.64) that we have

$$C \ge \int_{\Gamma_0} \rho_a^{5/3} - \int_{\Gamma_0} \widetilde{V_\infty^a}(x)\rho_a(x)\,dx$$

$$\ge \|\rho_a\|_{L^{5/3}(\Gamma_0)}^{5/3} - \|\widetilde{V_\infty^a}\|_{L^{5/2}(\Gamma_0)}\,\|\rho_a\|_{L^{5/3}(\Gamma_0)}.$$

Thus $\|\rho_a\|_{L^{5/3}(\Gamma_0)}$ and $\|\widetilde{V_\infty^a}\|_{L^{5/2}(\Gamma_0)}$ are bounded independently of a as $a \to 0^+$. Then, returning to (2.63), the three positive terms in (2.63) are bounded too.

So, μ_a has to converge to 1 as $a \to 0^+$ (thus yielding (2.59)) and $\sqrt{\rho_a}$ is bounded in $H^1(\Gamma_0)$. Up to a subsequence, we may assume that $\sqrt{\rho_a}$ converges to some function $\sqrt{\bar\rho} \in H^1(\Gamma_0)$, the convergence being weak in $H^1(\Gamma_0)$ and strong in $L^p(\Gamma_0)$, for all $1 \le p < 6$ (by Rellich's theorem).

Hence, we may pass to the weak limit in (2.63) and we obtain

$$\liminf_{a \to 0} I_{\mathrm{per}}^a(\mu_a) \geq E_{\mathrm{per}}^0(\overline{\rho}) + \frac{M}{2} \geq I_{\mathrm{per}}^0(1) + \frac{M}{2}.$$

Then, because of (2.63), $\overline{\rho}$ is simply ρ_0, the unique minimizer of $I_{\mathrm{per}}^0(1)$, the convergence of $\sqrt{\rho_a}$ to $\sqrt{\rho_0}$ is in fact valid for the whole sequence ρ_a (not only for a subsequence) and is a strong convergence in $H^1(\Gamma_0)$.

This concludes the proof of our claim. $\diamond$

Remark 2.32 The proof of Proposition 2.30 provides the additional information that

$$\frac{4\pi \, (\mu_a - 1)}{a^2} \longrightarrow \theta_0 \qquad \text{as} \quad a \to 0^+, \tag{2.65}$$

with θ_0 being the Lagrange multiplier associated with the Euler–Lagrange equation satisfied by ρ_0.

Indeed, we may write down the Euler–Lagrange equation satisfied by $u_a \equiv \sqrt{\rho_a}$ on $\mathbf{R}^3$ as follows:

$$-\Delta u_a + \tfrac{5}{3} u_a^{7/3} - \widetilde{V_\infty^a} \, u_a + \left(\int_{\Gamma_0} \widetilde{V_\infty^a}(x - y)\rho_a(y) \, \mathrm{d}y \right) u_a = 0.$$

We multiply it by u_a, integrate over Γ_0 (recall that u_a satisfies either the Neumann or the periodic boundary condition on $\partial \Gamma_0$), and finally use the equation obtained in this way to rewrite $I_{\mathrm{per}}^a(\mu_a)$ as

$$\begin{aligned}
I_{\mathrm{per}}^a(\mu_a) &= -\tfrac{2}{3} \int_{\Gamma_0} \rho_a^{5/3} - \tfrac{1}{2} \iint_{\Gamma_0 \times \Gamma_0} \rho_a(x) V_\infty^a(x - y)\rho_a(y) \, \mathrm{d}x \, \mathrm{d}y + \tfrac{1}{2} U_\infty^a \\
&= -\tfrac{2}{3} \int_{\Gamma_0} \rho_a^{5/3} - \tfrac{1}{2} \iint_{\Gamma_0 \times \Gamma_0} \rho_a(x) \widetilde{V_\infty^a}(x - y)\rho_a(y) \, \mathrm{d}x \, \mathrm{d}y \\
&\qquad + \tfrac{1}{2} \left(U_\infty^a - \frac{4\pi}{a^2} \right) + \frac{2\,\pi}{a^2} \, (1 - \mu_a^2).
\end{aligned} \tag{2.66}$$

If we perform the same computation with ρ_0, we obtain, in the same way,

$$I_{\mathrm{per}}^0(1) = -\tfrac{2}{3} \int_{\Gamma_0} \rho^{5/3} - \tfrac{1}{2} \iint_{\Gamma_0 \times \Gamma_0} \rho_0(x) G(x - y)\rho_0(y) \, \mathrm{d}x \, \mathrm{d}y - \theta_0 + \frac{M}{2}. \tag{2.67}$$

To conclude, we let $a \to 0^+$ in (2.66) and we compare with (2.67). $\diamond$

If we now look at (2.65), we see that if we had some information on the sign of θ_0, we could infer the position of μ_a with respect to 1 at least for a small enough. Unfortunately, this sign is strongly dependent on the coefficients we set in front of the von Weiszäcker term $\int_{\Gamma_0} |\nabla \sqrt{\rho}|^2$ or the Thomas–Fermi term $\int_{\Gamma_0} \rho^{5/3}$, as we shall see later on.

Therefore, it is not possible to decide whether or not $\mu_a \leq 1$. Then, the non-compactness of ρ_Λ (in the sense that $\frac{1}{|\Lambda|} \int_{\Gamma(\Lambda)^c} \rho_\Lambda(x) \, \mathrm{d}x$ may not go to zero as $\Lambda \to \infty$) is an intrinsic difficulty in the Yukawa case. In the same spirit, we remark that no matter what the sign of $\mu_a - 1$ is, if we choose when Λ is fixed a molecular system composed of $|\Lambda|$ nuclei of unit charge and $c|\Lambda|$ electrons (with $c > \mu_a$), we obtain in the limit only $min(c, \mu_a) = \mu_a$ electrons per cell. Thus some charge has necessarily escaped at infinity in this case.

Let us now turn to the discussion of the position of the absolute charge μ_a with respect to 1.

We illustrate on the four following examples the fact that μ_a can take arbitrary values in $]0, +\infty[$.

Example 1 If $a \to +\infty$, $\mu_a \to 0^+$. Then, for a large enough, $\mu_a < 1$.

We know from Proposition 4.4 that

$$I_{per}^a(\mu_a) < \tfrac{1}{2}U_\infty^a, \tag{2.68}$$

where

$$U_\infty^a = \sum_{y \in \mathbf{Z}^3 \setminus \{0\}} \frac{\exp(-a|y|)}{|y|}.$$

It is easily seen that

$$U_\infty^a \longrightarrow 0 \qquad \text{as } a \to +\infty,$$

since, for example,

$$0 \le U_\infty^a \le \sum_{y \in \mathbf{Z}^3} \exp(-a|y|) \le \exp\left(-\frac{a}{2}\right) \sum_{y \in \mathbf{Z}^3} \exp\left(-\frac{|y|}{2}\right),$$

as soon as $a \ge 1$.

However, we also have

$$I_{per}^a(\mu_a) \ge \|\rho_a\|_{L^{5/3}(\Gamma_0)}^{5/3} - \|V_a\|_{L^{5/2}(\Gamma_0)} \|\rho_a\|_{L^{5/3}(\Gamma_0)}, \tag{2.69}$$

as in Proposition 2.30.

It just remains to notice that Hölder's inequality gives

$$\|V_a\|_{L^{5/2}(\Gamma_0)} \le \|V_a\|_{L^1(\Gamma_0)}^\alpha \|V_a\|_{L^{3-\epsilon}(\Gamma_0)}^{1-\alpha}$$

$$\le \left(\frac{4\pi}{a^2}\right)^\alpha \|V_1\|_{L^{3-\epsilon}(\Gamma_0)}^{1-\alpha},$$

for some $0 < \alpha < 1$, for all $a \ge 1$, and for any $0 < \epsilon < \tfrac{1}{2}$; therefore,

$$\|V_a\|_{L^{5/2}(\Gamma_0)} \longrightarrow 0 \qquad \text{as } a \to +\infty.$$

This, together with (2.68) and (2.69), is enough to imply that we have

$$\|\rho_a\|_{L^{5/3}(\Gamma_0)} \longrightarrow 0 \qquad \text{as } a \to +\infty.$$

Our claim follows, since we have, by Hölder's inequality,

$$\mu_a = \|\rho_a\|_{L^1(\Gamma_0)} \leq \|\rho_a\|_{L^{5/3}(\Gamma_0)}.$$

$$\diamondsuit$$

The last three examples are based upon the convergence of the periodic Yukawa potential to the periodic Coulomb potential as $a \to 0^+$. More precisely, according to Remark 2.32, we now investigate the sign of θ_0.

For this purpose, we reintroduce the dependence of the periodic Coulomb energy on the coefficients in the front of the von Weiszäcker and the Thomas–Fermi terms by writing

$$E_{\text{per}}^{\text{TFW}}(\rho) = A \int_{\Gamma_0} |\nabla\sqrt{\rho}|^2 + c_1 \int_{\Gamma_0} \rho^{5/3} - \int_{\Gamma_0} G(x)\rho(x)\,\mathrm{d}x + \tfrac{1}{2}D_G(\rho,\rho),$$

for any $A, c_1 \geq 0$. From now on, we forget the constant M, which plays no role in what follows.

We then investigate the behaviour of θ_0 as $A \to +\infty$ (Example 2), $A \to 0^+$ (Example 3) and $c_1 \to 0^+$ (Example 4).

We will use the subscript A or c_1, when necessary, to recall the parameter we are working with.

Example 2 In this example, we prove that

$$\theta_0 \longrightarrow -\tfrac{5}{3}c_1 \qquad \text{as} \quad A \to +\infty.$$

Then, because of (2.65), we see that μ_a is strictly less than 1 when A is large enough and a is small enough.

We use the characteristic function of Γ_0 as a test function for $E_{\text{per}}^{\text{TFW}}(\rho)$ to show that

$$I_A^{\text{TFW}}(1) \leq E_{\text{per}}^{\text{TFW}}(1_{\Gamma_0}) = c_1.$$

Thus the corresponding minimizers ρ_A of $I_A^{\text{TFW}}(1)$ form a bounded sequence (ρ_A) in $L^1 \cap L^{5/3}(\Gamma_0)$. Then, we deduce the following bound:

$$A \int_{\Gamma_0} |\nabla\sqrt{\rho_A}|^2 \leq C,$$

for some positive constant C independent of A.

Thus, in particular,

$$\int_{\Gamma_0} |\nabla\sqrt{\rho_A}|^2\,\mathrm{d}x \longrightarrow 0 \qquad \text{as} \quad A \to +\infty.$$

Hence, from the Poincaré–Wirtinger inequality (and since $\int_{\Gamma_0} \rho_A = 1$), we deduce that

$$\rho_A \to 1_{\Gamma_0} \qquad \text{strongly in} \quad L^p(\Gamma_0),$$

for all $1 \leq p < 3$. Then, we may pass to the limit in $I_A^{\text{TFW}}(1)$ to deduce

$$c_1 \geq \liminf_{A \to +\infty} I_A^{\text{TFW}}(1) \geq c_1,$$

and the equality case in all the above inequalities show that, in fact,

$$A \int_{\Gamma_0} |\nabla \sqrt{\rho_A}|^2 \, dx \longrightarrow 0 \qquad \text{as} \quad A \to +\infty.$$

At this stage, we use the formulation (2.67) of the energy based on the Euler–Lagrange equation in order to conclude. ◇

Example 3 If we now let A go to zero, we expect to find as a limit the periodic Thomas–Fermi energy (that is to say, without the von Weizsäcker term). We know from Lieb and Simon [40] that in the Thomas–Fermi case the Lagrange multiplier θ^{TF} is negative. Thus we recover another situation where θ_0 is negative.

Let ρ be in $C^\infty(\bar{\Gamma}_0)$ such that $\int_{\Gamma_0} \rho(x) \, dx = 1$. Then ρ is a test function for $E_{\text{per}}^{\text{TFW}}$, so that

$$\limsup_{A \to 0^+} I_A^{\text{TFW}}(1) \leq \limsup_{A \to 0^+} E_{\text{per}}^{\text{TFW}}(\rho) = E_{\text{per}}^{\text{TF}}(\rho).$$

As such functions ρ form a dense subset of $L^1 \cap L^{5/3}(\Gamma_0)$, we obtain

$$\limsup_{A \to 0^+} I_A^{\text{TFW}}(1) \leq I^{\text{TF}}(1).$$

On the other hand, we obviously have

$$\liminf_{A \to 0^+} I_A^{\text{TFW}}(1) \geq I^{\text{TF}}(1).$$

Thus,

$$\lim_{A \to 0^+} I_A^{\text{TFW}}(1) = I^{\text{TF}}(1).$$

Now let ρ_A be the minimizer of $I_A^{\text{TFW}}(1)$. Then, from the following inequality,

$$I^{\text{TF}}(1) \leq E_{\text{per}}^{\text{TFW}}(\rho_A) + A \int_{\Gamma_0} \left|\nabla \sqrt{\rho_A}\right|^2 = I_A^{\text{TFW}}(1),$$

together with the analogue of (2.69) for $I_A^{\text{TFW}}(1)$, we deduce that (ρ_A) is a minimizing sequence of $I^{\text{TF}}(1)$ and that

$$A \int_{\Gamma_0} |\nabla \sqrt{\rho_A}|^2 \, dx \longrightarrow 0 \qquad \text{as} \quad A \to 0^+.$$

However, any minimizing sequence of $E_{\text{per}}^{\text{TF}}(1)$ is compact in $L^1 \cap L^{5/3}(\Gamma_0)$ (this is a strictly convex minimization problem) and converges to the unique minimizer ρ_0 of $E_{\text{per}}^{\text{TF}}(1)$.

As in the preceding example, we conclude by comparing the formulations of the energies $I_A^{\mathrm{TFW}}(1)$ and $I_{\mathrm{per}}^{\mathrm{TF}}(1)$, based upon the Euler–Lagrange equations, as $A \to 0^+$. $\diamond$

In view of these examples, it is reasonable to wonder whether θ_0 is always negative. This is not the case, as we show in our final example.

Example 4 We let the constant c_1 go to zero, and we prove that the charge of the absolute minimum, denoted by μ_{c_1}, goes to infinity. Therefore, the Lagrange multiplier θ_{c_1}, corresponding to the minimization problem with the charge constraint equal to 1, is positive for c_1 small enough.

We argue by contradiction, assuming that the charge of the absolute minimum μ_{c_1} is bounded by some positive constant C which does not depend on c_1. We call $\rho_{c_1} \equiv u_{c_1}^2$ the corresponding minimizer.

We argue as in Example 2, to show that

$$\limsup_{c_1 \to 0^+} I_{c_1}(\mu_{c_1}) \leq I(C),$$

where

$$I(C) = \inf\left\{ A \int_{\Gamma_0} |\nabla\sqrt{\rho}|^2 - \int_{\Gamma_0} G(x)\rho(x)\,\mathrm{d}x + \tfrac{1}{2}D_G(\rho,\rho) \Big/ \right.$$
$$\left. \sqrt{\rho} \in H_{\mathrm{per}}^1(\Gamma_0),\ \int_{\Gamma_0} \rho(x)\,\mathrm{d}x \leq C \right\}.$$

On the other hand, with the help of the Hölder and Sobolev inequalities, we have

$$I_{c_1}(\mu_{c_1}) \geq A \int_{\Gamma_0} |\nabla\sqrt{\rho_{c_1}}|^2 - \int_{\Gamma_0} G(x)\rho_{c_1}(x)\,\mathrm{d}x$$

$$\geq A \int_{\Gamma_0} |\nabla\sqrt{\rho_{c_1}}|^2\,\mathrm{d}x - \|G\|_{L^2(\Gamma_0)}\,\|\rho_{c_1}\|_{L^2(\Gamma_0)}$$

$$\geq A \int_{\Gamma_0} |\nabla\sqrt{\rho_{c_1}}|^2\,\mathrm{d}x - \|G\|_{L^2(\Gamma_0)}\,\|\rho_{c_1}\|_{L^1(\Gamma_0)}^{1/4}\|\rho_{c_1}\|_{L^3(\Gamma_0)}^{3/4}$$

$$\geq A\|\nabla\sqrt{\rho_{c_1}}\|_{L^2(\Gamma_0)}^2 - C\,\|G\|_{L^2(\Gamma_0)}\,\|\nabla\sqrt{\rho_{c_1}}\|_{L^2(\Gamma_0)}^{3/2}. \qquad (2.70)$$

Hence, $\int_{\Gamma_0} |\nabla\sqrt{\rho_{c_1}}|^2\,\mathrm{d}x$ is bounded uniformly with respect to c_1, since we may bound from above the left-hand side of (2.70) uniformly in c_1 (compare, for example, with $I_1(C)$, for $c_1 \leq 1$).

It is thus possible to extract a subsequence (still denoted by ρ_{c_1}) converging strongly in $L^p(\Gamma_0)$ to $\bar\rho$ for all $1 \leq p < 3$ (because of Rellich's theorem), and such that $\nabla\sqrt{\rho_{c_1}}$ converges weakly to $\nabla\sqrt{\bar\rho}$ in $L^2(\Gamma_0)$.

Then, passing to the (weak) limit in $I_{c_1}(\mu_1)$ yields that

$$
\begin{aligned}
I(C) &\geq \limsup_{c_1 \to 0^+} I_{c_1}(\mu_1) \\
&\geq \liminf_{c_1 \to 0^+} I_{c_1}(\mu_1) \\
&\geq A \int_{\Gamma_0} |\nabla \sqrt{\bar\rho}|^2 - \int_{\Gamma_0} G(x)\bar\rho(x)\, dx + \tfrac{1}{2} D_G(\bar\rho, \bar\rho) \\
&\geq I(C).
\end{aligned}
$$

Thus, all the above inequalities are in fact equalities, from which we recover the strong convergence of $\nabla \sqrt{\rho_{c_1}}$ to $\nabla \sqrt{\bar\rho}$ in $L^2(\Gamma_0)$. Next, we deduce that $\bar\rho$ is the (unique) minimizer of the strictly convex minimization problem $I(C)$. Moreover, we may pass to the limit in the Euler–Lagrange equation satisfied by $u_{c_1} = \sqrt{\rho_{c_1}}$ on $\mathbf{R}^3$, namely

$$
-A\Delta u_{c_1} + \tfrac{5}{3} c_1 u_{c_1}^{7/3} + G u_{c_1} + \left(\int_{\Gamma_0} G(x - y)\rho_{c_1}(y)\, dy \right) u_{c_1} = 0,
$$

and we obtain

$$
-A\Delta \bar u - G \bar u + \left(\int_{\Gamma_0} G(x - y)\bar\rho(y)\, dy \right) \bar u = 0 \qquad \text{in } \mathcal{D}'(\mathbf{R}^3), \tag{2.71}
$$

where $\bar u = \sqrt{\bar\rho}$ is a non-negative function in $H^1_{\text{per}}(\Gamma_0)$.

In order to reach a contradiction, we now go back to the minimization problem $I(C)$.

We may first notice that $I(C)$ is indeed a well defined minimization problem, that the minimizing sequences are bounded in $H^1_{\text{per}}(\Gamma_0)$ (apply the inequality analogous to (2.70) to $I(C)$) and, therefore, that $I(C)$ is achieved by a unique minimizer $\bar\rho$.

We now check that

$$
I(C) < 0, \tag{2.72}
$$

and thus $\int_{\Gamma_0} \bar\rho(x)\, dx > 0$ and $\bar\rho$ is a positive function on $\mathbf{R}^3$.

Indeed, let ρ be such that $\sqrt{\rho} \in H^1_{\text{per}}(\Gamma_0)$ and $\int_{\Gamma_0} \rho(x)\, dx = 1$.

Then, for every $0 < \lambda \leq C$,

$$
I(C) \leq \lambda \left[\int_{\Gamma_0} |\nabla \sqrt{\rho}|^2 - \int_{\Gamma_0} G(x)\rho(x)\, dx \right] + \frac{\lambda^2}{2} D_G(\rho, \rho).
$$

The quantity inside the brackets is negative, choosing for $\sqrt{\rho}$ the first eigenfunction of $-\Delta - G$ on $H^1_{\text{per}}(\Gamma_0)$.

Our claim follows, choosing λ small enough.

The final step towards the contradiction consists of dividing (2.71) by $\overline{u}$ (which is positive) and then integrating on Γ_0 (use the fact that by the definition of $I(C)$, $\overline{u}$ satisfies the Neumann or periodic boundary condition on $\partial\Gamma_0$).

A straightforward calculation yields

$$\int_{\Gamma_0} \frac{|\nabla\overline{u}|^2}{|\overline{u}|^2}\,\mathrm{d}x = 0.$$

Thus, $\overline{u}$ is a positive constant on Γ_0 and we reach a contradiction with (2.71) or (2.72). $\diamondsuit$

2.7 Extensions

This last section of this chapter is devoted to various extensions of the results we have obtained above. As announced in Subsection 1.5.2 of Chapter 1, our arguments hold in situations other than the academic situation of a point nucleus at the centre of a cubic unit cell of unit volume in the standard TFW model.

Let us now examine a few other cases, which we consider as the most interesting extensions. Of course, most of these extensions (and some others) will also be considered in the Coulomb case. On this subject, we refer the reader to the next chapter.

2.7.1 *Smeared out nuclei*

In this subsection, we indicate the modifications of our results in the case when we incorporate, in the various energies considered so far, interaction terms corresponding to smeared out nuclei. Let us already mention that the modification of point nuclei into smeared out nuclei may be made in any of the extensions of the following subsections, and that, consequently, the results described in these subsections remain unchanged.

For the sake of simplicity, we only investigate the case when the 'shape' of each nucleus is described through a smooth probability density m, on Γ_0, with compact support in the interior of Γ_0, as prescribed in the so-called $(Y - m)$ program defined in Chapter 1 and in Section 2.1 of the present chapter. However, it is possible to take, for instance, a finite number of point nuclei per cell, or mixing between point nuclei and smeared out nuclei in the cell. All our arguments go through *mutatis mutandis*. On the contrary, it is worth noticing that in the Coulomb case that we shall study in Chapters 3, 5, and 6, the argument we use strongly depends on the 'shape' of the nucleus (point, smeared, and so on).

In all the energies and minimization problems we have considered so far in this chapter, we replace the attractive potential created by the point nuclei,

$$V_\Lambda = \sum_{k\in\Lambda} \delta(x - k) \star V,$$

by the attractive potential created by smeared out nuclei,

$$V_\Lambda^m = \sum_{k \in \Lambda} m(x - k) \star V.$$

Likewise, the nuclear repulsion energy,

$$\tfrac{1}{2} \sum_{y \neq z \in \Lambda} V(y - z),$$

is changed into its analogue in the smeared out case,

$$\tfrac{1}{2} \sum_{y \neq z \in \Lambda} \iint m(x + y)m(t + z)V(x - t + y - z)\mathrm{d}x\,\mathrm{d}t.$$

In the thermodynamic limit, the energy that is likely to be obtained is

$$E_{\mathrm{per}}^m(\rho) = \int_{\Gamma_0} |\nabla\sqrt{\rho}|^2 + \int_{\Gamma_0} \rho^{5/3} - \int_{\Gamma_0} \rho V_\infty^m$$
$$+ \tfrac{1}{2} \iint_{\Gamma_0 \times \Gamma_0} \rho(x)\rho(y)V_\infty(x - y)\mathrm{d}x\,\mathrm{d}y$$
$$+ \tfrac{1}{2} \sum_{k \neq 0 \in \mathbf{Z}^3} \iint m(x)m(y)V(x - y + k)\,\mathrm{d}x\,\mathrm{d}y,$$

where

$$V_\infty^m(x) = \sum_{k \in \mathbf{Z}^3} (m \star V)(x - k).$$

Then, it is easily seen that our whole analysis goes through (in fact, we may even simplify the proofs in that case; see the details in the Coulomb case in Chapter 3), *except* on the following point.

When m does not share the symmetries of the cell (here a unit cube), it is not possible to recover the periodic boundary condition for the minimizer of $I_{\mathrm{per}}^m(\lambda)$ from the uniqueness of the minimizer together with symmetries invariance arguments (see Section 2.1). This means that the periodic problem defined on $H_{\mathrm{per}}^1(\Gamma_0)$ as in Section 2.1 (formula (2.4)) cannot be identified with the same problem defined on $H^1(\Gamma_0)$. This causes the following slight modification in our argument. We prove that actually $\tilde{\rho}$ belongs to $H_{\mathrm{per}}^1(\Gamma_0)$, so that we catch the lower limit of $\frac{I_\Lambda^m}{|\Lambda|}$ as for $\frac{I_\Lambda}{|\Lambda|}$. Indeed, the uniform L^∞ bound for ρ_Λ gives that, for every x in $\mathbf{R}^3$, and for every k in $\mathbf{Z}^3$,

$$|\widetilde{\rho_\Lambda}(x + k) - \widetilde{\rho_\Lambda}(x)| = \frac{1}{|\Lambda|}\left|\sum_{y \in \Lambda} \rho_\Lambda(x + y + k) - \sum_{z \in \Lambda} \rho_\Lambda(x + z)\right|$$
$$= \frac{1}{|\Lambda|}\left|\sum_{y \in \Lambda + k} \rho_\Lambda(x + y) - \sum_{z \in \Lambda} \rho_\Lambda(x + z)\right|$$

$$\leq \frac{1}{|\Lambda|} \left| \sum_{y \in (\Lambda+k)\setminus\Lambda} \rho_\Lambda(x+y) + \sum_{z \in \Lambda\setminus(\Lambda+k)} \rho_\Lambda(x+z) \right|$$

$$\leq 2 \frac{|\Lambda^{|k|}|}{|\Lambda|} \, \|\rho_\Lambda\|_{L^\infty(\mathbf{R}^3)}.$$

Thus, the bound from above goes to zero as Λ goes to infinity, by the definition of a Van Hove sequence, and this bound is uniform with respect to $x \in \mathbf{R}^3$, so that

$$\|\widetilde{\rho_\Lambda}(\cdot + k) - \widetilde{\rho_\Lambda}\|_{L^\infty(\mathbf{R}^3)} \longrightarrow 0 \qquad \text{as} \quad \Lambda \longrightarrow \infty.$$

Indeed, it is possible to define $\tilde{\rho}$ as the limit of $\widetilde{\rho_\Lambda}$ on the whole space $\mathbf{R}^3$ (almost everywhere or in H^1_{loc}-weak or in L^p_{loc}-strong, for all $1 \leq p < +\infty$), since

$$\widetilde{\rho_\Lambda}(x + k) = \widetilde{\rho_{\Lambda+k}}(x), \qquad \text{for all } x \in \mathbf{R}^3, \, k \in \mathbf{Z}^3.$$

(Recall that this comes from the uniqueness of the minimizing density.)

2.7.2 *Replacing $\rho^{5/3}$ by some other function*

It is soon noticed in Remark 2.3 that if the exponent p in the Thomas–Fermi term is changed into some $p \in]\frac{5}{3}, 3]$, then our whole analysis goes through straightforwardly.

If this exponent is now $p > 3$, it is necessary to add in the definition of the minimization problem that ρ must belong to $L^p(\mathbf{R}^3)$, since the fact that $\sqrt{\rho} \in H^1(\mathbf{R}^3)$ does not necessarily imply this for $p > 3$. Apart from this slight modification, our arguments apply again to this case.

Likewise, we are not obliged to keep a power function ρ^p. Indeed, if we replace the term ρ^p by a function $j(\rho)$ and denote $F(u) = j(\rho = u^2)$, we may see that the properties on F that we now detail are sufficient to conclude.

The function F is non-negative, even, C^2, satisfies $F(0) = F'(0) = 0$, and is convex with respect to $u^2 = \rho$. In addition, in order to have a well defined minimization problem, we may, for instance, suppose a bound of the kind

$$F(t) \leq \alpha t^2 + \beta t^6, \qquad \forall t \geq 0.$$

Next, applying our argument requires that the function F is flat enough in the neighbourhood of 0 and super-quadratic at infinity. This may, for instance, be ensured if the following two conditions are fulfilled.

To obtain a minimum at Λ fixed and to use the argument of Proposition 2.2, we assume the existence of some exponent $p \geq \frac{7}{3}$ such that

$$[F'(t)]_+ = O(t^p), \qquad \text{for } t \text{ small.} \tag{2.73}$$

To prove the uniform L^∞ bound in Proposition 2.5, we assume that there exists some exponent $p > 1$ and some constant $c > 0$ such that

$$F'(t) \geq ct^p, \qquad \text{for } t \text{ large.}$$

A case when one of the above conditions, namely (2.73), is not fulfilled is the case when one takes a term of the form ρ^p for some exponent $1 < p < \frac{5}{3}$. This leads to an interesting situation.

Indeed, at first, we may prove the existence of some critical charge number λ_c, depending on a (the exponent involved in the definition of the Yukawa potential) and on the total charge of the nuclei $|\Lambda|$, such that the infimum $I_\Lambda(\lambda)$ is not achieved for $\lambda > \lambda_c$. Unfortunately, we have failed until now in comparing λ_c and $|\Lambda|$, uniformly in a. Therefore we cannot go through our argument, since we do not know of the existence of a neutral molecule for all Λ.

However, there is an interesting point: in this case, it is possible to show, using estimations à la Véron (see [57]), that the compactness holds for the sequence ρ_Λ (if it exists), in the sense that

$$\frac{1}{|\Lambda|} \int_{\Gamma(\Lambda)^c} \rho_\Lambda(x)\,dx \stackrel{\Lambda \to \infty}{\longrightarrow} 0.$$

This therefore suggests that if we succeed in proving the existence of a minimum, then we shall obtain a thermodynamic limit with one electron per cell (in the spirit of what is the case in the Coulomb case; see the following chapters). The difficulty we experience with proving the existence of a minimum in this case leads to another question that goes beyond the particular case we are considering here. If there is no minimum to the minimization problem, can we, however, prove that the energy has a thermodynamic limit? This is an interesting extension of our work that is worth looking at. We have only preliminary results in this direction. In particular, in the Yukawa case that we have treated in this chapter, such an extension is easy (using the notion of almost minimizing sequences, following an idea by Ekeland—see, for instance, [20]), but we hope to complete as soon as possible these preliminary results with more comprehensive ones in the Coulomb setting.

2.7.3 Default and excess of charge

First, it is clear that if we take any positive ion at Λ fixed instead of considering the neutral molecule, then our argument also applies. In particular, if the negative charge per cell is small enough, then we can ensure that we do not lose charge in the thermodynamic limit.

In Remark 2.29, we have noticed that if the number of electrons considered at Λ fixed is large enough, namely of the type $c|\Lambda|$ with $c > \mu_0$, then we necessarily lose charge while passing to the thermodynamic limit. Physically, this means that although we may build any stable negative ion (recall that in Proposition 2.2 we have seen that, when Λ is fixed, a minimum exists whatever the negative charge is), we cannot expect to build a crystal (in this setting) where the number of electrons per cell is arbitrarily large.

So this raises the following question: From the physical standpoint, where does the excess negative charge escape? We do not know how to answer this. However, it is worth mentioning that the situation is different in the Coulomb case (see Chapter 6.)

2.7.4   *Other shapes of periodic cells*

Our last extension is the case when the periodic cell is not a cube. It is an easy modification of the above work to prove that all our arguments still hold in this case. The point is that the exponential fall-off at infinity of the Yukawa potential allows us to work with any periodic tiling of the space $\mathbf{R}^3$. The situation we shall encounter in the Coulomb case is again radically different. We shall apply another strategy of proof when the lattice is not cubic. On this question, we refer the reader to Chapter 6.

3

CONVERGENCE OF THE ENERGY FOR THE THOMAS–FERMI–VON WEIZSÄCKER MODEL

3.1 Introduction

In this chapter, the problem under consideration is the existence of the thermo-dynamic limit for the energy in Thomas–Fermi–von Weizsäcker type molecular models. We again consider a set of $|\Lambda|$ nuclei of unit charges located on a set Λ of points of integer coordinates in $\mathbf{R}^3$. Denoting by I_Λ the energy of the neutral molecular system and by ρ_Λ its electronic ground state, we investigate the be-haviour of both I_Λ and ρ_Λ as the set Λ goes to the periodic lattice $\mathbf{Z}^3$. For the (quite substantial!) physical background of this mathematical problem, we refer the reader to Chapter 1 and to the references mentioned therein from solid state physics, thermodynamics, and statistical mechanics.

In Chapter 2, we have proved the convergence of the energy per unit volume in the case when the interaction potential between the electrons and the nuclei (and the electrons and the nuclei between themselves) is the Yukawa potential

$$\frac{e^{-a|x-y|}}{|x-y|}.$$

We shall come back to this setting in Chapter 4 for the proof of the convergence of the electronic density.

We now want to show the existence of a limit for the energy per unit volume when the interaction potential is the Coulomb potential

$$\frac{1}{|x-y|}.$$

The Coulomb case seems to be much more difficult than the Yukawa case: when $a > 0$ is small, the Yukawa potential does look like the Coulomb potential when the charges are close to each other, but its short-range nature differs radically from the long-range nature of the Coulomb potential and greatly simplifies most of the proofs. To deal with the Coulomb case, we need to be more careful. In order to illustrate the new difficulty arising in the Coulomb case, let us just men-tion the following fact. When the interaction potential is short-ranged, each of the terms, electron–electron repulsion, nucleus–nucleus repulsion, and electron–nucleus attraction, grows like $|\Lambda|$, no matter what the ratio *total negative charge over total positive charge* is. On the contrary, in the $(Cb - \delta)$ case, each of the above terms is of order $|\Lambda|^{5/3}$, and it is a consequence of the global (asymptotic) neutrality that they 'cancel' each other to yield an energy that behaves like $|\Lambda|$.

However, as will become clear below, part of our analysis here will rely upon the analysis made in Chapter 2 in the Yukawa case, but we shall also give proofs that do not rely so much on the Yukawa approximation.

Let us now briefly recall, for the reader's convenience, the models we are studying here.

When Λ is fixed, the minimization problem

$$
E_\Lambda^{\mathrm{TFW}}(\rho) = \int_{\mathbf{R}^3} |\nabla\sqrt{\rho}|^2 + \int_{\mathbf{R}^3} \rho^{5/3} - \int_{\mathbf{R}^3} \left(\sum_{k\in\Lambda} \frac{1}{|x-k|} \right) \rho(x)\,\mathrm{d}x
$$
$$
+ \frac{1}{2} \int\!\!\!\int_{\mathbf{R}^3\times\mathbf{R}^3} \frac{\rho(x)\rho(y)}{|x-y|}\,\mathrm{d}x\,\mathrm{d}y, \tag{3.1}
$$

$$
I_\Lambda^{\mathrm{TFW}} = \inf \left\{ E_\Lambda^{\mathrm{TFW}}(\rho) + \frac{1}{2} \sum_{y\neq z\in\Lambda} \frac{1}{|y-z|}; \right.
$$
$$
\left. \rho \geq 0, \ \sqrt{\rho}\in H^1(\mathbf{R}^3), \int_{\mathbf{R}^3}\rho = |\Lambda| \right\}, \tag{3.2}
$$

has a unique minimizing density, denoted by ρ_Λ (see Benguria, Brézis, and Lieb [6]), such that $u_\Lambda = \sqrt{\rho_\Lambda}$ is a solution to

$$
-\triangle u_\Lambda + \left[\tfrac{5}{3}\rho_\Lambda^{2/3} - \Phi_\Lambda\right]u_\Lambda = -\theta_\Lambda u_\Lambda,
$$

where $\theta_\Lambda > 0$ and

$$
\Phi_\Lambda = \sum_{k\in\Lambda} \frac{1}{|x-k|} - \rho_\Lambda \star \frac{1}{|x|}. \tag{3.3}
$$

In the thermodynamic limit, we hope (and it is actually the case), in view of the results obtained by Lieb and Simon on the Thomas–Fermi model in [40], that the problem (3.1)–(3.3) converges to the following periodic problem:

$$
I_{\mathrm{per}}^{\mathrm{TFW}} = \inf \left\{ E_{\mathrm{per}}^{\mathrm{TFW}}(\rho); \rho \geq 0, \ \sqrt{\rho}\in H_{\mathrm{per}}^1(\Gamma_0), \int_{\Gamma_0}\rho = 1 \right\}, \tag{3.4}
$$

$$
E_{\mathrm{per}}^{\mathrm{TFW}}(\rho) = \int_{\Gamma_0} |\nabla\sqrt{\rho}|^2 + \int_{\Gamma_0} \rho^{5/3} - \int_{\Gamma_0} \rho(x)G(x)\,\mathrm{d}x
$$
$$
+ \frac{1}{2} \int\!\!\!\int_{\Gamma_0\times\Gamma_0} \rho(x)\rho(y)G(x-y)\,\mathrm{d}x\,\mathrm{d}y, \tag{3.5}
$$

where $H_{\mathrm{per}}^1(\Gamma_0)$ is the subset of the functions in $H^1(\Gamma_0)$ satisfying the periodic boundary conditions on the boundary of Γ_0 and G is the periodic potential defined by

$$-\Delta G = 4\pi\left(-1 + \sum_{y\in\mathbf{Z}^3}\delta(\cdot - y)\right), \tag{3.6}$$

$$\int_{\Gamma_0} G = 0 \tag{3.7}$$

(in other words, G is the periodic Green function for the Laplacian with periodic boundary conditions on $\partial\Gamma_0$). As in Chapter 1, we shall denote

$$M = \lim_{x\to 0} G(x) - \frac{1}{|x|}. \tag{3.8}$$

In addition to the above $(Cb - \delta)$ model (3.1)– (3.8), we shall also consider the following $(Cb - m)$ model:

$$E_\Lambda^m(\rho) = \int_{\mathbf{R}^3}|\nabla\sqrt{\rho}|^2 + \int_{\mathbf{R}^3}\rho^{5/3} - \int_{\mathbf{R}^3}V_\Lambda^m(x)\rho(x)\mathrm{d}x$$
$$+\frac{1}{2}\int\int_{\mathbf{R}^3\times\mathbf{R}^3}\frac{\rho(x)\rho(y)}{|x-y|}\mathrm{d}x\,\mathrm{d}y, \tag{3.9}$$

$$I_\Lambda^m = \inf\left\{E_\Lambda^{\mathrm{TFW}}(\rho) + \frac{1}{2}\sum_{z\neq t\in\Lambda}\int\int m(x+z)V(x+z-y-t)m(y+t)\mathrm{d}x\,\mathrm{d}y;\right.$$
$$\left.\rho\geq 0,\ \sqrt{\rho}\in H^1(\mathbf{R}^3),\ \int_{\mathbf{R}^3}\rho = |\Lambda|\right\},$$

with

$$V_\Lambda^m = \sum_{k\in\Lambda}m(x-k)\star\frac{1}{|x|},$$

together with its thermodynamic limit

$$I_{\mathrm{per}}^m = \inf\left\{E_{\mathrm{per}}^m(\rho); \rho\geq 0, \sqrt{\rho}\in H_{\mathrm{per}}^1(\Gamma_0), \int_{\Gamma_0}\rho = 1\right\}, \tag{3.10}$$

with the energy:

$$E_{\mathrm{per}}^m(\rho) = \int_{\Gamma_0}|\nabla\sqrt{\rho}|^2 + \int_{\Gamma_0}\rho^{5/3} - \int_{\Gamma_0}\rho(x)G_m(x)\mathrm{d}x$$

$$+\frac{1}{2}\iint_{\Gamma_0 \times \Gamma_0} \rho(x)\rho(y)G(x-y)\,dx\,dy, \tag{3.11}$$

where G_m is the periodic potential solution to

$$-\triangle G_m = 4\pi\Big(-1 + \sum_{y \in \mathbf{Z}^3} m(\cdot - y)\Big), \tag{3.12}$$

that satisfies

$$\int_{\Gamma_0} G_m = 0, \tag{3.13}$$

which is also given by

$$G_m = G \star m.$$

Before presenting the main result of this chapter, we want to make some remarks on the definitions (3.4) and (3.10) of the periodic minimization problems obtained by the thermodynamic limit. In fact, in the Coulomb case or when the smearing m shares the symmetry of the unit cube, it is equivalent to set the limit minimization problem on $H^1(\Gamma_0)$ (that is, without the periodic boundary conditions). Indeed, the energy functionals defined by (3.5) and (3.11) are strictly convex with respect to ρ. Therefore, there is a unique density that minimizes these functionals on $H^1(\Gamma_0)$ (with charge constraint 1). Hence, uniqueness implies that the minimizers also share the symmetry of the unit cube, and thus are periodic. However, since we shall also prove results without symmetry assumptions on m or, in the point nuclei, with several nuclei by unit cell or, even, with a non-cubic primitive cell, we have made the choice to introduce the models in the most general framework.

   We shall prove in this chapter the following theorem, announced in the preface and in Chapter 1.

Theorem 3.1 (Convergence of the energy in the Coulomb case) *Consider the $(Cb - \delta)$ (respectively, $(Cb - m)$) model. Assume that the sequence (Λ) satisfies the condition*

$$(c) \qquad\qquad \frac{|\Lambda^h|}{|\Lambda|} Log(|\Lambda^h|) \overset{\Lambda \to \infty}{\longrightarrow} 0$$

in addition to the conditions (a) and (b) of Definition 1 of Chapter 1. Also assume in the $(Cb-m)$ case that the 'shape' of the nucleus m shares the symmetry of the unit cube. Then

$$\lim_{\Lambda \to \infty} \frac{1}{|\Lambda|} I_\Lambda^{\mathrm{TFW}} = I_{\mathrm{per}}^{\mathrm{TFW}} + \frac{M}{2},$$

where $I_{\mathrm{per}}^{\mathrm{TFW}}$ is defined by (3.4)–(3.7), and $M = \lim_{x \to 0} G(x) - \frac{1}{|x|}$ (respectively, $= I_{\mathrm{per}}^m + \frac{M}{2}$, where I_{per}^m is defined by (3.11)– (3.13), and $M = \iint_{\Gamma_0 \times \Gamma_0} m(x)m(y) [G(x-y) - \frac{1}{|x-y|}]dx\,dy).$

Remark 3.2 In fact, we shall provide in Chapter 6 another strategy of proof that allows to state the same result for any Van Hove sequence (that means without the additional hypothesis (c)) and without the symmetry assumption on the density m.

This chapter is organized as follows.

As in Chapter 2, we first establish, in Section 3.2, some estimates on the energy and the density. In particular, we show, among other useful estimates, that

$$\left| \frac{I_\Lambda^{\text{TFW}}}{|\Lambda|} \right| \leq C \tag{3.14}$$

and that

$$\|\rho_\Lambda\|_{L^\infty} \leq C, \tag{3.15}$$

for some positive constants C that do not depend on Λ. Let us already mention that (3.15) will also be useful in the rest of our work dealing with the convergence and the periodicity of the limit density.

The bounds (3.14) and (3.15), like most of the other estimates we prove in Section 3.2, hold both in the smeared nuclei case (($Cb - m$) program), and the point nuclei case (($Cb - \delta$) program), but their proofs are simpler in the first case (at least as long as we restrict ourselves to smooth densities m, which is by definition the case in the ($Cb-m$) program) because we may write the interaction terms of the energy (3.9) in terms of

$$\int_{\mathbf{R}^3} |\nabla \Phi_\Lambda|^2,$$

where $\Phi_\Lambda = (m_\Lambda - \rho_\Lambda) \star \frac{1}{|x|}$ (see the details in Section 3.2 below).

In Section 3.3, we show that the cloud of electrons roughly remains around the positions of the nuclei Λ while the latter gets wider: no charge escapes at infinity in the sense that

$$\lim_{\Lambda \longrightarrow \infty} \frac{1}{|\Lambda|} \int_{\Gamma(\Lambda)} \rho_\Lambda = 1, \tag{3.16}$$

which can be seen as a compactness issue. We shall in fact give several proofs of (3.16).

The proofs given in Sections 3.3.1 and 3.3.2 are based upon the following remark: if (3.16) does not hold, then the negative electronic charge that remains in the neighbourhood of the nuclei cannot compensate for the positive charge of these nuclei, which causes the energy I_Λ^{TFW} to explode faster than $|\Lambda|$ (in fact, like $|\Lambda|^{5/3}$), which contradicts (3.14). This argument will hold both in the cases of point nuclei and smeared nuclei.

As an alternative proof for the point nuclei case, we consider in Section 3.3.3 a more particular sequence of domains going to infinity, namely a sequence of homothetic cubes, and present a somewhat different proof of (3.16) in this case. In Section 3.3.4, we generalize this proof, based on a scaling argument, in order to deal with an arbitrary sequence (Λ), again in the point nuclei case.

Eventually, in Section 3.3.5, we shall give a proof of (3.16) via a simple argument based on an evaluation of the charge outside $\Gamma(\Lambda)$.

It is worth noticing that, in the Yukawa case, we have exhibited cases when (3.16) does not hold (see Section 2.6 in Chapter 2).

In Section 3.4, we obtain lower bounds for $\frac{I_\Lambda^{\mathrm{TFW}}}{|\Lambda|}$. Here again, we obtain this lower limit by two different strategies of proof. The first one (Section 3.4.4) uses the fact that the Yukawa model goes to the Coulomb model when a goes to zero, in view of Section 2.6 of Chapter 2. Our second strategy of proof (Section 3.4.2 for the case of smeared nuclei and Section 3.4.3 for the point nuclei case) is the continuation of the $\sim$-transform idea that we introduced in Chapter 2, and uses in a fundamental way the effective potential Φ_Λ. Moreover, in Section 3.4.4, we shall in particular present an adaptation of our proof to the Yukawa case, therefore providing another determination of a sharp lower bound for the energy per cell in the Yukawa case (already made in Chapter 2 using different arguments).

Section 3.5 is devoted to an upper bound of $\frac{I_\Lambda^{\mathrm{TFW}}}{|\Lambda|}$. The idea is, as in the Yukawa case, to build a sequence of electronic densities with the help of the periodic density minimizing the periodic minimization problem on the unit cell Γ_0 (3.4) to which we hope to converge in the thermodynamic limit. This allows us to conclude the proof of the theorem stated above (Corollary 3.24). The symmetry assumption plays a crucial role in this strategy of proof and, in fact, it is the only point where it plays a role. It is also for the proof of that step that the additional assumption (c) on the Van Hove sequence is made necessary. Besides, in the smeared nuclei case, we prove the convergence of the density in an average sense ($\sim$-transform sense); see Corollaries 3.25 and 3.26.

In Section 3.6, we investigate some extensions of our results. We first replace the cubic cell of unit volume by a cube of different volume and investigate the existence of a cube with optimal size. We have to wait until Chapter 6 before dealing with the case of non-cubic cells. Next, we add to the TFW energy function the Dirac correction term for the exchange energy. In this new setting, when the parameter appearing in front of the Dirac term is small enough, the convergence of the energy per unit volume still holds. For both types of extensions, we do not detail the proofs but only indicate, for the sake of simplicity, where and how our previous proofs have to be modified.

In Chapter 5, we shall come back to the TFW model with Coulomb potential in order to investigate the convergence of the electronic density. As a corollary of our result of convergence of the densities, and as announced in Remark 3.2 above, we shall give in Chapter 6 another proof of the convergence of the energy per unit volume. This proof does not make any use of the $\sim$-transform idea that we use in this chapter and will cover more general situations (finite number of nuclei per cell, non-cubic cells, and so on).

3.2 Preliminaries: a priori estimates

Bearing in mind that our objective is to pass to the limit as Λ goes to infinity, it is clear that we are primarily interested in finding uniform bounds (with respect to Λ) on the various quantities involved in the problem: the energy per unit cell, the effective potential Φ_Λ, the density ρ_Λ, and the Lagrange multiplier θ_Λ. Obtaining such bounds is our first purpose in this section. Next, from these bounds, we deduce in Proposition 3.13 and Corollary 3.15 preliminary convergence results on the densities and estimates from below for some terms of the energy.

We begin our study by the $(Cb - m)$ case: in this case, the estimates are somewhat easier to obtain. Next, we turn to the $(Cb - \delta)$ case.

3.2.1 The case of smeared nuclei

In this special case, it is convenient to rewrite the functional of energy E_Λ^m in different ways. Let $m_\Lambda = \sum_{y \in \Lambda} m(\cdot - y)$, so that

$$V_\Lambda^m = m_\Lambda \star \frac{1}{|x|}$$

and

$$\Phi_\Lambda^m = (m_\Lambda - \rho_\Lambda) \star \frac{1}{|x|}$$

satisfies

$$-\triangle \Phi_\Lambda^m = 4\pi \left[m_\Lambda - \rho_\Lambda \right] \qquad \text{a.e. on } \mathbf{R}^3; \qquad (3.17)$$

while

$$U_\Lambda^m = D^{Cb}(m_\Lambda, m_\Lambda) - |\Lambda| D^{Cb}(m, m).$$

Therefore, the expression

$$\frac{U_\Lambda^m}{2} - \int_{\mathbf{R}^3} V_\Lambda^m(x)\rho_\Lambda(x)\, dx + \tfrac{1}{2}D^{Cb}(\rho_\Lambda, \rho_\Lambda)$$

is equivalent either to

$$\tfrac{1}{2}D^{Cb}(m_\Lambda - \rho_\Lambda, m_\Lambda - \rho_\Lambda) - \frac{|\Lambda|}{2}D^{Cb}(m, m) \qquad (3.18)$$

or to

$$\frac{1}{8\pi} \int_{\mathbf{R}^3} |\nabla \Phi_\Lambda^m|^2\, dx - \frac{|\Lambda|}{2}D^{Cb}(m, m) . \qquad (3.19)$$

Using the expression (3.18), it is easy to prove the following.

Lemma 3.3 $\dfrac{I_\Lambda^m}{|\Lambda|}$ *is bounded independently of* Λ, *for every* $\Lambda \subset \mathbf{Z}^3$.

Proof of Lemma 3.3 For the bound from below, we just use (3.18) to obtain

$$\frac{I_\Lambda^m}{|\Lambda|} \geq -\tfrac{1}{2} D^{Cb}(m, m),$$

since the other terms arising in E_Λ^m are non-negative.

On the other hand, if we apply m_Λ as a test function for E_Λ^m, we see that we have

$$\frac{I_\Lambda^m}{|\Lambda|} \leq \frac{E_\Lambda^m(m_\Lambda)}{|\Lambda|} \leq \int_{\Gamma_0} |\nabla \sqrt{m}|^2 + \int_{\Gamma_0} m^{5/3} - \tfrac{1}{2} D^{Cb}(m, m).$$

Our claim follows. $\diamondsuit$

As a consequence of Lemma 3.3, we deduce the following bounds on ρ_Λ and Φ_Λ^m.

Proposition 3.4 (Estimates in the $(Cb-m)$ case) *There exist various positive constants C, independent of $\Lambda \subset \mathbf{Z}^3$ such that, for all $\Lambda \subset \mathbf{Z}^3$, the following estimates hold:*

(i) $\dfrac{1}{|\Lambda|} \displaystyle\int_{\mathbf{R}^3} |\nabla \sqrt{\rho_\Lambda}|^2 \mathrm{d}x \leq C$;

(ii) $\|\rho_\Lambda\|_{L^p(\mathbf{R}^3)} \leq C\, |\Lambda|^{\frac{1}{p}}$, *for all* $1 \leq p \leq \tfrac{5}{3}$;

(iii) $\dfrac{1}{|\Lambda|} \displaystyle\int_{\mathbf{R}^3} |\nabla \Phi_\Lambda^m|^2 \, \mathrm{d}x \leq C$;

(iv) $\tfrac{1}{|\Lambda|} D^{Cb}(m_\Lambda - \rho_\Lambda, m_\Lambda - \rho_\Lambda) \leq C$;

(v) $0 < \theta_\Lambda \leq C$;

(vi) $0 \leq \dfrac{1}{|\Lambda|} \displaystyle\int_{\mathbf{R}^3} \Phi_\Lambda^m \, \rho_\Lambda \, \mathrm{d}x \leq C$;

(vii) $\dfrac{1}{|\Lambda|} \left| U_\Lambda^m - \displaystyle\int_{\mathbf{R}^3} V_\Lambda^m \rho_\Lambda \, \mathrm{d}x \right| \leq C$;

where ρ_Λ is the minimizer of I_Λ^m and θ_Λ is the corresponding Lagrange multiplier.

Proof of Proposition 3.4 (i)–(iv) are immediately deduced from Lemma 3.3 and the expressions for the energy given by (3.18) and (3.19).

The bound on the Lagrange multiplier θ_Λ is shown in [53] and is a consequence of the following pointwise inequality taken from [53]: there exists a positive constant C_0, such that, for all $\Lambda \subset \mathbf{Z}^3$ and for every x in $\mathbf{R}^3$,

$$\tfrac{10}{9} \rho_\Lambda(x)^{2/3} \leq \Phi_\Lambda^m(x) + [C_0 - \theta_\Lambda]_+ . \tag{3.20}$$

This yields the additional information that $0 < \theta_\Lambda \leq C_0$, for every $\Lambda \subset \mathbf{Z}^3$. In fact, the original inequality, as found in the above reference, applies for point nuclei. However, the proof extends to our case in a straightforward way.

In order to conclude, we write the Euler–Lagrange equation satisfied by $u_\Lambda \equiv \sqrt{\rho_\Lambda}$, namely

$$-\Delta u_\Lambda + \tfrac{5}{3} u_\Lambda^{7/3} - \Phi_\Lambda^m u_\Lambda + \theta_\Lambda u_\Lambda = 0. \qquad (3.21)$$

Next, we multiply it by u_Λ and integrate over $\mathbf{R}^3$. This yields (vi). Finally, (vii) follows, since it corresponds to the remaining term in the definition of $\frac{I_\Lambda^m}{|\Lambda|}$. $\diamondsuit$

In fact, pursuing our analysis of (3.20) and (3.21) a step further, it is even possible to derive uniform L^∞ bounds on ρ_Λ and Φ_Λ^m.

Proposition 3.5 $((Cb-m)$ **case**) *There exist positive constants C independent of Λ such that, for all $\Lambda \subset \mathbf{Z}^3$:*
(i) $\|\rho_\Lambda\|_{L^\infty(\mathbf{R}^3)} \le C$; *and*
(ii) $\|\Phi_\Lambda^m\|_{L^\infty(\mathbf{R}^3)} \le C$.

Remark 3.6 As a direct consequence of (i) above, we deduce that the estimates (ii) in Proposition 3.4 hold for every $1 \le p \le +\infty$.

Proof of Proposition 3.5 Our strategy of proof is as follows. We begin by proving that Φ_Λ^m is bounded (uniformly with respect to Λ) in some average sense (see (3.26) below), and next that it is bounded in $L^\infty(\mathbf{R}^3)$. It is then straightforward to deduce (i) from (ii) and (3.20).
The point is actually to prove that Φ_Λ^m is bounded from above, since a bound from below follows from (3.20). Indeed, (3.20) immediately yields

$$\Phi_\Lambda^m \ge -C_0 \quad \text{on } \mathbf{R}^3.$$

Finally, we deduce from our bound on Φ_Λ^m a bound on ρ_Λ using (3.20). Equivalently, we may use a comparison argument soon introduced in Chapter 2, and made precise in a very general setting in Lemma 4.11 of Chapter 4.
Let $\omega \ge 0$ belong to $\mathcal{D}(B_1)$, with $\int_{\mathbf{R}^3} \omega^2 = 1$. Let $\delta > 0$. We set

$$\omega_{\delta,y} = \frac{1}{\delta^{3/2}} \omega \left(\frac{\cdot - y}{\delta} \right),$$

for every $y \in \mathbf{R}^3$.
We first claim that the smallest eigenvalue $\lambda_1(\Omega)$ of $-\Delta + \left(\tfrac{5}{3} u_\Lambda^{4/3} - \Phi_\Lambda^m \right) + \theta_\Lambda$, on any bounded open subset Ω of $\mathbf{R}^3$ with Dirichlet boundary condition on $\partial\Omega$, is positive.
Indeed, let v be the associated eigenfunction, which thus satisfies

$$\begin{cases} -\Delta v + \left(\tfrac{5}{3} u_\Lambda^{4/3} - \Phi_\Lambda^m \right) v + \theta_\Lambda v = \lambda_1(\Omega) v \quad \text{in } \Omega, \\[2mm] v > 0 \text{ in } \Omega, \ v = 0 \quad \text{on } \partial\Omega. \end{cases}$$

The positivity of v follows from Harnack's inequality, which holds in our case since $\frac{5}{3}u_\Lambda^{4/3} - \Phi_\Lambda^m + \theta_\Lambda$ belongs to $L^p_{loc}(\mathbf{R}^3)$ for some $p > \frac{3}{2}$ (see [56] or [24]). Comparing with (3.21), we deduce

$$-\int_{\partial\Omega} \frac{\partial v}{\partial n} \cdot u_\Lambda = \lambda_1 \int_\Omega u_\Lambda v,$$

and we conclude, since $\dfrac{\partial v}{\partial n} \leq 0$ on $\partial\Omega$.

Since $\lambda_1(B(y;\delta)) > 0$, we then have

$$\int_{\mathbf{R}^3} |\nabla\omega_{\delta,y}|^2 \, dx + \int_{\mathbf{R}^3} \left(\tfrac{5}{3}u_\Lambda^{4/3} - \Phi_\Lambda^m\right) \omega_{\delta,y}^2 \, dx + \theta_\Lambda > 0$$

or, equivalently, skipping all the subscripts Λ from now on in order to simplify the notation,

$$\Phi^m \star \omega_\delta^2 \leq \tfrac{5}{3}\left(u^{4/3} \star \omega_\delta^2\right) + C + \frac{C}{\delta^2}, \qquad (3.22)$$

almost everywhere on $\mathbf{R}^3$, where we used the bound on θ_Λ (Proposition 3.4(v)). Here and below, C denotes various positive constants that are independent of Λ and δ.

To simplify the notation, we indicate with a subscript δ the result of a convolution with ω_δ^2, so that (3.22) becomes

$$\Phi_\delta^m \leq \tfrac{5}{3}(u^{4/3})_\delta + C + \frac{C}{\delta^2}. \qquad (3.23)$$

We now deduce from (3.17) satisfied by Φ_δ^m that

$$-\Delta\Phi_\delta^m = 4\pi \left[\mu_\delta - (u^2)_\delta\right], \qquad (3.24)$$

where

$$\mu_\delta = m_\Lambda \star \omega_\delta^2.$$

Besides, as $\displaystyle\int_{\mathbf{R}^3} \omega_{\delta,y}^2 = 1$, we deduce from (3.23) and Jensen's inequality that

$$(u^2)_\delta \geq (u^{4/3})_\delta^{3/2}$$

$$\geq \left(\Phi_\delta^m - C - \frac{C}{\delta^2}\right)_+^{3/2}.$$

Going back to (3.24), the above inequality yields the following inequality:

$$-\Delta\Phi_\delta^m + \left(\Phi_\delta^m - C - \frac{C}{\delta^2}\right)_+^{3/2} \leq C_0, \qquad (3.25)$$

since μ_δ is uniformly bounded on $\mathbf{R}^3$.

We claim that (3.25) implies

$$\Phi_\delta^m \leq C + \frac{C}{\delta^2} \quad \text{on } \mathbf{R}^3. \tag{3.26}$$

Indeed, let $\Omega = \left\{ x \in \mathbf{R}^3 / \Phi_\delta^m - C - \frac{C}{\delta^2} > 0 \right\}$. Since Φ_δ^m, like Φ_Λ^m, is a continuous function on $\mathbf{R}^3$, going to zero at infinity, Ω is an open and bounded subset of $\mathbf{R}^3$. Moreover, $\varphi_\delta \equiv \Phi_\delta^m - C - \frac{C}{\delta^2}$ vanishes on $\partial\Omega$.

Besides, the constant function $C_0^{2/3}$ satisfies

$$-\Delta h + h_+^{3/2} = C_0 \quad \text{on} \quad \mathbf{R}^3$$

and is strictly larger than φ_δ on $\partial\Omega$. Then, in view of the maximum principle, we deduce that

$$\varphi_\delta \leq C_0^{2/3} \quad \text{on} \quad \Omega.$$

However, the same inequality holds on Ω^c, by definition of Ω, and therefore $\varphi_\delta \leq C_0^{2/3}$ on $\mathbf{R}^3$. Hence, we recover (3.26).

Our aim now is to deduce a similar bound for Φ_Λ^m from (3.26). Recall that it is enough to find a bound from above for Φ_Λ^m, since we already have a bound from below. We first point out that if Φ_Λ^m is non-positive on $\mathbf{R}^3$, we conclude because of (3.20) and the uniform bound from below for Φ_Λ^m just recalled. Thus, we may assume that there exists $x_\Lambda \in \mathbf{R}^3$ satisfying

$$\Phi_\Lambda^m(x_\Lambda) = \|\Phi_\Lambda^{m\,+}\|_{L^\infty(\mathbf{R}^3)} > 0$$

(such a point exists since Φ_Λ^m is a continuous function going to zero at infinity).

We then intend to apply a comparison principle, recalling that from (3.17) satisfied by Φ_Λ^m, we deduce that

$$-\Delta \Phi_\Lambda^{m\,+} \leq C_0 \quad \text{a.e. on} \quad \mathbf{R}^3.$$

This is in fact the same constant C_0 as the one appearing in (3.25). In order to simplify the notation, we suppose, without loss of generality, that $x_\Lambda = 0$ and, once more, we skip the subscript Λ everywhere.

We introduce the notation

$$\fint_{B_\delta} = \frac{1}{\frac{4}{3}\pi\delta^3} \int_{B_\delta}$$

and

$$\fint_{S_\delta} = \frac{1}{4\pi\delta^2} \int_{S_\delta}$$

where S_δ is the sphere centered at 0 with radius δ.

We now choose ω such that $\omega \equiv 1$ on $B_{\frac{1}{2}}$. Then, we use (3.26) (with 2δ instead of δ to be precise) to obtain

$$\fint_{B_\delta} \Phi^m \leq C + \frac{C}{\delta^2}.$$

Indeed, (3.26) holds for $\Phi_\delta^{m\,+}$ as well (with possibly different positive constants), since Φ_δ^m, like Φ_Λ^m, is bounded from below by $-C_0$ and $\int \omega_\delta^2 \, dx = 1$. Then, there exists $\delta' \in \left(\frac{\delta}{2}; \delta\right)$, possibly depending on Λ, such that

$$\fint_{S_{\delta'}} \Phi^{m\,+} \leq C + \frac{C}{\delta^2}, \tag{3.27}$$

for different positive constants C.

Now, we construct two positive functions, Φ_1 and Φ_2, determined by

$$\begin{cases} -\Delta\Phi_1 = 0 & \text{in } B_{\delta'}, \\ \Phi_1 = \Phi^{m\,+} & \text{on } \partial B_{\delta'} = S_{\delta'}, \end{cases}$$

and

$$\begin{cases} -\Delta\Phi_2 = C_0 & \text{in } B_{\delta'}, \\ \Phi_2 = 0 & \text{on } \partial B_{\delta'}. \end{cases}$$

It is easy to compute explicitly Φ_2, which is given by

$$\Phi_2(x) = \frac{C_0}{6}(\delta'^2 - |x|^2) \qquad \text{for all } x \in B_{\delta'}.$$

In particular,

$$\Phi_2(0) = \frac{C_0}{6}\delta'^2 \leq \frac{C_0}{6}\delta^2. \tag{3.28}$$

In addition, from the mean value property of harmonic functions, we obtain

$$\Phi_1(0) = \fint_{S_{\delta'}} \Phi^{m\,+} \leq C + \frac{C}{\delta^2}, \tag{3.29}$$

because of (3.27). We may then conclude, since

$$-\Delta\left(\Phi^{m\,+} - (\Phi_1 + \Phi_2)\right) \leq 0 \qquad \text{on } B_{\delta'},$$

together with

$$\Phi^{m\,+} - (\Phi_1 + \Phi_2) = 0 \qquad \text{on } \partial B_{\delta'},$$

imply

$$\Phi^{m\,+} \leq \Phi_1 + \Phi_2 \qquad \text{on } \quad B_{\delta'}.$$

In particular, this inequality holds at 0 and we conclude with the help of (3.28) and (3.29). $\diamond$

We now state, in the following Lemma 3.7 and Proposition 3.8, the analogous properties in the case of point nuclei.

3.2.2   *The point nuclei case*

Lemma 3.7 *Let Λ be a Van Hove sequence. Then, $\dfrac{I_\Lambda^{\mathrm{TFW}}}{|\Lambda|}$ is bounded independently of $\Lambda \subset \mathbf{Z}^3$.*

Proof of Lemma 3.7   In order to obtain the bound from below, it suffices to notice that

$$\frac{I_\Lambda^{\mathrm{TFW}}}{|\Lambda|} \geq \frac{I_\Lambda^{\mathrm{TF}}}{|\Lambda|},$$

and to refer the reader to [40], where such a lower bound is shown to exist for the Thomas–Fermi model (without the monotonicity assumption on the sequence Λ).

To prove the bound from above, we use the technical results given by Lemma 2.24 in Section 2.4 of Chapter 2.

We take a positive function $\rho_0 \in \mathcal{D}(\overset{\circ}{\Gamma_0})$, sharing the symmetries of the unit cube and such that $\int_{\Gamma_0} \rho_0 \, dx = 1$ and we define $\bar{\rho}_\Lambda = \sum_{y \in \Lambda} \rho_0(\cdot - y)$. Then, $\bar{\rho}_\Lambda \in \mathcal{D}(\Gamma(\Lambda))$ and $\int_{\mathbf{R}^3} \bar{\rho}_\Lambda(x) \, dx = |\Lambda|$, thus $\bar{\rho}_\Lambda$ is a test function for I_Λ^{TFW}. Therefore, we have

$$
\begin{aligned}
I_\Lambda^{\mathrm{TFW}} &\leq E_\Lambda^{\mathrm{TFW}}(\bar{\rho}_\Lambda) + \tfrac{1}{2} U_\Lambda^{Cb} \\
&= |\Lambda| \left(\int_{\Gamma_0} |\nabla \sqrt{\rho_0}|^2 \, dx + \int_{\Gamma_0} \rho_0^{5/3} \, dx \right) \\
&\quad - \sum_{y,z \in \Lambda} \int_{\Gamma_0} \frac{\rho_0(x)}{|x - z + y|} \, dx + \tfrac{1}{2} U_\Lambda^{Cb} \\
&\quad + \tfrac{1}{2} \iint_{\Gamma_0 \times \Gamma_0} \rho_0(x)\rho_0(y) \left(\sum_{z,t \in \Lambda} \frac{1}{|x - y + z - t|} \right) dx \, dy.
\end{aligned}
$$

Now, we claim that the function defined by

$$f(u) = \frac{1}{|u|} - \int_{\Gamma_0} \frac{\rho_0(x)}{|x - u|} \, dx$$

satisfies $|f(u)| \leq \dfrac{C}{|u|^4}$ for $|u|$ large.

Then, the series $\sum_{y \in \mathbf{Z}^3} |f(x - y)|$ converges. This point follows from the symmetry properties of ρ_0 and is proved in detail in Section 3.5.

Finally, we mimic the proof of Proposition 3.22 (it is in fact slightly simpler) with the help of Lemma 2.24(i) (Chapter 2, Section 2.4), to conclude that

$$I_\Lambda^{\mathrm{TFW}} \leq |\Lambda| \left(E_{\mathrm{per}}^{\mathrm{TFW}}(\rho_0) + \frac{M}{2} \right) + o(|\Lambda|),$$

which proves our claim. $\diamond$

Proposition 3.8 (Estimates in the $(Cb - \delta)$ case) *There exist various positive constants C, independent of $\Lambda \subset \mathbf{Z}^3$ such that, for all $\Lambda \subset \mathbf{Z}^3$, the following estimates hold:*

(i) $\dfrac{1}{|\Lambda|} \displaystyle\int_{\mathbf{R}^3} |\nabla \sqrt{\rho_\Lambda}|^2 \, \mathrm{d}x \leq C$;

(ii) $\|\rho_\Lambda\|_{L^p(\mathbf{R}^3)} \leq C \, |\Lambda|^{\frac{1}{p}}$, *for all* $1 \leq p \leq \frac{5}{3}$;

(iii) $0 < \theta_\Lambda \leq C$;

(iv) $0 \leq \dfrac{1}{|\Lambda|} \displaystyle\int_{\mathbf{R}^3} \Phi_\Lambda \, \rho_\Lambda \, \mathrm{d}x \leq C$;

(v) $\dfrac{1}{|\Lambda|} \left| U_\Lambda^{Cb} - \displaystyle\int_{\mathbf{R}^3} V_\Lambda \, \rho_\Lambda \, \mathrm{d}x \right| \leq C$;

where ρ_Λ is the minimizer of I_Λ^{TFW}, θ_Λ its Lagrange multiplier, and Φ_Λ is defined as in (3.3) by $V_\Lambda - \rho_\Lambda \star \frac{1}{|x|}$.

Proof of Proposition 3.8 For all $\gamma > 0$, we may define $I_\Lambda^{\mathrm{TF}}(\gamma)$, the Thomas–Fermi problem we obtain when we put a multiplicative parameter γ in front of the term $\displaystyle\int_{\mathbf{R}^3} \rho^{5/3}$ in the definition of the Thomas–Fermi functional given by (1.5) in Chapter 1. Then, all the proofs done in [40] go through and, in particular, $\frac{I_\Lambda^{\mathrm{TF}}(\gamma)}{|\Lambda|}$ is bounded independently of $\Lambda \subset \mathbf{Z}^3$.

Since, for any $0 < \gamma < 1$, ρ_Λ is a test function for $I_\Lambda^{\mathrm{TF}}(\gamma)$, we may write that

$$\frac{I_\Lambda^{\mathrm{TFW}}}{|\Lambda|} = E_\Lambda^{\mathrm{TFW}}(\rho_\Lambda)$$

$$\geq \frac{I_\Lambda^{\mathrm{TF}}(\gamma)}{|\Lambda|} + \frac{(1-\gamma)}{|\Lambda|} \int_{\mathbf{R}^3} \rho_\Lambda^{5/3} \, \mathrm{d}x + \frac{1}{|\Lambda|} \int_{\mathbf{R}^3} |\nabla \sqrt{\rho_\Lambda}|^2 \, \mathrm{d}x.$$

Then, using Lemma 3.7, we may bound from above $\dfrac{I_\Lambda^{\mathrm{TFW}}}{|\Lambda|} - \dfrac{I_\Lambda^{\mathrm{TF}}(\gamma)}{|\Lambda|}$ by some positive constant independent of $\Lambda \subset \mathbf{Z}^3$. This bound yields (i) and

$$\frac{1}{|\Lambda|} \int_{\mathbf{R}^3} \rho_\Lambda^{5/3} \, \mathrm{d}x \leq C,$$

from which we infer (ii) with the help of the Hölder inequalities.

(iii)–(v) are shown as in Proposition 3.4 above.   ◇

Remark 3.9 None of the three terms, $\frac{U_\Lambda^{Cb}}{|\Lambda|}$, $\frac{1}{|\Lambda|} \int_{\mathbf{R}^3} V_\Lambda \rho_\Lambda$, and $\frac{1}{|\Lambda|} D^{Cb}(\rho_\Lambda, \rho_\Lambda)$ is bounded. More precisely, since $\frac{U_\Lambda^{Cb}}{|\Lambda|} = O(|\Lambda|^{2/3})$, we deduce from (iv) and (v) that each term is of the same order $|\Lambda|^{2/3}$ as Λ goes to infinity.

We state and prove in the following proposition the analogue of Proposition 3.5 in the case of point nuclei.

Proposition 3.10 (($Cb - \delta$) **case**) *There exists a positive constant C indepen-dent of Λ such that, for all $\Lambda \subset \mathbf{Z}^3$:*

$$\|\rho_\Lambda\|_{L^\infty(\mathbf{R}^3)} \leq C.$$

Remark 3.11 (ii) in Proposition 3.8 therefore holds for $1 \leq p \leq +\infty$.

We group in the following proposition some estimates on Φ_Λ that we obtain in the course of the proof of the Proposition 3.10 and that will be very useful in the rest of our work.

Proposition 3.12 *There exist positive constants C independent of Λ such that:*

(i) $\|\Phi_\Lambda\|_{L^\infty(\Gamma(\Lambda)^c)} \leq C$;

(ii) $\|\Phi_\Lambda\|_{L^p_{\mathrm{unif}}(\mathbf{R}^3)} \leq C,$ *for all $1 \leq p < 3$*
       in the sense that
       $$\sup_{x\in\mathbf{R}^3} \|\Phi_\Lambda\|_{L^p(x+B_1)} \leq C;$$

and, in particular,
       (iii) $\|\Phi_\Lambda\|_{L^p(\Gamma(\Lambda))} \leq C\,|\Lambda|^{1/p}.$

Proof of Proposition 3.10 and Proposition 3.12 The main difference with the ($Cb - m$) case comes from the fact that the effective potential Φ_Λ admits singularities at each point of Λ. Hence, of course, we have to modify the argument we used in the proof of Proposition 3.5. However, we shall obtain uniform bounds for Φ_Λ in $L^p_{\mathrm{unif}}(\mathbf{R}^3)$, for every $1 \leq p < 3$. Next, we shall recover uniform L^∞ bounds for ρ_Λ with a comparison argument.

We begin the argument as in the proof of Proposition 3.5 and we keep the same notation. We shall only indicate the necessary modifications.

The definition of the measure μ_δ arising in (3.24) has to be replaced by

$$\mu_\delta = \left(\sum_{y\in\Lambda} \delta_y\right) \star \omega_\delta^2,$$

which is simply

$$\mu_\delta(x) = \sum_{y\in\Lambda} \omega_\delta^2(x-y)$$

$$= \frac{1}{\delta^3}\sum_{y\in\Lambda} \omega^2\left(\frac{x-y}{\delta}\right). \tag{3.30}$$

Recall now that $supp\ \omega \subset B_1$; then, we assume from now on that $\frac{\sqrt{3}}{2\sqrt{2}} < \delta < 1$, so that the right-hand side of (3.30) is either 0 if $x \notin \cup_{y\in\Lambda}(y+B_\delta)$, or $\frac{1}{\delta^3}\omega_\delta^2(\frac{x-[x]}{\delta})$

if $[x]$ denotes the point of Λ such that $x \in [x] + B_\delta$. In addition, the lower bound we impose on δ ensures that

$$\Gamma(\Lambda) \subset \bigcup_{y \in \Lambda} (y + B_\delta).$$

In any case, one has that

$$\mu_\delta \leq \frac{C}{\delta^3},$$

for some positive constant C independent of δ and Λ.

At this stage, we may copy the proof of Proposition 3.5 to deduce

$$\Phi_\delta^+ \leq C + \frac{C}{\delta^2} \qquad \text{on } \mathbf{R}^3,$$

from which follows the existence of $\frac{1}{2}(\frac{\sqrt{3}}{2\sqrt{2}} + \delta) < \delta' < \delta$ such that

$$\fint_{x+S_{\delta'}} \Phi^+ \leq C + \frac{C}{\delta^2} \qquad \text{for every } x \in \mathbf{R}^3 \tag{3.31}$$

(the choice for the bound from below for δ' will become clear later on).

We first use (3.31) for x in Λ to estimate Φ_Λ in the neighbourhood of its singularities. To fix ideas, we argue for $x = 0$.

In B_δ, it is clear that Φ^+ satisfies

$$-\Delta \Phi^+ \leq 4\pi \, \delta_0$$

(this is where we use the assumption $\delta < 1$). Then, we construct a positive function Φ_1 satisfying

$$\begin{cases} -\Delta \Phi_1 = 0 & \text{in } B_{\delta'}, \\ \quad \Phi_1 = \Phi^+ & \text{on } \partial B_{\delta'} = S_{\delta'}, \end{cases}$$

while it is easy to check that the function $\Phi_2 = \frac{1}{|x|} - \frac{1}{\delta'}$ is the positive solution to

$$\begin{cases} -\Delta \Phi_2 = 4\pi \, \delta_0 & \text{in } B_{\delta'}, \\ \quad \Phi_2 = 0 & \text{on } \partial B_{\delta'}. \end{cases}$$

Hence,

$$\begin{cases} -\Delta(\Phi_1 + \Phi_2) = 4\pi \, \delta_0 & \text{in } B_{\delta'}, \\ \quad \Phi_1 + \Phi_2 = \Phi^+ & \text{on } \partial B_{\delta'}. \end{cases}$$

Then, we apply the maximum principle to write

$$\Phi^+ \leq \Phi_1 + \Phi_2 \qquad \text{in } B_{\delta'}.$$

Next, because of (3.31) and the definition of Φ_2,

$$\Phi^+ \le C + \frac{C}{\delta^2} + \frac{C}{|x|} \qquad \text{in } B_{\delta'}, \tag{3.32}$$

for some various constants C independent of δ and Λ. Indeed, from the mean value property of harmonic functions together with (3.31) and the positivity of Φ^+, we have, for every x in $B_{\delta'}$

$$\Phi_1(x) \le \fint_{S_{\delta'}} \Phi^+ \le C + \frac{C}{\delta^2}.$$

We now set

$$r_0 = \tfrac{1}{2}\left(\frac{\sqrt{3}}{2\sqrt{2}} + \delta \right).$$

Then r_0 does not depend on Λ. In addition,

$$\Gamma(\Lambda) \subset \bigcup_{y \in \Lambda} (y + B_{r_0}),$$

since $\frac{\sqrt{3}}{2\sqrt{2}} < r_0$ and, for every y in Λ, we deduce from (3.32) that

$$\Phi^+(x) \le \frac{C}{|x - y|} \qquad \text{in} \quad y + B_{r_0}, \tag{3.33}$$

for some positive constant C that depends neither on Λ nor on y.

At this stage, we just copy the proof we made in the Yukawa case (see Proposition 2.5 in Chapter 2) to recover a uniform L^∞ bound for ρ_Λ on $\bigcup_{y \in \Lambda}(y + B_{r_0})$.

Moreover, from (3.33), we immediately obtain claim (iii) of Proposition 3.12, once we have recalled that, as in the $(Cb - m)$ case, we already know a uniform L^∞ bound for Φ_Λ^- coming from (3.20).

To complete the proof, our aim now is to obtain some bound for ρ_Λ and Φ_Λ on $\mathbf{R}^3 \setminus \left(\bigcup_{y \in \Lambda}(y + B_{r_0}) \right)$. In fact, away from the singularities, more precisely on $\mathbf{R}^3 \setminus \left(\bigcup_{y \in \Lambda}(y + B_\delta) \right)$, for any $\delta > 0$, we may obtain a uniform L^∞ bound on Φ_Λ which provides a uniform L^∞ bound on ρ_Λ with the help of (3.20).

Let us first point out that, since we already know a uniform L^∞ bound for Φ_Λ^- on $\mathbf{R}^3$, it is sufficient to obtain a uniform bound from above for Φ_Λ^+.

On $\mathbf{R}^3 \setminus \Lambda$, Φ_Λ satisfies $-\triangle\Phi_\Lambda \le 0$. We assume from now on that the cut-off function ω we introduced in the proof of the Proposition 3.5 is a radial function. Then, we claim that for all $0 < \delta < 1$ and for all x, say, in $\mathbf{R}^3 \setminus \left(\bigcup_{y \in \Lambda}(y + B_\delta) \right)$,

$$\Phi(x) \le (\Phi \star \omega_\delta^2)(x) = \Phi_\delta(x)$$

(once more we skip the subscript Λ everywhere). This is a standard fact, but we briefly present the proof for the sake of completeness.

Indeed,

$$
\begin{aligned}
(\Phi \star \omega_\delta^2)(x) &= \int_{\mathbf{R}^3} \Phi(x - y)\, \omega_\delta^2(y)\, \mathrm{d}y \\
&= \int_0^{+\infty} r^2 \mathrm{d}r \int_{S_1} \Phi(x - r\sigma)\, \omega_\delta^2(r\sigma)\, \mathrm{d}\sigma \\
&= \int_0^\delta \omega_\delta^2(r)\, r^2 \mathrm{d}r \int_{S_1} \Phi(x - r\sigma)\, \mathrm{d}\sigma \\
&\geq \Phi(x) \int_0^\delta \omega_\delta^2(r) |S_r|\, \mathrm{d}r,
\end{aligned}
$$

since ω is a radial function with compact support in B_δ, and Φ is subharmonic on $\mathbf{R}^3 \setminus \Lambda$. And, we conclude because $\int_{\mathbf{R}^3} \omega_\delta^2 \, \mathrm{d}x = 1$.

However, we have already shown that

$$
\Phi_\delta \leq \Phi_\delta^+ \leq C + \frac{C}{\delta^2} \qquad \text{on } \mathbf{R}^3,
$$

and thus, for any $0 < \delta < 1$,

$$
|\Phi_\Lambda| \leq C + \frac{C}{\delta^2} \qquad \text{on} \quad \mathbf{R}^3 \setminus \left(\bigcup_{y \in \Lambda} (y + B_\delta) \right), \tag{3.34}
$$

for some constants C that are independent of Λ and δ, using the uniform L^∞ bound for Φ_Λ^-.

Finally, (3.34) together with (3.20) provides the uniform L^∞ bound for ρ_Λ.

The claim (i) in Proposition 3.12 is a simple consequence of (3.34), while (ii) follows upon collecting (3.33) and (3.34).   ◇

For the sake of simplicity, we forget from now on the superscript m or TFW, since the results we state until the end of this section are valid in both cases.

From the estimates we have already obtained, we may derive strong convergence results for the sequence $(\tilde{\rho}_\Lambda)$ defined in Chapter 1, Definition 3, for example, which will be our main idea to derive a lower bound on $\frac{I_\Lambda}{|\Lambda|}$ as Λ goes to infinity. More precisely, we have the following:

Proposition 3.13 ($(Cb - m)$ **and** $(Cb - \delta)$ **cases**) *The sequence* $\sqrt{\tilde{\rho}_\Lambda}$ *is bounded in* $H^1(\Gamma_0)$ *and in* $L^p(\Gamma_0)$, *for all* $1 \leq p \leq +\infty$.

Then, up to a subsequence, we may assume that $\sqrt{\tilde{\rho}_\Lambda}$ *converges to some nonnegative function* $\sqrt{\tilde{\rho}}$ *strongly in* $L^p(\Gamma_0)$, *for all* $1 \leq p < +\infty$, *almost everywhere on* $\mathbf{R}^3$, *and that* $\nabla \sqrt{\tilde{\rho}_\Lambda}$ *converges to* $\nabla \sqrt{\tilde{\rho}}$ *weakly in* $L^2(\Gamma_0)^3$.

Remark 3.14 (1) We establish in Section 3.4, from the L^p bounds on ρ_Λ, that $\tilde{\rho}$ satisfies the periodic boundary conditions on $\partial\Gamma_0$.

(2) *A posteriori*, we shall be able to prove that the limit function $\tilde{\rho}$ is defined in a unique way as the minimizer of I_{per}. Hence, the whole sequence $(\tilde{\rho}_\Lambda)$ is in

fact convergent and its limit does not depend on the sequence (Λ) we consider. By the way, we shall establish a similar result for the sequences $\tilde{\Phi}_\Lambda^m - \int_{\Gamma_0} \tilde{\Phi}_\Lambda^m$ in the $(Cb - m)$ case, and, respectively, $\tilde{\Phi}_\Lambda - \int_{\Gamma_0} \tilde{\Phi}_\Lambda$ in the $(Cb - \delta)$ case (see Corollary 3.25 and Corollary 3.26 for the precise statements).

As the conclusion of this section, we are able to pass to the weak lower limit in some terms of the energy functional, with the help of the $\sim$-transform. The following corollary has to be seen as the first step towards the identification of a lower bound for the energy per cell involving the expected periodic problem (with the $\sim$-transform method) that we prove in Section 3.4.

Corollary 3.15 *For any Van Hove sequence* (Λ), *we have*

$$\liminf_{\Lambda \to \infty} \frac{1}{|\Lambda|} \int_{\mathbf{R}^3} |\nabla \sqrt{\rho_\Lambda}|^2 \, dx \geq \int_{\Gamma_0} |\nabla \sqrt{\tilde{\rho}}|^2 \, dx,$$

$$\liminf_{\Lambda \to \infty} \frac{1}{|\Lambda|} \int_{\mathbf{R}^3} \rho_\Lambda^p \, dx \geq \int_{\Gamma_0} \tilde{\rho}^p \, dx \qquad for \ all \ 1 \leq p < +\infty.$$

Proof of Proposition 3.13 and Corollary 3.15 Recall that $\tilde{\rho}_\Lambda = \frac{1}{|\Lambda|} \sum_{y \in \Lambda} \rho_\Lambda(\cdot + y)$. Then, we use the convexity of the function $t \mapsto t^p$, for $1 \leq p < +\infty$, to obtain

$$\frac{1}{|\Lambda|} \int_{\mathbf{R}^3} \rho_\Lambda^p(x) \, dx \geq \frac{1}{|\Lambda|} \int_{\Gamma(\Lambda)} \rho_\Lambda^p \, dx$$

$$= \frac{1}{|\Lambda|} \sum_{y \in \Lambda} \int_{\Gamma_0} \rho_\Lambda(x + y)^p \, dx$$

$$\geq \int_{\Gamma_0} \left(\frac{1}{|\Lambda|} \sum_{y \in \Lambda} \rho_\Lambda(x + y) \right)^p dx.$$

Thus,

$$\frac{1}{|\Lambda|} \int_{\mathbf{R}^3} \rho_\Lambda^p \, dx \geq \int_{\Gamma_0} \tilde{\rho}_\Lambda^p \, dx.$$

For the same reason, we have

$$\frac{1}{|\Lambda|} \int_{\mathbf{R}^3} |\nabla \sqrt{\rho_\Lambda}|^2 \, dx \geq \frac{1}{|\Lambda|} \int_{\Gamma(\Lambda)} |\nabla \sqrt{\rho_\Lambda}|^2 \, dx \geq \int_{\Gamma_0} |\nabla \sqrt{\tilde{\rho}_\Lambda}|^2 \, dx.$$

We then complete the proof of Proposition 3.13, with the help of Propositions 3.4, 3.5, 3.8, and 3.10, and Rellich's theorem. $\diamond$

In the forthcoming Section 3.3, we prove that the electronic density ρ_Λ is essentially concentrated on $\Gamma(\Lambda)$, in the following sense:

$$\frac{1}{|\Lambda|} \int_{\Gamma(\Lambda)^c} \rho_\Lambda^q(x)\, dx \longrightarrow 0 \qquad \text{as } \Lambda \to \infty,$$

for all $1 \leq q < +\infty$.

3.3   Proof of compactness

In this section, we intend to show that

$$\frac{1}{|\Lambda|} \int_{\Gamma(\Lambda)} \rho_\Lambda(x)\, dx \longrightarrow 1 \qquad \text{as} \quad \Lambda \to \infty, \qquad (3.35)$$

or, equivalently, that

$$\frac{1}{|\Lambda|} \int_{\Gamma(\Lambda)^c} \rho_\Lambda(x)\, dx \longrightarrow 0 \qquad \text{as} \quad \Lambda \to \infty.$$

This is what we call 'compactness for the sequence ρ_Λ'. By the way, (3.35) provides a rough estimate of the 'size' of large neutral molecules; more precisely, we learn from it that the volume is of the order of $|\Lambda|$ for a large neutral molecule with total nuclear charge $|\Lambda|$.

A direct consequence of (3.35) is that

$$\int_{\Gamma_0} \tilde{\rho}_\Lambda(x)\, dx \longrightarrow 1 \qquad \text{as} \quad \Lambda \to \infty.$$

Then, $\int_{\Gamma_0} \tilde{\rho}(x)\, dx = 1$ and there is no loss of 'charge' for the limit periodic problem, a different situation from the Yukawa case (see Chapter 2).

We provide different proofs of this claim.

The first one is specific to smeared out nuclei.

The last four ones apply to both cases, but we only develop the argument in the case of point nuclei, for the sake of brevity.

The second one is based on the quadratic nature of the inter-electronic repulsion term. This quadratic nature allows us to compare the minimum ρ_Λ with some well chosen test function (that actually plays the role of the smeared out nuclei density of the first proof), and to obtain, in a similar fashion to the first proof, the desired compactness.

The third one is based upon a scaling argument that makes sense only when, for the sequence (Λ) we consider, $\Gamma(\Lambda)$ is the cube centered at the origin with volume $|\Lambda|$.

The fourth one has to be seen as a generalization of the scaling argument above to any arbitrary sequence Λ.

Finally, the fifth one is based on an estimation of the charge outside $\Gamma(\Lambda)$. The bounds on the Lagrange multiplier and on the effective potential Φ_Λ that we have obtained in the preceding section (Propositions 3.8 and 3.10) play a crucial role in this strategy of proof.

3.3.1   *Smeared out nuclei*

Our proof relies on the following inequality:

$$\left| \int_{\mathbf{R}^3} (m_\Lambda - \rho_\Lambda) h \right| \leq \frac{1}{(2\pi)^3} \, D^{Cb}(m_\Lambda - \rho_\Lambda, m_\Lambda - \rho_\Lambda)^{1/2} \, \|\nabla h\|_{L^2(\mathbf{R}^3)}, \quad (3.36)$$

which holds for any function h in $H^1(\mathbf{R}^3)$ as soon as $m_\Lambda - \rho_\Lambda$ belongs to $L^{6/5}(\mathbf{R}^3)$. This inequality is easy to prove with the help of the Fourier transform and, in fact, it is simply the standard Cauchy–Schwarz inequality in the duality $H^{-1} \times H^1$ ($(D(\cdot, \cdot))^{1/2}$ is a norm on the dense subspace $L^{6/5}(\mathbf{R}^3)$ of $H^{-1}(\mathbf{R}^3)$).

We define $\Lambda^1 = \{x \in \Gamma(\Lambda)^c \, / d(x; \Gamma(\Lambda)) \leq 1\}$. We introduce a function h_Λ in $H^1(\mathbf{R}^3)$ such that $0 \leq h_\Lambda \leq 1$ almost everywhere on $\mathbf{R}^3$, $h_\Lambda \equiv 1$ on $\Gamma(\Lambda)$, $h_\Lambda \equiv 0$ on $\Gamma(\Lambda)^c \cap (\Lambda^1)^c$, and

$$\int_{\mathbf{R}^3} |\nabla h_\Lambda|^2 \, dx \leq C|\Lambda^1|,$$

where, here and below, C denotes various positive constants independent of Λ.

Next, from Proposition 3.4(iv), we know that

$$D^{Cb}(m_\Lambda - \rho_\Lambda, m_\Lambda - \rho_\Lambda) \leq C|\Lambda|,$$

for some positive constant C that is independent of Λ.

Then, (3.36) can be written as

$$\left| \int_{\Gamma(\Lambda)} (m_\Lambda - \rho_\Lambda) \, dx + \int_{\Lambda^1} \rho_\Lambda \, h_\Lambda \, dx \right| \leq C \, |\Lambda|^{\frac{1}{2}} \, |\Lambda^1|^{\frac{1}{2}},$$

since $supp(m_\Lambda) \subset \Gamma(\Lambda)$.

Now, we notice that

$$\int_{\Gamma(\Lambda)} m_\Lambda \, dx = |\Lambda|$$

and that, from Hölder's inequality and Proposition 3.4(ii),

$$0 \leq \int_{\Lambda^1} \rho_\Lambda \, h_\Lambda \, dx$$

$$\leq \int_{\Lambda^1} \rho_\Lambda \, dx$$

$$\leq \|\rho_\Lambda\|_{L^{5/3}(\mathbf{R}^3)} \, |\Lambda^1|^{\frac{2}{5}}$$

$$\leq C \, |\Lambda|^{\frac{3}{5}} \, |\Lambda^1|^{\frac{2}{5}}.$$

Gathering all these inequalities, we obtain

$$0 \leq 1 - \frac{1}{|\Lambda|} \int_{\Gamma(\Lambda)} \rho_\Lambda(x) \, dx \leq C \left[\left(\frac{|\Lambda^1|}{|\Lambda|} \right)^{\frac{1}{2}} + \left(\frac{|\Lambda^1|}{|\Lambda|} \right)^{\frac{2}{5}} \right]. \tag{3.37}$$

Finally, we conclude, with the help of the Van Hove condition, letting Λ go to infinity. $\diamondsuit$

Remark 3.16 In fact, since we know that ρ_Λ is bounded in L^∞ independently of Λ, we may bound $\int_{\Lambda^1} \rho_\Lambda$ from above by $|\Lambda^1|$, and therefore improve (3.37) a little by

$$0 \leq 1 - \frac{1}{|\Lambda|} \int_{\Gamma(\Lambda)} \rho_\Lambda(x) \, dx \leq C \left[\left(\frac{|\Lambda^1|}{|\Lambda|} \right)^{\frac{1}{2}} + \frac{|\Lambda^1|}{|\Lambda|} \right].$$

In addition, if $\Gamma(\Lambda) = |\Lambda|^{1/3} \cdot \Gamma_0$ is the cube centered at 0, with volume $|\Lambda|$, we have already pointed out that $\frac{|\Lambda^1|}{|\Lambda|} = O(|\Lambda|^{-1/3})$. Hence, with this special choice for the sequence (Λ), we obtain the additional information that

$$\frac{1}{|\Lambda|} \int_{\Gamma(\Lambda)^c} \rho_\Lambda(x) \, dx \leq C |\Lambda|^{-\frac{1}{6}}.$$

We give now four alternative proofs of compactness that work both in the case of smeared and point nuclei.

3.3.2 Convexity argument

We now give a simple proof in the case of point nuclei. We use the same test function, $\bar{\rho}_\Lambda = \sum_{y \in \Lambda} \rho_0(\cdot - y)$, for $\rho_0 \in \mathcal{D}(\mathbf{R}^3)$, $\int_{\Gamma_0} \rho_0 = 1$, as in the proof of Lemma 3.7 to show that $\dfrac{I_\Lambda^{\mathrm{TFW}}}{|\Lambda|}$ is bounded from above independently of Λ.

Next, we deduce from the convexity of the functional $\rho \longrightarrow \int |\nabla \sqrt{\rho}|^2 + \int \rho^{5/3}$ that we have

$$E_\Lambda^{\mathrm{TFW}} \left(\frac{\rho_\Lambda + \bar{\rho}_\Lambda}{2} \right) \leq \tfrac{1}{2} E_\Lambda^{\mathrm{TFW}}(\rho_\Lambda) + \tfrac{1}{2} E_\Lambda^{\mathrm{TFW}}(\bar{\rho}_\Lambda) + \tfrac{1}{2} D^{Cb} \left(\frac{\rho_\Lambda + \bar{\rho}_\Lambda}{2}, \frac{\rho_\Lambda + \bar{\rho}_\Lambda}{2} \right)$$
$$- \tfrac{1}{4} D^{Cb}(\rho_\Lambda, \rho_\Lambda) - \tfrac{1}{4} D^{Cb}(\bar{\rho}_\Lambda, \bar{\rho}_\Lambda)$$
$$= \tfrac{1}{2} E_\Lambda^{\mathrm{TFW}}(\rho_\Lambda) + \tfrac{1}{2} E_\Lambda^{\mathrm{TFW}}(\bar{\rho}_\Lambda)$$
$$- \tfrac{1}{8} D^{Cb}(\rho_\Lambda - \bar{\rho}_\Lambda, \rho_\Lambda - \bar{\rho}_\Lambda). \tag{3.38}$$

In addition, since $\displaystyle\int_{\mathbf{R}^3} \frac{\rho_\Lambda + \bar{\rho}_\Lambda}{2} \, dx = |\Lambda|$, we know that

$$E_\Lambda^{\mathrm{TFW}}(\rho_\Lambda) \leq E_\Lambda^{\mathrm{TFW}} \left(\frac{\rho_\Lambda + \bar{\rho}_\Lambda}{2} \right),$$

by the definition of ρ_Λ. Then, comparing with (3.38), we obtain

$$\tfrac{1}{4} D^{Cb}(\rho_\Lambda - \bar{\rho}_\Lambda, \rho_\Lambda - \bar{\rho}_\Lambda) \leq E_\Lambda^{\mathrm{TFW}}(\bar{\rho}_\Lambda) - E_\Lambda^{\mathrm{TFW}}(\rho_\Lambda)$$

$$\leq C \left| \Lambda \right|,$$

from Lemma 3.7 and its proof.

Next, it just remains to apply the proof given in Section 3.3.1 to $\rho_\Lambda - \bar{\rho}_\Lambda$ instead of $m_\Lambda - \rho_\Lambda$ to conclude. $\diamond$

The convexity of the functional seems to be the key point of the above argument. In fact, we shall see in the sequel of our work that this proof may be adapted to various other situations, even situations where the energy is not convex, provided that the non-convex term is well controlled.

In the three following sections, we get rid of any convexity assumption on the energy functional, and then highlight the key reason why the compactness should hold.

3.3.3 Scaling argument

When we are dealing with a sequence of cubes centered at 0, in the sense that $\Gamma(\Lambda) = |\Lambda|^{1/3}.\Gamma_0$, for all the subsets Λ considered, there is a very simple proof of (3.35) that we now describe.

We use the notation

$$\mu_\Lambda = \sum_{y \in \Lambda} \delta_y,$$

and we set

$$\hat{\rho}_\Lambda = \rho_\Lambda \left(|\Lambda|^{1/3}. \right),$$

and

$$\hat{\mu}_\Lambda = \frac{1}{|\Lambda|} \sum_{y \in \Lambda} \delta_{\frac{y}{|\Lambda|^{1/3}}},$$

so that

$$\int_{\Gamma_0} \hat{\rho}_\Lambda(x) \, dx = \frac{1}{|\Lambda|} \int_{\Gamma(\Lambda)} \rho_\Lambda(x) \, dx,$$

and

$$\hat{\mu}_\Lambda \longrightarrow \mathbf{1}_{\Gamma_0} \qquad \text{as } \Lambda \to \infty,$$

weakly in the sense of measures (or of distributions).

Let us now set

$$A_\Lambda = - \int_{\mathbf{R}^3} V_\Lambda \, \rho_\Lambda + \tfrac{1}{2} D^{Cb}(\rho_\Lambda, \rho_\Lambda) + \tfrac{1}{2} U_\Lambda^{Cb}.$$

We rewrite A_Λ with the above scaling as

$$A_\Lambda = |\Lambda|^{5/3} \left\{ -D^{Cb}(\hat\rho_\Lambda, \hat\mu_\Lambda) + \tfrac{1}{2} D^{Cb}(\hat\rho_\Lambda, \hat\rho_\Lambda) + \frac{1}{2|\Lambda|^2} \sum_{\substack{y,z\in\Lambda \\ y\neq z}} \frac{1}{\left| \frac{y}{|\Lambda|^{1/3}} - \frac{z}{|\Lambda|^{1/3}} \right|} \right\}.$$

From the bounds we already have on (ρ_Λ), we know that $(\hat\rho_\Lambda)$ is a bounded sequence of $L^1 \cap L^\infty(\mathbf{R}^3)$. Then, up to a subsequence, we may assume that $\hat\rho_\Lambda$ converges to $\hat\rho$ weakly in $L^p(\mathbf{R}^3)$, for all $1 < p < +\infty$ and in $L^\infty - \star$ weak. Therefore, we may pass to the limit in $|\Lambda|^{5/3} A_\Lambda$ using these weak convergences, as $\Lambda \to \infty$, and we obtain

$$\liminf_{\Lambda\to\infty} \frac{A_\Lambda}{|\Lambda|^{5/3}} \geq \tfrac{1}{2} D^{Cb}(\hat\rho - \mathbf{1}_{\Gamma_0}, \hat\rho - \mathbf{1}_{\Gamma_0}) \geq 0.$$

Indeed, we recognize in the last term of $\frac{A_\Lambda}{|\Lambda|^{5/3}}$ a Riemann sum on $\Gamma_0 \times \Gamma_0$, thus converging to $\int_{\Gamma_0\times\Gamma_0} \int \frac{1}{|x-y|} \, dx \, dy$.

We now recall that $\frac{A_\Lambda}{|\Lambda|}$ is bounded from above independently of Λ, from Proposition 3.8(iv), (v). Then, $\frac{A_\Lambda}{|\Lambda|^{5/3}}$ has to go to zero as Λ goes to infinity; hence, we deduce that

$$\hat\rho = \mathbf{1}_{\Gamma_0}.$$

But, on the one hand,

$$\int_{\Gamma_0} \hat\rho_\Lambda \, dx \longrightarrow 1 = \int_{\Gamma_0} \hat\rho \, dx,$$

since $\hat\rho_\Lambda$ converges to $\hat\rho$ weakly in $L^{5/3}(\mathbf{R}^3)$; while, on the other hand,

$$\int_{\Gamma_0} \hat\rho_\Lambda(x) \, dx = \frac{1}{|\Lambda|} \int_{\Gamma(\Lambda)} \rho_\Lambda(x) \, dx.$$

And so we conclude.   ◇

Remark 3.17 (1) From this proof, we first learn that the compactness is due to the fact that the sum of the Coulombic interactions terms, namely

$$\tfrac{1}{2} \frac{U_\Lambda^{Cb}}{|\Lambda|} - \frac{1}{|\Lambda|} \int_{\mathbf{R}^3} V_\Lambda \, \rho_\Lambda \, dx + \frac{1}{2|\Lambda|} D^{Cb}(\rho_\Lambda, \rho_\Lambda) \,, \tag{3.39}$$

is bounded independently of Λ. Therefore, we should be able to apply the same argument to any suitable energy functional containing a sum of terms of this kind (the Hartree–Fock functional, for instance). The general argument that we develop hereafter is based on the same idea, but holds for any admissible choice of sequence Λ.

(2) By the way, the above scaling shows that each term in (3.39) is exactly of order $|\Lambda|^{5/3}$, as claimed several times before.

3.3.4 General argument

The key point of our argument consists of proving that, as soon as we know that the sum of terms arising in (3.39) is bounded independently of Λ, we may also bound

$$\frac{1}{|\Lambda|} D^{Cb}(\mathbf{1}_{\Gamma(\Lambda)} - \rho_\Lambda, \mathbf{1}_{\Gamma(\Lambda)} - \rho_\Lambda). \tag{3.40}$$

Then, we may once more copy the proof given in Section 3.3.1 to conclude, applying the inequality (3.36) to $\mathbf{1}_{\Gamma(\Lambda)} - \rho_\Lambda$ instead of $m_\Lambda - \rho_\Lambda$.

Let us now return to the proof of this claim. As in [40], we define

$$f(x) = \frac{1}{|x|} - \int_{\Gamma_0} \frac{dy}{|x - y|}$$

and

$$f_\Lambda(x) = \sum_{k \in \Lambda} \left(\frac{1}{|x - k|} - \int_{\Gamma_0} \frac{dy}{|x - k - y|} \right).$$

It is convenient to rewrite f_Λ as

$$f_\Lambda = V_\Lambda - \mathbf{1}_{\Gamma(\Lambda)} \star \frac{1}{|x|}. \tag{3.41}$$

Besides, it is proved in [40] (see also (3.80) in Section 3.5) that

$$|f(x)| \leq \frac{C}{|x|^4} \tag{3.42}$$

almost everywhere on $\mathbf{R}^3$, for some positive constant C and that f_Λ converges to the periodic potential $G + d$, for some real constant d independent of Λ, uniformly on compact subsets of $\mathbf{R}^3 \setminus \mathbf{Z}^3$. Moreover, for any compact subset K of $\mathbf{R}^3$, $f_\Lambda - \sum_{k \in \Lambda \cap K} \frac{1}{|x - k|}$ converges uniformly on K to $G + d - \sum_{k \in \mathbf{Z}^3 \cap K} \frac{1}{|x - k|}$ (see [40]).

We make use of this last convergence result to show that

$$\frac{U_\Lambda^{Cb}}{|\Lambda|} - \frac{1}{|\Lambda|} D^{Cb}(\mathbf{1}_{\Gamma(\Lambda)}, \mathbf{1}_{\Gamma(\Lambda)}) \tag{3.43}$$

is bounded independently of Λ, which is the first step in recovering (3.40) from (3.39).

Indeed, on the one hand, we write

$$\frac{U_\Lambda^{Cb}}{|\Lambda|} - \frac{1}{|\Lambda|} \int_{\Gamma(\Lambda)} V_\Lambda \, dx$$

$$= \frac{1}{|\Lambda|} \sum_{\substack{y,z \in \Lambda \\ y \neq z}} \frac{1}{|y - z|} - \frac{1}{|\Lambda|} \sum_{y,z \in \Lambda} \int_{\Gamma_0} \frac{dx}{|x + z - y|}$$

$$= \frac{1}{|\Lambda|} \sum_{z \in \Lambda} \lim_{\substack{x \to z \\ x \neq z}} \left[f_\Lambda(x) - \frac{1}{|x - z|} \right].$$

Besides, we express the last identity, with the help of the $\sim$-transform, as

$$\lim_{\substack{x \to 0 \\ x \neq 0}} \left[\tilde{f}_\Lambda(x) - \frac{1}{|x|} \right].$$

Finally, appealing to the technical Lemma 2.24 given in Section 2.4 of Chapter 2, we may check that this quantity converges to

$$\lim_{\substack{x \to 0 \\ x \neq 0}} \left[G(x) - \frac{1}{|x|} + d \right] = M + d,$$

by the definition of M, as $\Lambda \to \infty$, and thus is bounded.

Hence, at this stage, we have proved that

$$\frac{U_\Lambda^{Cb}}{|\Lambda|} - \frac{1}{|\Lambda|} \int_{\Gamma(\Lambda)} V_\Lambda \, dx$$

is bounded independently of Λ. In order to establish (3.43), we now prove that, on the other hand,

$$\frac{1}{|\Lambda|} \int_{\Gamma(\Lambda)} V_\Lambda \, dx - \frac{1}{|\Lambda|} D^{Cb}(\mathbf{1}_{\Gamma(\Lambda)}, \mathbf{1}_{\Gamma(\Lambda)})$$

is bounded too. However, because of (3.41), this quantity is simply $\frac{1}{|\Lambda|} \int_{\Gamma(\Lambda)} f_\Lambda \, dx$, and it is easy to deduce from (3.42) that this is of order $O(1)$ as Λ goes to infinity (see [40]). By the way, we may even show that it converges to $\int_{\Gamma_0} (G(x) + d) \, dx = d$.

Finally, it remains to check that

$$\frac{1}{|\Lambda|} \int_{\mathbf{R}^3} V_\Lambda \, \rho_\Lambda \, dx - \frac{1}{|\Lambda|} D^{Cb}(\mathbf{1}_{\Gamma(\Lambda)}, \rho_\Lambda)$$

is indeed of order $O(1)$ as Λ goes to infinity. For this purpose, we rewrite this expression as

$$\frac{1}{|\Lambda|} \int_{\mathbf{R}^3} f_\Lambda \, \rho_\Lambda \, dx.$$

Next, we check as in [40] that

$$\|f_\Lambda\|_{L^p(\mathbf{R}^3)} \leq C \, |\Lambda|^{1/p} \qquad \text{for all } 1 \leq p < 3.$$

We conclude with the help of the Proposition 3.8, together with Hölder's in-equalities.

Our claim follows, collecting all these estimates.   ◇

Remark 3.18 This proof and the scaling argument follow the same pattern; indeed, if we apply the scaling given in Section 3.3.3 to the quantity

$$\frac{1}{|\Lambda|} D^{Cb}(\mathbf{1}_{\Gamma(\Lambda)} - \rho_\Lambda, \mathbf{1}_{\Gamma(\Lambda)} - \rho_\Lambda),$$

we find that

$$|\Lambda|^{2/3} D^{Cb}(\mathbf{1}_{\Gamma_0} - \hat{\rho}_\Lambda, \mathbf{1}_{\Gamma_0} - \hat{\rho}_\Lambda).$$

The fact that it is bounded yields the convergence of $\hat{\rho}_\Lambda$ to $\mathbf{1}_{\Gamma_0}$ as Λ goes to infinity, and hence the compactness of ρ_Λ.

3.3.5 *An argument via the estimation of the outside charge*

The last proof of the compactness that we develop now relies very much upon the bound on the Lagrange multiplier due to Solovej (see [53] and Proposition 3.8(iii)). In particular, we used this bound in the course of the proof of Proposition 3.12 to ensure that Φ_Λ was bounded in $L^\infty \left(\mathbf{R}^3 \setminus \bigcup_{y \in \Lambda} y + B_\delta \right)$, for every $0 < \delta < 1$, independently of Λ.

Now, we recall that Φ_Λ satisfies

$$-\Delta \Phi_\Lambda = -4\pi \, \rho_\Lambda \qquad \text{on } \mathbf{R}^3 \setminus \Lambda. \tag{3.44}$$

We integrate this equation by parts over $\Gamma(\Lambda)^c$ and we obtain

$$\int_{\Gamma(\Lambda)^c} \rho_\Lambda = \frac{1}{4\pi} \int_{\partial \Gamma(\Lambda)} \frac{\partial \Phi_\Lambda}{\partial n} \tag{3.45}$$

(this calculation makes sense since, for example, Φ_Λ is C^1 in the vicinity of $\partial \Gamma(\Lambda)$). But, since, from Proposition 3.10, we know that

$$\|\rho_\Lambda\|_{L^\infty(\mathbf{R}^3)} \leq C,$$

we also have from (3.44) that, in particular,

$$\|\Delta \Phi_\Lambda\|_{L^\infty(\mathbf{R}^3 \setminus \bigcup_{y \in \Lambda} y + B_\delta)} \leq C,$$

for every $0 < \delta < 1$, where, here and below, C denotes various positive constants independent of Λ. Thus, from standard elliptic regularity results, the two facts that $|\Delta \Phi_\Lambda| \leq C$ and $|\Phi_\Lambda| \leq C$ imply that $\nabla \Phi_\Lambda$ is bounded in $L^\infty(\mathbf{R}^3 \setminus \bigcup_{y \in \Lambda} y + B_\delta)$ independently of Λ. In addition, Φ_Λ being C^1 in the neighbourhood of $\partial \Gamma(\Lambda)$, the uniform L^∞ bound on $\nabla\Phi_\Lambda$ holds pointwise there. Then, using this estimate and (3.45), we obtain

$$\int_{\Gamma(\Lambda)^c} \rho_\Lambda \leq C \, |\partial\Gamma(\Lambda)|.$$

Besides, for any Van Hove sequence,

$$|\partial\Gamma(\Lambda)| = o(|\Lambda|),$$

since, for example,

$$|\partial\Gamma(\Lambda)| \leq 12 \, \sharp \left\{ y \in \Lambda; dist(y; \partial\Gamma(\Lambda)) \leq \tfrac{1}{2} \right\},$$

and the proof is complete.   ◇

Before turning to the following section, let us make some comparisons with the Thomas–Fermi case.

3.3.6   Compactness in the TF case

The proofs of compactness given in the TFW setting obviously carry through, since, as soon as we know that the energy per unit cell is bounded, we know that the sum of the electrostatic terms (3.39) is bounded; next, we reproduce the proof given in Section 3.3.4. However, it seems to us noteworthy to propose an alternative proof which only relies upon the properties of the minimizing TF density, denoted by ρ_Λ^{TF}, and the corresponding effective potential, Φ_Λ. Indeed, in that case, the Euler–Lagrange equation reads

$$\tfrac{5}{3} \rho_\Lambda^{TF\,2/3} = \Phi_\Lambda, \tag{3.46}$$

while Φ_Λ is the solution to

$$-\Delta\Phi_\Lambda = 4\pi \left[\sum_{y \in \Lambda} \delta_y - \rho_\Lambda^{TF} \right],$$

going to zero at infinity. From these two equations, it follows that Φ_Λ also satisfies

$$0 \leq \Phi_\Lambda \leq C \sum_{y \in \Lambda} \frac{1}{|x-y|^4} \qquad \text{a.e. on } \mathbf{R}^3, \tag{3.47}$$

as proved by Lieb and Simon in [40]. From (3.47), one then obtains

$$\lim_{\Lambda \to \infty} \frac{1}{|\Lambda|} \int_{\Gamma(\Lambda)^c} \Phi_\Lambda^p = 0, \qquad \text{for all} \quad 1 \leq p < +\infty \tag{3.48}$$

(we refer the reader to [40] for a proof of this claim).

The compactness of ρ_Λ^{TF} follows, gathering together (3.46) and (3.48).

3.4   Lower bound for the energy

In this section, we are going to obtain a lower bound for $\dfrac{I_\Lambda^{\mathrm{TFW}}}{|\Lambda|}$ (respectively, $\dfrac{I_\Lambda^m}{|\Lambda|}$) in terms of $I_{\mathrm{per}}^{\mathrm{TFW}} + \dfrac{M}{2}$ (respectively, $I_{\mathrm{per}}^m + \dfrac{M}{2}$).

Our first strategy consists of comparing the Coulomb model with the Yukawa model with a parameter a, next letting a go to 0^+.

Next, we develop alternative proofs that we call direct, since they do not rely upon the approximation of the Coulomb potential by a Yukawa potential, both in the case of smeared out and point nuclei.

3.4.1   Comparison proof

In order to simplify the notation, in this chapter, the Yukawa-type expressions are referred to with the superscript a, while the Coulombic ones use the superscript Cb; in addition, we skip the superscript TFW everywhere. Moreover, as there should be no ambiguity, we forget the subscript m in the subsection devoted to the $(Cb - m)$ case.

3.4.1.1   *Smeared out nuclei*   In the following, ρ is a positive function such that $\int_{\mathbf{R}^3} \rho = |\Lambda|$, $\sqrt{\rho} \in H^1(\mathbf{R}^3)$ and Λ is an arbitrary Van Hove sequence of $\mathbf{Z}^3$. Then, $I_\Lambda^{Cb} + \frac{|\Lambda|}{2} D^{Cb}(m, m)$ is, by definition and by using the formulation (3.18), the infimum of the functional defined by

$$E_\Lambda^{Cb}(\rho) = \int_{\mathbf{R}^3} |\nabla \sqrt{\rho}|^2 + \int_{\mathbf{R}^3} \rho^{5/3} + \tfrac{1}{2} D^{Cb}(m_\Lambda - \rho, m_\Lambda - \rho),$$

over all such functions ρ.

We obtain a similar expression for

$$I_\Lambda^a + \frac{|\Lambda|}{2} D^a(m, m).$$

Hence, it is straightforward to check that

$$\frac{I_\Lambda^{Cb}}{|\Lambda|} + \tfrac{1}{2} D^{Cb}(m, m) \geq \frac{I_\Lambda^a}{|\Lambda|} + \tfrac{1}{2} D^a(m, m). \tag{3.49}$$

Indeed, we have

$$D^a(\mu, \mu) \leq D^{Cb}(\mu, \mu)$$

for any function μ for which these quantities make sense, since, with the help of the Fourier transform,

$$D^a(\mu, \mu) = \int_{\mathbf{R}^3} \frac{|\hat{\mu}(\xi)|^2}{a^2 + |\xi|^2} \, \mathrm{d}\xi \leq \int_{\mathbf{R}^3} \frac{|\hat{\mu}(\xi)|^2}{|\xi|^2} \, \mathrm{d}\xi = D^{Cb}(\mu, \mu),$$

where

$$\hat{\mu}(\xi) = \int_{\mathbf{R}^3} \mu(x) e^{-2i\pi (x,\xi)} \, dx.$$

Now, one passes to the limit in (3.49) as $\Lambda \to \infty$ and obtains

$$\liminf_{\Lambda \to \infty} \frac{I_\Lambda^{Cb}}{|\Lambda|} \geq \liminf_{\Lambda \to \infty} \frac{I_\Lambda^a}{|\Lambda|} + \tfrac{1}{2} D^a(m,m) - \tfrac{1}{2} D^{Cb}(m,m)$$

$$= I_{\mathrm{per}}^a(\mu_a) + \tfrac{1}{2} D^a(m,m) - \tfrac{1}{2} D^{Cb}(m,m), \qquad (3.50)$$

where μ_a is the charge of the absolute minimum of $\lambda \mapsto I_{\mathrm{per}}^a(\lambda)$ (see Section 2.5 of Chapter 2).

From Section 2.6 of Chapter 2, we know that

$$\lim_{a \to 0^+} I_{\mathrm{per}}^a(\mu_a) = I_{\mathrm{per}}^{Cb} + \frac{M}{2},$$

while

$$0 \leq D^{Cb}(m,m) - D^a(m,m)$$

$$\leq \iint_{\Gamma_0 \times \Gamma_0} m(x) \left[\frac{1}{|x-y|} - \frac{e^{-a|x-y|}}{|x-y|} \right] m(y) \, dx \, dy,$$

$$\leq a,$$

since

$$0 \leq 1 - e^{-at} \leq at \qquad \text{for all } t \geq 0.$$

Our claim follows immediately from (3.50) letting a go to 0^+.   ◇

3.4.1.2 *Point nuclei* We have to compare

$$\tfrac{1}{2} D^{Cb}(\rho,\rho) - \int_{\mathbf{R}^3} V_\Lambda^{Cb}(x)\rho(x) \, dx + \tfrac{1}{2} U_\Lambda^{Cb}$$

with the corresponding Yukawa expression. We rewrite each term in the above sum with the help of the Fourier transform in the following way:

$$\tfrac{1}{2} \int_{\mathbf{R}^3} \frac{|\hat{\rho}(\xi)|^2}{|\xi|^2} \, d\xi - \sum_{y \in \Lambda} \int_{\mathbf{R}^3} \frac{e^{2i\pi (y,\xi)}}{|\xi|^2} \hat{\rho}(\xi) \, d\xi$$

$$+ \tfrac{1}{2} \sum_{\substack{y,z \in \Lambda \\ y \neq z}} \int_{\mathbf{R}^3} \frac{e^{2i\pi (y-z,\xi)}}{|\xi|^2} \, d\xi.$$

Next, we notice that this quantity is also

$$\frac{1}{2} \lim_{R \to +\infty} \left[\int_{B_R} \frac{\left| \sum_{y \in \Lambda} e^{2i\pi (y,\xi)} - \hat{\rho}(\xi) \right|^2}{|\xi|^2} \, d\xi - |\Lambda| \int_{B_R} \frac{d\xi}{|\xi|^2} \right]. \qquad (3.51)$$

Now, we make two observations. On the one hand, the first integral is obviously bounded below, for every $R > 0$, by

$$\int_{B_R} \frac{\left| \sum_{y \in \Lambda} e^{2i\pi (y,\xi)} - \hat{\rho}(\xi) \right|^2}{a^2 + |\xi|^2} \, d\xi.$$

On the other hand, it is easy to check that

$$\lim_{R \to +\infty} \int_{B_R} \left(\frac{1}{|\xi|^2} - \frac{1}{a^2 + |\xi|^2} \right) d\xi = 2\pi^2 a.$$

Collecting these two facts and comparing with (3.51), we obtain that

$$\frac{1}{2} \lim_{R \to +\infty} \left[\int_{B_R} \frac{\left| \sum_{y \in \Lambda} e^{2i\pi (y,\xi)} - \hat{\rho}(\xi) \right|^2}{|\xi|^2} \, d\xi - |\Lambda| \int_{B_R} \frac{d\xi}{|\xi|^2} \right]$$

$$\geq \frac{1}{2} \lim_{R \to +\infty} \left[\int_{B_R} \frac{\left| \sum_{y \in \Lambda} e^{2i\pi (y,\xi)} - \hat{\rho}(\xi) \right|^2}{a^2 + |\xi|^2} \, d\xi - |\Lambda| \int_{B_R} \frac{d\xi}{a^2 + |\xi|^2} \right] - 2\pi^2 a |\Lambda|$$

$$= \frac{1}{2} D^a(\rho, \rho) - \int_{\mathbf{R}^3} V_\Lambda^a(x) \rho(x) \, dx + \frac{1}{2} U_\Lambda^a - \pi^2 a |\Lambda|.$$

Thus we have proved that

$$\frac{I_\Lambda^{Cb}}{|\Lambda|} \geq \frac{I_\Lambda^a}{|\Lambda|} - \pi^2 a.$$

We conclude as in the preceding case, letting Λ go to infinity, and then letting a go to 0^+. $\diamond$

3.4.2 Direct proof for the $(Cb - m)$ program

Before entering the details of the proof, we first show a preliminary result whose goal is to legitimate the use of the $\sim$-transform trick. Indeed, the forthcoming lemma will ensure that the $\sim$-transform forces the sequences of the so-transformed quantities (densities, effective potentials, and so on) to become asymptotically periodic.

Lemma 3.19 *Let (Λ) be a Van Hove sequence and let h_Λ be a sequence in $L^p(\mathbf{R}^3)$ for some $1 \le p \le +\infty$, such that*

$$\|h_\Lambda\|_{L^p(\mathbf{R}^3)} \le C|\Lambda|^{\frac{1}{p}}, \qquad (3.52)$$

for some constant C independent of Λ.

Suppose that $\tilde{h}_\Lambda$ converges to some function $\tilde{h}$ a.e. on $\mathbf{R}^3$. Then, $\tilde{h}$ is a periodic function on $\mathbf{R}^3$.

Proof of Lemma 3.19 We begin with the simpler case when $p = +\infty$. Then, for every k in $\mathbf{Z}^3$, we have that

$$\tilde{h}_\Lambda(\cdot + k) - \tilde{h}_\Lambda \longrightarrow \tilde{h}(\cdot + k) - \tilde{h}$$

as $\Lambda \to \infty$, almost everywhere on $\mathbf{R}^3$.

Besides, for almost every x in $\mathbf{R}^3$,

$$
\begin{aligned}
|\tilde{h}_\Lambda(x &+ k) - \tilde{h}_\Lambda(x)| \\
&= \frac{1}{|\Lambda|} \left| \sum_{y \in \Lambda} h_\Lambda(x + y + k) - \sum_{z \in \Lambda} h_\Lambda(x + z) \right| \\
&= \frac{1}{|\Lambda|} \left| \sum_{y \in \Lambda + k} h_\Lambda(x + y) - \sum_{z \in \Lambda} h_\Lambda(x + z) \right| \\
&\le \frac{1}{|\Lambda|} \left| \sum_{\substack{y \in (\Lambda + k) \\ y \notin \Lambda}} h_\Lambda(x + y) \right| + \frac{1}{|\Lambda|} \left| \sum_{\substack{y \in \Lambda \\ y \notin (\Lambda + k)}} h_\Lambda(x + y) \right| \\
&\le \|h_\Lambda\|_{L^\infty(\mathbf{R}^3)} \frac{1}{|\Lambda|} \Big(vol\big((\Lambda + k) \setminus \Lambda\big) + vol\big(\Lambda \setminus (\Lambda + k)\big) \Big) \\
&\le 2 \|h_\Lambda\|_{L^\infty(\mathbf{R}^3)} \frac{|\Lambda^{|k|}|}{|\Lambda|}.
\end{aligned}
$$

We conclude by letting $\Lambda \to \infty$, using (3.52) and the definition of a Van Hove sequence.

When $p < +\infty$, the proof is not really more sophisticated. We proceed as above, to write that

$$
\begin{aligned}
\|\tilde{h}_\Lambda(\cdot &+ k) - \tilde{h}_\Lambda\|_{L^1(\Gamma_0)} \\
&\le \frac{1}{|\Lambda|} \sum_{\substack{y \in (\Lambda + k) \\ y \notin \Lambda}} \|h_\Lambda(\cdot + y)\|_{L^1(\Gamma_0)} + \frac{1}{|\Lambda|} \sum_{\substack{y \in \Lambda \\ y \notin (\Lambda + k)}} \|h_\Lambda(\cdot + y)\|_{L^1(\Gamma_0)} \\
&\le \frac{C}{|\Lambda|} |\Lambda^{|k|}|^{\frac{1}{p'}} |\Lambda|^{\frac{1}{p}}, \qquad \text{when } 1 < p, \quad \text{with } \frac{1}{p'} + \frac{1}{p} = 1
\end{aligned}
$$

or $\leq C \dfrac{|\Lambda^{|k|}|}{|\Lambda|},$ \qquad when $p = 1,$

because of Hölder's inequalities and (3.52). \hfill $\Diamond$

Now, we emphasize the fact that the uniqueness of the minimizer of I_Λ^m yields that

$$\text{for every } k \in \mathbf{Z}^3, \qquad \rho_{\Lambda+k} = \rho_\Lambda(\cdot - k). \tag{3.53}$$

Then, we may apply the same analysis made on $\tilde{\rho}_\Lambda$ over Γ_0 to any Γ_k, k in $\mathbf{Z}^3$; so that we obtain by this process $\tilde{\rho}$ as the limit of $\tilde{\rho}_\Lambda$, a.e. on $\mathbf{R}^3$, and as the strong limit of $\tilde{\rho}_\Lambda$ in $L_{\text{loc}}^p(\mathbf{R}^3)$, for all $1 \leq p < +\infty$. In addition, $\tilde{\rho}$ is a periodic function on $\mathbf{R}^3$ because of Lemma 3.19.

Let us now define

$$\tilde{\Phi}_\Lambda^m = \frac{1}{|\Lambda|} \sum_{y \in \Lambda} \Phi_\Lambda^m(\cdot + y).$$

Then, $\tilde{\Phi}_\Lambda^m$ is the unique solution to the Poisson equation

$$-\triangle \tilde{\Phi}_\Lambda^m = 4\pi\, [\tilde{m}_\Lambda - \tilde{\rho}_\Lambda] \qquad \text{on } \mathbf{R}^3, \tag{3.54}$$

going to zero at infinity.

Moreover, as we know from Propositions 3.4 and 3.5 that

$$\left| \frac{1}{|\Lambda|} \int_{\Gamma(\Lambda)} \Phi_\Lambda^m \,\mathrm{d}x \right| \leq C$$

and

$$\frac{1}{|\Lambda|} \int_{\Gamma(\Lambda)} |\nabla \Phi_\Lambda^m|^2 \,\mathrm{d}x \leq C$$

(where C is independent of Λ), we deduce that

$$\left| \int_{\Gamma_0} \tilde{\Phi}_\Lambda^m \,\mathrm{d}x \right| \leq C$$

and

$$\int_{\Gamma_0} |\nabla \tilde{\Phi}_\Lambda^m|^2 \,\mathrm{d}x \leq C.$$

Hence, by the Poincaré–Wirtinger inequality, $(\tilde{\Phi}_\Lambda^m)$ is bounded in $H^1(\Gamma_0) \cap L^\infty(\mathbf{R}^3)$, and thus up to a subsequence converges to $\tilde{\Phi}^m \in H^1(\Gamma_0) \cap L^\infty(\mathbf{R}^3)$, weakly in $H^1(\Gamma_0)$, a.e. on Γ_0 and strongly in $L^p(\Gamma_0)$, for all $1 \leq p < +\infty$ (from Rellich's theorem and then Hölder's inequalities).

In addition, we deduce from (3.53) that, for any k in $\mathbf{Z}^3$,

$$\Phi^m_{\Lambda+k} = \Phi^m_\Lambda(\cdot - k),$$

since, by definition,

$$\Phi^m_\Lambda = (m_\Lambda - \rho_\Lambda) \star \frac{1}{|x|}.$$

Then, we may repeat the argument we applied to $\tilde{\rho}_\Lambda$ to check that $\tilde{\Phi}^m$ is a periodic function on $\mathbf{R}^3$, which satisfies, in addition

$$-\triangle\tilde{\Phi}^m = 4\pi\,[\tilde{m} - \tilde{\rho}] \qquad \text{on } \mathbf{R}^3, \tag{3.55}$$

at least in the sense of distributions, $\tilde{m}$ being the periodic function equal to m on Γ_0 ((3.55) follows directly from (3.54), letting $\Lambda \to \infty$).

However, we know another solution to (3.55), namely

$$\bar{\Phi}(x) = \int_{\Gamma_0} G(x - y)[m(y) - \tilde{\rho}(y)]\,\mathrm{d}y. \tag{3.56}$$

(We remark that the fact that $\tilde{\Phi}^m$ is periodic implies from (3.54) that $\int_{\Gamma_0} \tilde{\rho}(x)\,\mathrm{d}x = 1$). Then, there is a constant c (possibly depending on the sequence Λ) such that

$$\tilde{\Phi}^m(x) = \bar{\Phi}(x) + c \tag{3.57}$$

(actually, $c = \int_{\Gamma_0} \tilde{\Phi}^m(x)\,\mathrm{d}x$, since we may verify that $\int_{\Gamma_0} \bar{\Phi}(x)\,\mathrm{d}x = 0$).

Now, we may go back to the expression for the energy given by (3.19) and write the following lower bounds on all these non-negative terms:

$$\liminf_{\Lambda\to\infty} \frac{I^m_\Lambda}{|\Lambda|} \geq \liminf_{\Lambda\to\infty} \left[\frac{1}{|\Lambda|} \int_{\mathbf{R}^3} |\nabla\sqrt{\rho_\Lambda}|^2 + \frac{1}{|\Lambda|} \int_{\mathbf{R}^3} \rho_\Lambda^{5/3} \right.$$
$$\left. + \frac{1}{8\pi|\Lambda|} \int_{\mathbf{R}^3} |\nabla\Phi^m_\Lambda|^2 \right] - \tfrac{1}{2} D^{Cb}(m, m)$$

$$\tag{3.58}$$

$$\geq \int_{\Gamma_0} |\nabla\sqrt{\tilde{\rho}}|^2 + \int_{\Gamma_0} \tilde{\rho}^{5/3}$$
$$+ \frac{1}{8\pi} \int_{\Gamma_0} |\nabla\tilde{\Phi}^m|^2 - \tfrac{1}{2} D^{Cb}(m, m). \tag{3.59}$$

Besides, from the periodic boundary condition satisfied by $\tilde{\Phi}^m$ and the fact that (3.55) implies the continuity of $\frac{\partial\tilde{\Phi}^m}{\partial n}$, we check that

$$\int_{\partial\Gamma_0} \frac{\partial\tilde{\Phi}^m}{\partial n}\,\tilde{\Phi}^m = 0.$$

Thus,

$$\int_{\Gamma_0} |\nabla \tilde{\Phi}^m|^2 = \int_{\Gamma_0} -\triangle \tilde{\Phi}^m \cdot \tilde{\Phi}^m$$

(the second integral makes sense since, because of (3.55), $\tilde{\Phi}^m \in H^2_{\mathrm{loc}}(\mathbf{R}^3)$).

Finally, the third term in (3.59) is written as

$$\frac{1}{2} \iint_{\Gamma_0 \times \Gamma_0} (\tilde{m}(x) - \tilde{\rho}(x)) G(x - y)(\tilde{m}(y) - \tilde{\rho}(y)) \, dx \, dy$$

$$= \tfrac{1}{2} D_G(m, m) - \int_{\Gamma_0} G_m(x) \tilde{\rho}(x) \, dx + \tfrac{1}{2} D_G(\tilde{\rho}, \tilde{\rho}),$$

with the help of (3.55) together with (3.56) and (3.57) (let us recall that $\int_{\Gamma_0} (\tilde{m} - \tilde{\rho}) = 0$).

Then, we recognize in (3.59) the expression of $E^m_{\mathrm{per}}(\tilde{\rho}) + \frac{M}{2}$, which is larger than $I^m_{\mathrm{per}} + \frac{M}{2}$, since $\tilde{\rho}$ is a test function for I^m_{per}. $\diamondsuit$

Remark 3.20 In fact, as already pointed out in the introduction to this chapter, when we assume that m shares the symmetry of the unit cube, the periodic minimization problem with periodic boundary conditions is equivalent to the minimization problem without boundary conditions. Thus, in that case, there is no need of Lemma 3.19 to ensure that $\tilde{\rho}$ is a test function for I^m_{per}. However, it is worth noticing that the argument developed above works if m is not symmetric or if the primitive cell is not cubic, and that the proof given in the forthcoming section (devoted to the point nuclei case) will also be valid if there are several point nuclei in the (not necessarily cubic) unit cell.

3.4.3   Direct proof for the $(Cb - \delta)$ program

We first recall that the energy per cell is given by

$$\frac{I^{\mathrm{TFW}}_\Lambda}{|\Lambda|} = \frac{1}{|\Lambda|} \int_{\mathbf{R}^3} |\nabla \sqrt{\rho_\Lambda}|^2 + \frac{1}{|\Lambda|} \int_{\mathbf{R}^3} \rho_\Lambda^{5/3}$$

$$+ \frac{1}{2|\Lambda|} U^{Cb}_\Lambda - \frac{1}{|\Lambda|} \int_{\mathbf{R}^3} V_\Lambda \, \rho_\Lambda + \frac{1}{2|\Lambda|} D^{Cb}(\rho_\Lambda, \rho_\Lambda). \qquad (3.60)$$

Because of Proposition 3.8, Proposition 3.10, and Lemma 3.19, we already know, as in the $(Cb - m)$ case above, that the sequence $\tilde{\rho}_\Lambda = \frac{1}{|\Lambda|} \sum_{y \in \Lambda} \rho_\Lambda(\cdot + y)$ converges to a positive periodic function $\tilde{\rho}$ on $\mathbf{R}^3$, almost everywhere on $\mathbf{R}^3$, strongly in $L^p_{\mathrm{loc}}(\mathbf{R}^3)$, for all $1 \le p < +\infty$, and $\nabla \sqrt{\tilde{\rho}_\Lambda}$ converges to $\nabla \sqrt{\tilde{\rho}}$ weakly in $H^1_{\mathrm{loc}}(\mathbf{R}^3)$. Then, we may pass to the weak limit and obtain lower bounds in the first two terms of (3.60), as in the smeared out nuclei case, with the help of the $\sim$-transform.

Thus, we now concentrate now our attention on the sum of the last three terms of (3.60).

In the compactness proof of Section 3.3.4, we have already pointed out that it was convenient to introduce the function

$$f_\Lambda(x) = \sum_{k \in \Lambda} \left(\frac{1}{|x - k|} - \int_{\Gamma_0} \frac{dy}{|x - k - y|} \right)$$

$$= V_\Lambda - 1_{\Gamma(\Lambda)} \star \frac{1}{|x|},$$

and to rewrite this sum as

$$\frac{1}{2|\Lambda|} \sum_{\substack{z \in \Lambda}} \lim_{\substack{x \to z \\ x \neq z}} \left[f_\Lambda(x) - \frac{1}{|x - z|} \right] + \frac{1}{2|\Lambda|} \int_{\Gamma(\Lambda)} f_\Lambda \, dx$$

$$- \frac{1}{|\Lambda|} \int_{\Gamma(\Lambda)} f_\Lambda \, \rho_\Lambda \, dx + \frac{1}{2|\Lambda|} D^{Cb}(1_{\Gamma(\Lambda)} - \rho_\Lambda, 1_{\Gamma(\Lambda)} - \rho_\Lambda). \quad (3.61)$$

We recall from Section 3.3.4 that f_Λ converges to the periodic potential $G + d$ uniformly on the compact subsets of $\mathbf{R}^3 \setminus \mathbf{Z}^3$, d being a constant independent of Λ. In Section 3.3.4, once more, we proved that the sum of the first two terms arising in (3.61) above converges to $\frac{M}{2} + d$ as Λ goes to infinity. We are not going to redo the proof of this fact here, but we only reproduce the main lines of the argument and we refer the reader to this section for the notation and the details.

Indeed, we rewrite the first term in (3.61) above as

$$\tfrac{1}{2} \lim_{\substack{x \to 0 \\ x \neq 0}} \left[\tilde{f}_\Lambda(x) - \frac{1}{|x|} \right].$$

Then, using the technical Lemma 2.24 given in Section 2.4 of Chapter 2, we show that the above quantity converges to $\frac{1}{2} \lim_{\substack{x \to 0 \\ x \neq 0}} \left[\tilde{G}(x) + d - \frac{1}{|x|} \right]$ as Λ goes to infinity; that is, $\frac{M+d}{2}$.

Besides, the second term of (3.61) is also $\frac{1}{2} \int_{\Gamma_0} \tilde{f}_\Lambda$, and thus converges to $\frac{1}{2} \int_{\Gamma_0} (G + d) = \frac{d}{2}$ as Λ goes to infinity.

We now prove that

$$\lim_{\Lambda \to \infty} \frac{1}{|\Lambda|} \int_{\mathbf{R}^3} f_\Lambda \, \rho_\Lambda = d + \int_{\Gamma_0} G(y) \, \tilde{\rho}(y) \, dy. \quad (3.62)$$

Indeed, on the one hand, we first state that

$$\frac{1}{|\Lambda|} \int_{\Gamma(\Lambda)^c} |f_\Lambda|^p \overset{\Lambda \to \infty}{\longrightarrow} 0,$$

for all $1 \leq p < +\infty$ (the proof is taken from [40] and we just redo it for the sake of consistency). Then, because of Proposition 3.8(ii) and Hölder's inequalities, we immediately deduce

$$\frac{1}{|\Lambda|} \int_{\Gamma(\Lambda)^c} f_\Lambda \, \rho_\Lambda \stackrel{\Lambda \to \infty}{\longrightarrow} 0. \qquad (3.63)$$

For this purpose, we recall from Section 3.3 (formula (3.42)) the existence of a positive constant C such that

$$|f(x)| \leq \frac{C}{|x|^4} \qquad \text{a.e. on } \mathbf{R}^3.$$

Hence, the function $f_\Lambda = \sum_{y \in \Lambda} f(\cdot - y)$ is bounded on $\Gamma(\Lambda)^c$ independently of Λ. Then, in view of Hölder's inequalities, it is sufficient to check that

$$\frac{1}{|\Lambda|} \int_{\Gamma(\Lambda)^c} |f_\Lambda| \stackrel{\Lambda \to \infty}{\longrightarrow} 0$$

or even that

$$\frac{1}{|\Lambda|} \int_{\Gamma(\Lambda)^c} \sum_{y \in \Lambda} \frac{1}{|x - y|^4} \, dx \stackrel{\Lambda \to \infty}{\longrightarrow} 0.$$

We thus fix $\frac{1}{2} < h$ and define the set

$$\Lambda^h = \{y \in \Lambda / d(y; \partial\Gamma(\Lambda)) \leq h\}.$$

By definition of a Van Hove sequence, its volume $|\Lambda^h|$ is of order $o(|\Lambda|)$ as $\Lambda \to \infty$. Besides,

$$\frac{1}{|\Lambda|} \int_{\Gamma(\Lambda)^c} \sum_{y \in \Lambda^h} \frac{1}{|x - y|^4} \, dx \leq \frac{|\Lambda^h|}{|\Lambda|} \int_{|x| \geq \frac{1}{2}} \frac{1}{|x|^4} \, dx,$$

while

$$\frac{1}{|\Lambda|} \int_{\Gamma(\Lambda)^c} \sum_{y \in \Lambda \backslash \Lambda^h} \frac{1}{|x - y|^4} \, dx \leq \int_{|x| \geq h} \frac{1}{|x|^4} \, dx = \frac{4\pi}{h}.$$

Gathering these two estimates, we recover our claim by letting Λ and then h go to infinity.

On the other hand, we define

$$f_{\Lambda^c}(x) = \sum_{y \in \mathbf{Z}^3 \backslash \Lambda} f(x - y)$$

$$= G(x) + d - f_\Lambda, \qquad (3.64)$$

where $d \in \mathbf{R}$ is a constant independent of Λ (see Section 3.3.4 for the details) and we now prove that we have

$$\|f_{\Lambda^c}\|_{L^1(\Gamma(\Lambda))} = o(|\Lambda|).$$

For the same reasons as above, we rather show instead that

$$\frac{1}{|\Lambda|} \int_{\Gamma(\Lambda)} \sum_{y \in \mathbf{Z}^3 \setminus \Lambda} \frac{dx}{|x - y|^4} \stackrel{\Lambda \to \infty}{\longrightarrow} 0.$$

We once more fix $h > \frac{1}{2}$ and define this time

$$\Lambda^h = \{x \in \Gamma(\Lambda)/d(x; \Lambda^c) \le h\}.$$

We first point out that we have

$$\|f_{\Lambda^c}\|_{L^\infty(\Gamma(\Lambda))} \le C,$$

for some constant C that is independent of Λ. This uniform bound allows us to write

$$\frac{1}{|\Lambda|} \int_{\Lambda^h} \sum_{y \in \mathbf{Z}^3 \setminus \Lambda} \frac{dx}{|x - y|^4} \le C \frac{|\Lambda^h|}{|\Lambda|}.$$

In addition,

$$\frac{1}{|\Lambda|} \int_{\Gamma(\Lambda) \setminus \Lambda^h} \sum_{y \in \mathbf{Z}^3 \setminus \Lambda} \frac{dx}{|x - y|^4} \le \frac{1}{|\Lambda|} \int_{\Gamma(\Lambda) \setminus \Lambda^h} \frac{1}{d(x; \Lambda^c)^{1/2}} \sum_{y \in \mathbf{Z}^3 \setminus \Lambda} \frac{dx}{|x - y|^{7/2}}$$

$$\le \frac{C}{\sqrt{h}},$$

since

$$\left\| \sum_{y \in \mathbf{Z}^3 \setminus \Lambda} \frac{1}{|x - y|^{7/2}} \right\|_{L^\infty(\Gamma(\Lambda))} \le C$$

(this infinite sum makes sense because $\frac{7}{2} > 3$). We reach the conclusion once more by letting $\Lambda \to \infty$ and then $h \to \infty$.

Thus, because of the uniform L^∞ bound on ρ_Λ given by Proposition 3.10, we also obtain

$$\frac{1}{|\Lambda|} \int_{\Gamma(\Lambda)} f_{\Lambda^c} \rho_\Lambda \stackrel{\Lambda \to \infty}{\longrightarrow} 0. \tag{3.65}$$

Finally, with the help of (3.63) and (3.65), together with the definition (3.64), we write

$$\frac{1}{|\Lambda|}\int_{\mathbf{R}^3}f_\Lambda\,\rho_\Lambda\,\mathrm{d}x = \frac{1}{|\Lambda|}\int_{\Gamma(\Lambda)}f_\Lambda\,\rho_\Lambda\,\mathrm{d}x + o(1)$$

$$= \frac{1}{|\Lambda|}\int_{\Gamma(\Lambda)}(G(x)+d)\,\rho_\Lambda(x)\,\mathrm{d}x + o(1)$$

$$= \int_{\Gamma_0}G(x)\,\tilde{\rho}_\Lambda(x)\,\mathrm{d}x + d\int_{\Gamma_0}\tilde{\rho}_\Lambda(x)\,\mathrm{d}x + o(1).$$

Hence, we recover (3.62) by letting Λ go to infinity since, from the compactness results of Section 3.3, we know that $\int_{\Gamma_0}\tilde{\rho}_\Lambda(x)\,\mathrm{d}x$ goes to 1.

According to our aim to bound from below $\frac{I_\Lambda^{\mathrm{TFW}}}{|\Lambda|}$ by $E_{\mathrm{per}}^{\mathrm{TFW}}(\tilde{\rho}) + \frac{M}{2}$ (which is itself bounded below by $I_{\mathrm{per}}^{\mathrm{TFW}} + \frac{M}{2}$), we concentrate on lower bounds of the last remaining term in (3.61), namely $\frac{1}{2|\Lambda|}D^{Cb}(\mathbf{1}_{\Gamma(\Lambda)}-\rho_\Lambda,\mathbf{1}_{\Gamma(\Lambda)}-\rho_\Lambda)$. More precisely, we intend to show that

$$\liminf_{\Lambda\to\infty}\frac{1}{|\Lambda|}D^{Cb}(\mathbf{1}_{\Gamma(\Lambda)}-\rho_\Lambda,\mathbf{1}_{\Gamma(\Lambda)}-\rho_\Lambda) \geq D_G(\tilde{\rho},\tilde{\rho}). \qquad (3.66)$$

This is where we take advantage of the way in which we brought the terms together in (3.61).

Indeed, we now set

$$g_\Lambda = \Phi_\Lambda - f_\Lambda$$

$$= (\mathbf{1}_{\Gamma(\Lambda)}-\rho_\Lambda)\star\frac{1}{|x|},$$

and we reformulate the term we are interested in as

$$\frac{1}{|\Lambda|}D^{Cb}(\mathbf{1}_{\Gamma(\Lambda)}-\rho_\Lambda,\mathbf{1}_{\Gamma(\Lambda)}-\rho_\Lambda) = \frac{1}{8\pi|\Lambda|}\int_{\mathbf{R}^3}|\nabla g_\Lambda|^2\,.$$

Then, we know from the proof of the compactness given in Section 3.3.4 that

$$\frac{1}{|\Lambda|}\int_{\mathbf{R}^3}|\nabla g_\Lambda|^2 \leq C \qquad (3.67)$$

(here and below, C denotes various positive constants that are independent of Λ). Our strategy now consists of reproducing the proof we made in the $(Cb-m)$ case, allowing g_Λ (that belongs to $L^\infty(\mathbf{R}^3)$) to play the role of Φ_Λ^m. Indeed, from the information we already know for f_Λ (collected in Section 3.3.4) and from Proposition 3.12, we see that we have

$$\frac{1}{|\Lambda|}\int_{\Gamma(\Lambda)}|\Phi_\Lambda| \leq C$$

and

$$\frac{1}{|\Lambda|} \int_{\Gamma(\Lambda)} |f_\Lambda| \leq C;$$

hence

$$\frac{1}{|\Lambda|} \int_{\Gamma(\Lambda)} |g_\Lambda| \leq C.$$

Therefore, defining, as usual, $\tilde{g}_\Lambda = \frac{1}{|\Lambda|} \sum_{y \in \Lambda} g_\Lambda(\cdot + y)$, we deduce from the above inequality that

$$\left| \int_{\Gamma_0} \tilde{g}_\Lambda \right| \leq C. \tag{3.68}$$

Moreover, (3.67) implies that

$$\int_{\Gamma_0} |\nabla \tilde{g}_\Lambda|^2 \leq C. \tag{3.69}$$

Then, because of (3.68) and (3.69), and from the Poincaré–Wirtinger inequality, $\tilde{g}_\Lambda$ is bounded in $H^1(\Gamma_0)$. Thus, up to a subsequence, we may assume, with the help of Rellich's theorem, that $\tilde{g}_\Lambda$ converges to some function $\tilde{g} \in H^1(\Gamma_0)$, weakly in $H^1(\Gamma_0)$, almost everywhere on $\mathbf{R}^3$ and strongly in $L^p(\Gamma_0)$, for all $1 \leq p < 6$. In addition, $\tilde{g}$ satisfies the equation

$$-\triangle\tilde{g} = 4\pi \left[1 - \tilde{\rho}\right] \qquad \text{on } \mathbf{R}^3 \tag{3.70}$$

(at least in the distribution sense) since, by definition of g_Λ and of the $\sim$-transform, $\tilde{g}_\Lambda$ is the solution to

$$-\triangle\tilde{g}_\Lambda = 4\pi \left[\tilde{1}_{\Gamma(\Lambda)} - \tilde{\rho}_\Lambda\right] \qquad \text{on } \mathbf{R}^3,$$

going to zero at infinity.

The point now consists of proving that $\tilde{g}$ is periodic. Indeed, suppose this is the case. We may construct explicitly a periodic solution to (3.70), namely

$$\bar{g}(x) = \int_{\Gamma_0} G(x - y) \left[1 - \tilde{\rho}(y)\right] dy.$$

However, since two periodic solutions to (3.70) are equal to an additive constant and $\int_{\Gamma_0} G(x - y) dy = 0$ for all x in $\mathbf{R}^3$ (G is periodic and $\int_{\Gamma_0} G(y) dy = 0$), we deduce the existence of a constant d' (possibly depending on the sequence Λ) such that

$$\tilde{g}(x) = d' - \int_{\Gamma_0} G(x - y) \tilde{\rho}(y) dy.$$

Then, we recover (3.66) by copying, step by step, the proof we made in the case of smeared out nuclei in Section 3.4.2, replacing Φ^m by g everywhere.

It is worth noticing at this stage that we have used several times in our argument the compactness of the sequence ρ_Λ. We want to emphasize that this is a crucial point to ensure the existence of the thermodynamic limit in the Coulomb case. Again, this is completely different from the Yukawa case (see Chapter 2).

We now go back to the proof of the periodicity of $\tilde{g}$. We already know that $\tilde{f}_\Lambda$ converges to the periodic function $G + d$; hence, we study the sequence $\tilde{\Phi}_\Lambda$ and its limit, as Λ goes to infinity, denoted by $\tilde{\Phi}$.

Unfortunately, we may not apply Lemma 3.19 directly, because an estimate similar to (3.52) is not available for Φ_Λ. However, we may modify the proof of Lemma 3.19, given in Section 3.4.2, in order to apply to any function h_Λ satisfying, instead of (3.52),

$$\sup_{k \in \mathbf{Z}^3} \|h_\Lambda\|_{L^1(\Gamma_k)} \le C \tag{3.71}$$

for some constant C independent of Λ. Of course, the sequence Φ_Λ enjoys this property, since we even have a uniform bound for Φ_Λ in $L^p_{\text{unif}}(\mathbf{R}^3)$, for all $1 \le p < 3$ (see Proposition 3.12). Indeed, this time, we may write, for all $k \in \mathbf{R}^3$,

$$\|\tilde{h}_\Lambda(\cdot + k) - \tilde{h}_\Lambda\|_{L^1(\Gamma_0)}$$
$$\le \frac{1}{|\Lambda|} \sum_{\substack{y \in (\Lambda+k) \\ y \notin \Lambda}} \|h_\Lambda\|_{L^1(\Gamma_y)} + \frac{1}{|\Lambda|} \sum_{\substack{y \in \Lambda \\ y \notin (\Lambda+k)}} \|h_\Lambda\|_{L^1(\Gamma_y)}$$
$$\le C \frac{|\Lambda^{|k|}|}{|\Lambda|},$$

in view of the assumption (3.71).

The bound from above goes to zero as Λ goes to infinity, by the definition of a Van Hove sequence. ◇

Remark 3.21 As $\tilde{f}_\Lambda$ converges to $G + d$, we deduce from the definition of g_Λ and the above proof that $\tilde{\Phi}_\Lambda - \int_{\Gamma_0} \tilde{\Phi}_\Lambda$ converges (up to a subsequence) to the periodic function

$$G(x) - \int_{\Gamma_0} G(x - y) \, \tilde{\rho}(y) \, \mathrm{d}y.$$

At the end of the following Section 3.5, we shall be able to identify $\tilde{\rho}$ with ρ_{per}, the unique minimizer of $I^{\text{TFW}}_{\text{per}}$. Then, finally, the limit of $\tilde{\Phi}_\Lambda - \int_{\Gamma_0} \tilde{\Phi}_\Lambda$ does not depend on Λ and the convergence concerns the whole sequence $\tilde{\Phi}_\Lambda$.

We conclude this section by giving another proof of the lower bound for the energy per cell in the Yukawa case, which draws its inspiration from the argument we developed above for the Coulomb case.

3.4.4 *Adaptation to the Yukawa case*

In this section, we adapt the proof we made in the Coulomb case in order to apply to the Yukawa case, therefore giving another proof of

$$\liminf_{\Lambda \to \infty} \frac{1}{|\Lambda|} D^a(\rho_\Lambda, \rho_\Lambda) = \liminf_{\Lambda \to \infty} \frac{1}{|\Lambda|} \iint_{\mathbf{R}^3 \times \mathbf{R}^3} \rho_\Lambda(x) \frac{e^{-a|x-y|}}{|x-y|} \rho_\Lambda(y) \, dx \, dy,$$

$$\geq \iint_{\Gamma_0 \times \Gamma_0} \tilde{\rho}(x) V_\infty^a(x-y) \tilde{\rho}(y) \, dx \, dy \qquad (3.72)$$

where $V_\infty^a(x) = \sum_{y \in \mathbf{Z}^3} V^a(x-y)$ and $V^a(x) = \dfrac{e^{-a|x|}}{|x|}$.

However, let us emphasize that the proof we give below is specific to the Yukawa potential, while the proof given in Section 2.3 of Chapter 2, holds if we replace V^a by any short-ranged potential $V \geq 0$ (for which the series $\sum_{y \in \mathbf{Z}^3} V$ $(\cdot - y)$ converges).

Let us now turn to the proof of (3.72). First of all, we define

$$h_\Lambda = \rho_\Lambda \star V^a.$$

Thus, h_Λ is the solution to

$$-\Delta h_\Lambda + a^2 h_\Lambda = 4\pi \rho_\Lambda, \qquad (3.73)$$

which goes to zero when $|x|$ goes to infinity. We multiply this equation by h_Λ and then integrate over $\mathbf{R}^3$, to obtain

$$\int_{\mathbf{R}^3} |\nabla h_\Lambda|^2 + a^2 \int_{\mathbf{R}^3} |h_\Lambda|^2 = 4\pi \int_{\mathbf{R}^3} \rho_\Lambda h_\Lambda = 4\pi D^a(\rho_\Lambda, \rho_\Lambda) \qquad (3.74)$$

(it is easily seen with the help of the Fourier transform that each term on the left-hand side above makes sense, since $\sqrt{\rho_\Lambda}$ belongs to $H^1(\mathbf{R}^3)$). However, we know from the bounds on the energy in the Yukawa case (see Proposition 2.5 in Chapter 2) that

$$\frac{1}{|\Lambda|} D^a(\rho_\Lambda, \rho_\Lambda) \leq C,$$

where, here and below, C denotes various positive constants that are independent of Λ. Then, we deduce from (3.74) that

$$\|h_\Lambda\|_{H^1(\mathbf{R}^3)} \leq C |\Lambda|^{1/2}. \qquad (3.75)$$

Next, we recall from Proposition 2.5 in Section 2.2 of Chapter 2, that ρ_Λ is bounded in $L^\infty(\mathbf{R}^3)$ independently of Λ. Thus, h_Λ also is bounded in $L^\infty(\mathbf{R}^3)$ (by $\|\rho_\Lambda\|_{L^\infty(\mathbf{R}^3)} \|V^a\|_{L^1(\mathbf{R}^3)}$).

Denoting now, as usual, $\tilde{h}_\Lambda = \frac{1}{|\Lambda|} \sum_{y \in \mathbf{Z}^3} h_\Lambda(\cdot + y)$, we deduce from (3.75) that $\tilde{h}_\Lambda$ is bounded in $H^1(\Gamma_0) \cap L^\infty(\mathbf{R}^3)$ independently of Λ; thus, $\tilde{h}_\Lambda$ converges (up to a subsequence) to $\tilde{h} \in H^1(\Gamma_0) \cap L^\infty(\mathbf{R}^3)$ as Λ goes to infinity, weakly in $H^1(\Gamma_0)$ and strongly in $L^p(\Gamma_0)$ for all $1 \leq p < +\infty$ (in view of Rellich's theorem and Hölder's inequalities). Moreover, with the help of Lemma 3.19, we know that $\tilde{h}$ is periodic.

In addition, applying the $\sim$-transform to each member of (3.73) and then passing to the weak limit as Λ goes to infinity, it is clear that

$$-\Delta \tilde{h} + a^2 \tilde{h} = 4\pi \tilde{\rho}. \tag{3.76}$$

Thus,

$$\tilde{h} = \tilde{\rho} \star V^a$$
$$= \int_{\Gamma_0} \tilde{\rho}(x - y) V_\infty^a(y) \, \mathrm{d}y. \tag{3.77}$$

It now just remains to use (3.74) and to proceed as in the Coulomb case, writing

$$\liminf_{\Lambda \to \infty} \frac{1}{|\Lambda|} D^a(\rho_\Lambda, \rho_\Lambda) = \liminf_{\Lambda \to \infty} \frac{1}{4\pi |\Lambda|} \left[\int_{\mathbf{R}^3} |\nabla h_\Lambda|^2 + a^2 \int_{\mathbf{R}^3} |h_\Lambda|^2 \right]$$
$$\geq \frac{1}{4\pi} \liminf_{\Lambda \to \infty} \left[\int_{\Gamma_0} |\nabla \tilde{h}_\Lambda|^2 + a^2 \int_{\Gamma_0} |\tilde{h}_\Lambda|^2 \right]$$
$$\geq \frac{1}{4\pi} \left[\int_{\Gamma_0} |\nabla \tilde{h}|^2 + a^2 \int_{\Gamma_0} |\tilde{h}|^2 \right]$$
$$\geq \iint_{\Gamma_0 \times \Gamma_0} \tilde{\rho}(x) V_\infty^a(x - y) \tilde{\rho}(y) \, \mathrm{d}x \, \mathrm{d}y$$

the last inequality holding true because of (3.76), (3.77), and the periodicity of $\tilde{h}$. Hence (3.72) follows.

Of course, there is no difficulty in obtaining a lower bound for the energy in the Yukawa case with smeared nuclei by adapting the proof given for the smeared nuclei in the Coulomb case (see Section 3.4.2) with the help of the above argument.

The following section will be devoted to the study of an upper bound for the energy per cell as Λ goes to infinity.

3.5   Upper bound for the energy

We are now going to address the problem of finding an upper bound for $\frac{1}{|\Lambda|} I_\Lambda^{\mathrm{TFW}}$ and $\frac{1}{|\Lambda|} I_\Lambda^m$. In fact, we are going to prove that it is asymptotically bounded from above by the periodic energy $I_{\mathrm{per}}^{\mathrm{TFW}} + \frac{M}{2}$ (respectively, $I_{\mathrm{per}}^m + \frac{M}{2}$) defined in Section 3.1.

We copy the proof used in Chapter 2 to deal with the Yukawa case. Our strategy of proof is here again to build a convenient sequence of functions whose energy approximates the infimum energy when Λ is fixed, and that converges to the periodic minimizing density when Λ goes to infinity.

The main result of this section is contained in the following proposition, where we have on purpose omitted the superscripts TFW or m:

Proposition 3.22 *We assume that the sequence Λ satisfies, for h fixed:*

$$\frac{|\Lambda^h|}{|\Lambda|} Log(|\Lambda^h|) \longrightarrow 0, \qquad as \; \Lambda \longrightarrow \infty.$$

Then,

$$\limsup_{\Lambda \to \infty} \frac{1}{|\Lambda|} I_\Lambda \leq I_{\text{per}} + \frac{M}{2},$$

where, according to the notation of the introduction to this chapter, the constant M is either

$$\lim_{x \to 0} \left[G(x) - \frac{1}{|x|} \right],$$

in the $(Cb - \delta)$ case, or

$$\iint_{\Gamma_0 \times \Gamma_0} m(x) \left[G(x - y) - \frac{1}{|x - y|} \right] m(y) \, dx \, dy$$

in the $(Cb - m)$ case, where we assume that m shares the same s symmetries as the unit cube.

Remark 3.23 The additional hypothesis made on the sequence (Λ) comes from our strategy of proof. Actually, we shall prove in Chapter 6 that the Van Hove condition is sufficient.

Proof of Proposition 3.22 We shall detail the proof only in the case of point nuclei, since, as usual, some simplifications are available in the case of smeared out nuclei. However, we shall indicate at the end of the proof how our argument has to be modified in order to apply to this particular framework.

When Λ is fixed, we are going to build a test function that will approach the ground state energy in a convenient way, in order to obtain an upper bound for this energy.

Fix $h > \frac{1}{2}$ independent of Λ; for instance, $h = 1$. We denote by $\overset{\circ}{\Lambda}$ the set of points in Λ that are at least at a distance h from $\partial\Gamma(\Lambda)$ and $\partial\Lambda = \Lambda \setminus \overset{\circ}{\Lambda}$. Next, we denote $\Gamma(\overset{\circ}{\Lambda}) = \bigcup_{y \in \overset{\circ}{\Lambda}} \Gamma_y$.

On $\Gamma(\overset{\circ}{\Lambda})$, we define the function $\bar{\rho}_\Lambda$ by

$$\bar{\rho}_\Lambda(x) = \rho_{\text{per}}(x),$$

where ρ_{per} is the minimum of the problem $I_{\text{per}}^{\text{TFW}}$. Thus, we have

$$\int_{\Gamma(\overset{\circ}{\Lambda})} \bar{\rho}_\Lambda = |\Gamma(\overset{\circ}{\Lambda})|.$$

Next, we extend the function $\bar{\rho}_\Lambda$ to $\mathbf{R}^3$ in such a way that we have

$$\begin{cases} \displaystyle\int_{\mathbf{R}^3} \bar{\rho}_\Lambda = |\Lambda|, \\[2mm] \displaystyle\int_{\Gamma_y} \bar{\rho}_\Lambda = 1, \qquad \text{for all } y \in \partial\Lambda \\[2mm] \sqrt{\bar{\rho}_\Lambda} \in H^1(\mathbf{R}^3), \\[2mm] \|\bar{\rho}_\Lambda\|_{L^{5/3}} \le C\,|\Lambda|^{3/5} \qquad \text{independently of } \Lambda, \\[2mm] \bar{\rho}_\Lambda = 0 \qquad \text{a.e. outside } \Gamma(\Lambda), \\[2mm] \displaystyle\int_{\Gamma(\partial\Lambda)} |\nabla\sqrt{\bar{\rho}_\Lambda}|^2 \le C|\Lambda^h|, \qquad \text{for some } C \text{ independent of } \Lambda. \end{cases}$$

(the existence of such a test function is established in Section 2.4 of Chapter 2).
We now claim that we have

$$\frac{1}{|\Lambda|}\left(E_\Lambda^{\mathrm{TFW}}(\bar{\rho}_\Lambda) + \tfrac{1}{2}U_\Lambda^{Cb} - E_{\overset{\circ}{\Lambda}}^{\mathrm{TFW}}(\bar{\rho}_\Lambda) - \tfrac{1}{2}U_{\overset{\circ}{\Lambda}}^{Cb}\right) \longrightarrow 0$$

and

$$\frac{1}{|\Lambda|}E_{\overset{\circ}{\Lambda}}^{\mathrm{TFW}}(\bar{\rho}_\Lambda) + \frac{1}{2|\Lambda|}U_{\overset{\circ}{\Lambda}}^{Cb} \longrightarrow I_{\mathrm{per}}^{\mathrm{TFW}} + \frac{M}{2}, \tag{3.78}$$

as Λ goes to infinity.

Both assertions are a consequence of the fact that, for a sequence satisfying the criteria of Definition 1, what happens on the 'boundary' of $\Gamma(\Lambda)$ can be neglected in front of what happens 'inside' $\Gamma(\Lambda)$. The proofs of these claims follow the same patterns as the corresponding proof in the Yukawa case, and rely once more very much upon the general technical Lemma 2.24 that we have stated in Section 2.4 of Chapter 2.

In view of the 'periodicity' of $\bar{\rho}_\Lambda$ over $\Gamma(\overset{\circ}{\Lambda})$, we have

$$E_{\overset{\circ}{\Lambda}}^{\mathrm{TFW}}(\bar{\rho}_\Lambda) + \tfrac{1}{2}U_{\overset{\circ}{\Lambda}}^{Cb}$$
$$= |\overset{\circ}{\Lambda}|\left(\int_{\Gamma_0} |\nabla\sqrt{\rho_{\mathrm{per}}}|^2\,\mathrm{d}x + \int_{\Gamma_0} \rho_{\mathrm{per}}^{5/3}\,\mathrm{d}x\right)$$
$$+ \left(\int_{\Gamma(\partial\Lambda)} |\nabla\sqrt{\bar{\rho}_\Lambda}|^2\,\mathrm{d}x + \int_{\Gamma(\partial\Lambda)} \bar{\rho}_\Lambda^{5/3}\,\mathrm{d}x\right)$$

$$- \int_{\Gamma_0} \sum_{y \in \overset{\circ}{\Lambda}} \sum_{z \in \overset{\circ}{\Lambda}} \frac{1}{|x+y-z|} \rho_{\text{per}}(x) \, dx$$

$$- \int_{\Gamma_0} \sum_{y \in \overset{\circ}{\Lambda}} \sum_{z \in \partial \Lambda} \frac{1}{|x+y-z|} \bar{\rho}_{\Lambda}(x+z) \, dx$$

$$+ \frac{1}{2} \iint_{\Gamma_0 \times \Gamma_0} \rho_{\text{per}}(x) \rho_{\text{per}}(y) \sum_{z \in \overset{\circ}{\Lambda}} \sum_{t \in \overset{\circ}{\Lambda}} \frac{1}{|x-y+z-t|} \, dx \, dy$$

$$+ \iint_{\Gamma_0 \times \Gamma_0} \sum_{z \in \overset{\circ}{\Lambda}} \sum_{t \in \partial \Lambda} \frac{1}{|x-y+z-t|} \rho_{\text{per}}(x) \bar{\rho}_{\Lambda}(y+t) \, dx \, dy$$

$$+ \frac{1}{2} \iint_{\Gamma_0 \times \Gamma_0} \sum_{z \in \partial \Lambda} \sum_{t \in \partial \Lambda} \frac{1}{|x-y+z-t|} \bar{\rho}_{\Lambda}(x+z) \bar{\rho}_{\Lambda}(y+t) \, dx \, dy$$

$$+ \frac{1}{2} U_{\overset{\circ}{\Lambda}}^{Cb}.$$

Since the Coulomb potential is long-range, we cannot use Lemma 2.24 directly as we did in Section 2.4 of Chapter 2, for the Yukawa case. More precisely, we have to group the various terms appearing in the decomposition of $E_{\Lambda}^{\text{TFW}}(\bar{\rho}_{\Lambda})$ in a convenient way.

Indeed, the series

$$\sum_{z \in \mathbf{Z}^3} \frac{1}{|x-z|}$$

does not converge. However, if we denote

$$f(u) = \frac{1}{|u|} - \int_{\Gamma_0} \frac{\rho_{\text{per}}(x)}{|x-u|} \, dx, \qquad (3.79)$$

we claim that the series

$$\sum_{z \in \mathbf{Z}^3} |f(x-z)|$$

is convergent. This comes from the fact that ρ_{per} shares the same symmetries as the unit cube, which yields $f(u) = O\left(\frac{1}{|u|^4}\right)$ as $|x| \to \infty$.

In addition, the series converges uniformly on compact subsets of $\mathbf{R}^3 \backslash \mathbf{Z}^3$. For this purpose, we expand $\frac{1}{|x-u|}$ for u large:

$$\frac{1}{|x-u|} = \frac{1}{|u|} \times [1 + \frac{\langle x, u \rangle}{|u|^2} - \frac{1}{2} \frac{|x|^2}{|u|^2} + \frac{3}{2} \frac{\langle x, u \rangle^2}{|u|^4}$$

$$- \frac{3}{2} \frac{|x|^2 \langle x, u \rangle}{|u|^4} + \frac{5}{2} \frac{\langle x, u \rangle^3}{|u|^6} + o\Big(\frac{|x|^3}{|u|^3}\Big)\Bigg] . \qquad (3.80)$$

Next, we notice that since ρ_{per} shares the same symmetries as the unit cube, it satisfies, in particular,

$$\rho_{\text{per}}(\varepsilon_1 x_1, \varepsilon_2 x_2, \varepsilon_3 x_3) = \rho_{\text{per}}(x_1, x_2, x_3)$$

for all $\varepsilon_i \in \{-1, 1\}$, and

$$\rho_{\text{per}}(x_{\sigma(1)}, x_{\sigma(2)}, x_{\sigma(3)}) = \rho_{\text{per}}(x_1, x_2, x_3),$$

for any permutation σ of $\{1, 2, 3\}$.

Then, all terms in (3.80) involving an odd power of $\langle x, u \rangle$ vanish when integrated against ρ_{per} on Γ_0. In particular, the term of order 2 with respect to $\frac{1}{|u|}$ cancels. Moreover,

$$\int_{\Gamma_0} x_i x_j \rho_{\text{per}}(x)\, \mathrm{d}x = \delta_{ij} \frac{1}{3} \int_{\Gamma_0} |x|^2 \rho_{\text{per}}(x)\, \mathrm{d}x,$$

for all $i, j \in \{1, 2, 3\}$, and thus

$$\int_{\Gamma_0} \langle x, u \rangle^2 \rho_{\text{per}}(x)\, \mathrm{d}x = \sum_{i=1}^{3} \int_{\Gamma_0} x_i^2 u_i^2 \rho_{\text{per}}(x)\, \mathrm{d}x$$

$$= \frac{1}{3} \sum_{i=1}^{3} \int_{\Gamma_0} |x|^2 |u_i|^2 \rho_{\text{per}}(x)\, \mathrm{d}x$$

$$= \frac{1}{3} |u|^2 \int_{\Gamma_0} |x|^2 \rho_{\text{per}}(x)\, \mathrm{d}x.$$

Then, if we compare with (3.80), we can easily see that the term of order 3 cancels too, whence, for u large,

$$|f(u)| \le \frac{C}{|u|^4}. \qquad (3.81)$$

(In fact, we even have $f(u) = O\left(\frac{1}{|u|^5}\right)$, since an odd power of $\langle x, u \rangle$ appears in the term of order $\frac{1}{|u|^4}$, but (3.81) suffices for our argument.)

We now can apply Lemma 2.24 in Chapter 2 to our function f. It just remains to identify the series with the function F, defined as

$$F(x) = G(x) - \int_{\Gamma_0} G(x - y) \rho_{\text{per}}(y)\, \mathrm{d}y + d, \qquad (3.82)$$

for some real constant d. For this purpose, we just point out that the series and the function F satisfy the same partial differential equation on $\mathbf{R}^3$, and then their difference is a harmonic and periodic function, and hence a constant.

With (i) and (ii) and (3.81), we treat all the terms where a double sum of the kind $\sum_{y\in\partial\Lambda}\sum_{z\in\overset{\circ}{\Lambda}}$ appears. Next, (iii) allows to deal with the terms

$$-\sum_{z\in\partial\Lambda}\sum_{t\in\partial\Lambda}\int_{\Gamma_t}\frac{1}{|x-z|}\bar{\rho}_\Lambda(x)\,\mathrm{d}x$$

$$+\tfrac{1}{2}\sum_{z\in\partial\Lambda}\sum_{t\in\partial\Lambda}\iint_{\Gamma_z\times\Gamma_t}\frac{\bar{\rho}_\Lambda(x)\bar{\rho}_\Lambda(y)}{|x-y|}\,\mathrm{d}x\,\mathrm{d}y$$

$$+\tfrac{1}{2}\sum_{z\in\partial\Lambda}\sum_{t\neq z\in\partial\Lambda}\frac{1}{|z-t|}, \qquad (3.83)$$

and shows that they behave like $o(|\Lambda|)$ as Λ goes to infinity. Then, (3.78) is shown. Indeed, we have

$$\frac{1}{|x-z|}-\int_{\Gamma_z}\frac{\bar{\rho}_\Lambda(y)}{|x-y|}\mathrm{d}y=\frac{1}{|x-z|}-\int_{\Gamma_0}\frac{\bar{\rho}_\Lambda(y+z)}{|x-y-z|}\mathrm{d}y$$

$$=\int_{\Gamma_0}\left(\frac{1}{|x-z|}-\frac{1}{|x-y-z|}\right)\bar{\rho}_\Lambda(y+z)\mathrm{d}y,$$

since $\int_{\Gamma_z}\bar{\rho}_\Lambda=1$ by (3.5); while, by the same kinds of arguments as above,

$$\int_{\Gamma_0}\left|\frac{1}{|x-z|}-\frac{1}{|x-y-z|}\right|\bar{\rho}_\Lambda(y+z)\mathrm{d}y\leq\frac{C}{|x-z|^2}.$$

It follows that

$$\left|\sum_{z\in\partial\Lambda}\sum_{t\in\partial\Lambda}\int_{\Gamma_t}\left(\frac{1}{|x-z|}-\int_{\Gamma_z}\frac{\bar{\rho}_\Lambda(y)}{|x-y|}\mathrm{d}y\right)\bar{\rho}_\Lambda(x)\mathrm{d}x\right|$$

$$\leq C\sum_{z\in\partial\Lambda}\sum_{t\in\partial\Lambda}\frac{1}{|z-t|^2}.$$

Using Lemma 2.24(iii) in Section 2.4 of Chapter 2, we obtain

$$-\tfrac{1}{2}\sum_{z\in\partial\Lambda}\sum_{t\in\partial\Lambda}\int_{\Gamma_t}\frac{1}{|x-z|}\bar{\rho}_\Lambda(x)\,\mathrm{d}x$$

$$+\tfrac{1}{2}\sum_{z\in\partial\Lambda}\sum_{t\in\partial\Lambda}\iint_{\Gamma_z\times\Gamma_t}\frac{\bar{\rho}_\Lambda(x)\bar{\rho}_\Lambda(y)}{|x-y|}\,\mathrm{d}x\,\mathrm{d}y=o(|\Lambda|).$$

The term

$$-\tfrac{1}{2}\sum_{z\in\partial\Lambda}\sum_{t\in\partial\Lambda}\int_{\Gamma_t}\frac{1}{|x-z|}\bar{\rho}_\Lambda(x)\,\mathrm{d}x+\tfrac{1}{2}\sum_{z\in\partial\Lambda}\sum_{t\neq z\in\partial\Lambda}\frac{1}{|z-t|}$$

in (3.83) can be treated in the same way.

Eventually, the properties (3.5) alone allow us to show that

$$\frac{1}{|\Lambda|} \sum_{y \in \partial \Lambda} \left(\int_{\Gamma_y} |\nabla \sqrt{\bar{\rho}_\Lambda}|^2 + \int_{\Gamma_y} \bar{\rho}_\Lambda^{5/3} \right) \longrightarrow 0.$$

Finally, the terms

$$|\mathring{\Lambda}| \left(\int_{\Gamma_0} |\nabla \sqrt{\rho}_{\text{per}}|^2 + \int_{\Gamma_0} \rho_{\text{per}}^{5/3} \right)$$

$$- \int_{\Gamma_0} \sum_{y \in \mathring{\Lambda}} \sum_{z \in \mathring{\Lambda}} \frac{\rho_{\text{per}}(x)}{|x + y - z|} \, dx + \frac{1}{2} U_{\mathring{\Lambda}}^{Cb}$$

$$+ \frac{1}{2} \iint_{\Gamma_0 \times \Gamma_0} \rho_{\text{per}}(x) \rho_{\text{per}}(y) \left[\sum_{z \in \mathring{\Lambda}} \sum_{t \in \mathring{\Lambda}} \frac{1}{|x - y + z - t|} \right] dx \, dy$$

give rise to the periodic energy $E_{\text{per}}^{\text{TFW}}(\rho_{\text{per}}) + \frac{M}{2}$.
    For this purpose, we check that

$$\frac{1}{2|\mathring{\Lambda}|} U_{\mathring{\Lambda}}^{Cb} - \frac{1}{2|\mathring{\Lambda}|} \int_{\Gamma_0} \sum_{y \in \mathring{\Lambda}} \sum_{z \in \mathring{\Lambda}} \frac{\rho_{\text{per}}(x)}{|x + y - z|} dx$$

$$= \frac{1}{2} \frac{1}{|\mathring{\Lambda}|} \sum_{y \in \mathring{\Lambda}} \sum_{z \neq y \in \mathring{\Lambda}} \left[\frac{1}{|y - z|} - \int_{\Gamma_0} \frac{\rho_{\text{per}}(x)}{|x + y - z|} dx \right]$$

$$= \frac{1}{2} \lim_{\substack{x \to 0 \\ x \neq 0}} \left[\frac{1}{|\mathring{\Lambda}|} \sum_{y \in \mathring{\Lambda}} \sum_{z \in \mathring{\Lambda}} f(x + y - z) - \frac{1}{|x|} \right]$$

converges to

$$\frac{1}{2} \lim_{\substack{x \to 0 \\ x \neq 0}} \left[F(x) - \frac{1}{|x|} \right] = \frac{M}{2} - \frac{1}{2} \int_{\Gamma_0} G(y) \rho_{\text{per}}(y) \, dy + \frac{d}{2},$$

as Λ goes to infinity, because of (3.82).
    Indeed, the proof of Lemma 2.24 shows, in addition, that

$$\sum_{y \in \mathring{\Lambda}} \sum_{z \in \mathring{\Lambda}} f(x + y - z) - \frac{1}{|x|}$$

converges uniformly to $\displaystyle\sum_{k \in \mathbf{Z}^3} f(x + k) - \frac{1}{|x|}$ on any compact subset K of $\mathbf{R}^3$ such that $K \cap \mathbf{Z}^3 = \{0\}$.

Besides,

$$\frac{1}{2\,|\overset{\circ}{\Lambda}|}\iint_{\Gamma_0\times\Gamma_0}\sum_{z,t\in\overset{\circ}{\Lambda}}\frac{\rho_{\mathrm{per}}(x)\rho_{\mathrm{per}}(y)}{|x-y+z-t|}\,dx\,dy-\frac{1}{2\,|\overset{\circ}{\Lambda}|}\int_{\Gamma_0}\sum_{y,z\in\overset{\circ}{\Lambda}}\frac{\rho_{\mathrm{per}}(x)}{|x+y-z|}\,dx$$

$$=-\frac{1}{2}\int_{\Gamma_0}\rho_{\mathrm{per}}(x)\frac{1}{|\overset{\circ}{\Lambda}|}\sum_{z\in\overset{\circ}{\Lambda}}\sum_{t\in\overset{\circ}{\Lambda}}f(x+y-z)\,dx$$

converges to $-\frac{1}{2}\int_{\Gamma_0}\rho_{\mathrm{per}}(x)F(x)\,dx$, which we identify with

$$-\frac{1}{2}\int_{\Gamma_0}\rho_{\mathrm{per}}(x)G(x)\,dx+\frac{1}{2}\iint_{\Gamma_0\times\Gamma_0}\rho_{\mathrm{per}}(x)G(x-y)\rho_{\mathrm{per}}(y)\,dx\,dy-\frac{d}{2}$$

with the help of (3.82).

Proposition 3.22 then follows in a straightforward way:

$$\limsup_{\Lambda\to\infty}\frac{1}{|\Lambda|}I_\Lambda^{\mathrm{TFW}}\le\limsup_{\Lambda\to\infty}\frac{1}{|\Lambda|}E_\Lambda^{\mathrm{TFW}}(\bar{\rho}_\Lambda)+\frac{U_\Lambda^{Cb}}{2\,|\Lambda|}=I_{\mathrm{per}}^{\mathrm{TFW}}+\frac{M}{2},$$

since $\int\bar{\rho}_\Lambda=|\Lambda|$. We only explain briefly the modifications to be made in our proof in order to treat the smeared nuclei case.

This time, we need an estimate of the term

$$D^{Cb}(m_\Lambda-\bar{\rho}_\Lambda,m_\Lambda-\bar{\rho}_\Lambda).$$

In fact, the computations are easier to perform, since there is no singularity at the nuclei. Our whole analysis goes through provided that we work with the function

$$f_m(x)=\int_{\Gamma_0}\frac{m(y)}{|x-y|}\,dy-\int_{\Gamma_0}\frac{\rho_{\mathrm{per}}(y)}{|x-y|}\,dy,$$

instead of f given by (3.79), the series associated to f_m then being identified with

$$G_m(x)-\int_{\Gamma_0}G(x-y)\rho_{\mathrm{per}}(y)\,dy+d',$$

for some real constant d'.

We want to emphasize at this stage that this is where the assumption on m sharing the symmetries of the unit cube plays a fundamental role. Otherwise, f_m does not behave like $\frac{1}{|x|^4}$ at infinity, which is a key point of our strategy (see Section 3.6 about this particular point). $\diamondsuit$

As the conclusion of this section, we are able to prove the following corollary, which is a direct consequence of Section 3.4 (we once more skip the superscripts TFW or m in order to simplify the notation).

Corollary 3.24 *We assume that the Van Hove sequence satisfies assumption (c) (and, in addition, that in the $(Cb - m)$ case, m shares the same symmetries as the unit cube). Then we have*

$$\lim_{\Lambda \to \infty} \frac{I_\Lambda}{|\Lambda|} = I_{\text{per}} + \frac{M}{2}.$$

At this stage, our goal to establish the existence of the thermodynamic limit for the TFW model is fulfilled for what concerns the energy per cell. However, we now want to understand what happens to the minimizing densities. As a consequence of the convergence of the energies given by Corollary 3.24, we shall first deduce in the following corollary some additional information on the sequence $\tilde\rho_\Lambda$:

Corollary 3.25 *($(Cb - m)$ case) Let ρ_{per} be the (unique) minimizer of I^m_{per}. Then, $\sqrt{\tilde\rho_\Lambda}$ converges to $\sqrt{\rho_{\text{per}}}$ strongly in $L^p_{\text{loc}}(\mathbf{R}^3)$, for all $1 \le p < \infty$ and $\nabla\sqrt{\tilde\rho_\Lambda}$ converges to $\nabla\sqrt{\rho_{\text{per}}}$ strongly in $L^2_{\text{loc}}(\mathbf{R}^3)^3$.*

In addition, the sequence $\tilde\Phi^m_\Lambda - \int_{\Gamma_0} \tilde\Phi^m_\Lambda \, dx$ converges to Φ^m_{per} defined by $G_m(x) - \int_{\Gamma_0} G(x - y)\rho_{\text{per}}(y) \, dy$, strongly in $H^1_{\text{loc}}(\mathbf{R}^3)$ and in $L^p_{\text{loc}}(\mathbf{R}^3)$, for all $1 \le p < \infty$.

Proof of Corollary 3.25 This corollary follows immediately from Corollary 3.24 and the direct proof of the lower bound for the energy per cell in the case of smeared nuclei (see Section 3.4.2), which yields

$$\lim_{\Lambda \to \infty} \frac{1}{|\Lambda|} \int_{\mathbf{R}^3} |\nabla\sqrt{\rho_\Lambda}|^2 = \lim_{\Lambda \to \infty} \frac{1}{|\Lambda|} \int_{\Gamma(\Lambda)} |\nabla\sqrt{\rho_\Lambda}|^2 \, dx$$

$$= \lim_{\Lambda \to \infty} \int_{\Gamma_0} |\nabla\sqrt{\tilde\rho_\Lambda}|^2 \, dx$$

$$= \int_{\Gamma_0} |\nabla\sqrt{\rho_{\text{per}}}|^2 \, dx, \qquad (3.84)$$

$$\lim_{\Lambda \to \infty} \frac{1}{|\Lambda|} \int_{\mathbf{R}^3} \rho^p_\Lambda \, dx = \lim_{\Lambda \to \infty} \frac{1}{|\Lambda|} \int_{\Gamma(\Lambda)} \rho^p_\Lambda \, dx$$

$$= \lim_{\Lambda \to \infty} \int_{\Gamma_0} \tilde\rho^p_\Lambda \, dx$$

$$= \int_{\Gamma_0} \rho^p_{\text{per}} \qquad \text{for all} \quad 1 \le p < \infty,$$

and

$$\lim_{\Lambda \to \infty} \frac{1}{|\Lambda|} \int_{\mathbf{R}^3} |\nabla\Phi^m_\Lambda|^2 \, dx = \lim_{\Lambda \to \infty} \frac{1}{|\Lambda|} \int_{\Gamma(\Lambda)} |\nabla\Phi^m_\Lambda|^2 \, dx$$

$$= \lim_{\Lambda \to \infty} \int_{\Gamma_0} |\nabla \tilde{\Phi}_\Lambda^m|^2 \, dx$$

$$= \int_{\Gamma_0} |\nabla \Phi_{\mathrm{per}}|^2 \, dx \ .$$

We may then conclude with the help of the Poincaré–Wirtinger inequality, together with Rellich's theorem. $\diamond$

We now state the analogous result in the case of point nuclei.

Corollary 3.26 (($Cb - \delta$) **case**) *Let ρ_{per} be the (unique) minimizer of $I_{\mathrm{per}}^{\mathrm{TFW}}$. $\sqrt{\tilde{\rho}_\Lambda}$ converges to $\sqrt{\rho_{\mathrm{per}}}$ strongly in $L^p_{\mathrm{loc}}(\mathbf{R}^3)$, for all $1 \leq p < \infty$ and $\nabla \sqrt{\tilde{\rho}_\Lambda}$ converges to $\nabla \sqrt{\rho_{\mathrm{per}}}$ strongly in $L^2_{\mathrm{loc}}(\mathbf{R}^3)^3$.*

Moreover, the sequence $\tilde{\Phi}_\Lambda - \int_{\Gamma_0} \tilde{\Phi}_\Lambda \, dx$ converges to Φ_{per} defined by $G(x) - \int_{\Gamma_0} G(x - y)\rho_{\mathrm{per}}(y) \, dy$, strongly in $L^p_{\mathrm{loc}}(\mathbf{R}^3)$ for all $1 \leq p < 3$.

The proof of Corollary 3.26 follows exactly the same pattern as the proof of Corollary 3.25; therefore, we skip it (the convergence results for $\tilde{\Phi}_\Lambda$ are proved in Section 3.4.3).

Let us point out that in the above two corollaries the limit functions do not depend on the sequence Λ and are uniquely determined by the limit periodic minimization problem; hence, the entire sequences converge to their limit (not only some subsequence).

Corollaries 3.25 and 3.26 are the first (weak) step to the proof of the periodicity of the sequence of densities when Λ goes to infinity that we shall solve completely in Chapter 5.

3.6   Some extensions

We give in this section a few extensions of the results we have just obtained.

Let us first of all say that the extensions that we have mentioned in the Yukawa case in Subsection 2.7.2 about other forms of functions than $\rho^{5/3}$ may also be considered in the Coulomb case, the results being the same. On the contrary, the situation when the system is not neutral (as in Subsection 2.7.3), or when the cell is not cubic (as in Subsection 2.7.4) have to be considered with more attention in the Coulomb case. We refer the reader to Chapter 6 for these extensions (see also the comments at the end of the next subsection).

We first consider here the case when the cubic cell is not of unit volume, then make a few comments on the shape of the nucleus, and then (Subsection 3.6.3) consider the very important extension of the TFDW model.

3.6.1   *Other sizes of elementary cubic cells*

We consider cubes of volume R^3, R being different from 1. It is straightforward to see that our whole analysis applies.

The periodic potential that appears satisfies

$$-\Delta G_R = 4\pi \left(\sum_{y \in \mathbf{Z}^3} \delta_{Ry} - \frac{1}{R^3} \right).$$

Of course, G_R is really defined up to a constant and we may normalize it by

$$G_R(x) - \frac{1}{|x|} \xrightarrow{x \to 0} 0.$$

Then, we obtain the relationship

$$G_R(x) = \frac{1}{R} G_1 \left(\frac{x}{R} \right), \tag{3.85}$$

with

$$G_1(x) = G(x) - M.$$

The periodic problem which is the thermodynamic limit becomes

$$I_{\text{per}}(1, R) = \inf \left\{ \int_{\Gamma_R} |\nabla u|^2 + |u|^{10/3} - G_R\, u^2 \right.$$
$$\left. + \frac{1}{2} \iint_{\Gamma_R \times \Gamma_R} u^2(x) G_R(x - y) u^2(y)\, dx\, dy; \int_{\Gamma_R} u^2 = 1 \right\}, \tag{3.86}$$

where we denote by $\Gamma_R = R \cdot \Gamma_0$ the cube of volume R^3 centered at 0 (Γ_1 is thus equal to what we have denoted by Γ_0 so far).

(a) On the above expression (3.86) of the periodic problem $I_{\text{per}}(1, R)$, it is easy to see that

$$\begin{cases} I_{\text{per}}(1, R) \xrightarrow{R \to 0} +\infty, \\ I_{\text{per}}(1, R) \xrightarrow{R \to +\infty} I^{\text{TFW}}(1), \end{cases} \tag{3.87}$$

where $I^{\text{TFW}}(1)$ is the usual TFW model for an atom of unit charge:

$$I^{\text{TFW}}(1) = \inf \left\{ \int_{\mathbf{R}^3} |\nabla \sqrt{\rho}|^2 - \int_{\mathbf{R}^3} \frac{1}{|x|} \rho + \int_{\mathbf{R}^3} \rho^{5/3} + \frac{1}{2} D(\rho, \rho); \right.$$
$$\left. \rho \geq 0,\ \sqrt{\rho} \in H^1(\mathbf{R}^3),\ \int_{\mathbf{R}^3} \rho = 1 \right\}.$$

We therefore recover the somewhat intuitive result that, when all the nuclei except the nucleus located at zero go to infinity, then the periodic model for the crystal goes to the atomic model.

For the first assertion of (3.87), we note that, using (3.85), we have

$$\left| \int G_R \, \rho \right| \leq \|G_R\|_{L^{5/2}(\Gamma_R)} \|\rho\|_{L^{5/3}(\Gamma_R)}$$

$$= R^{1/5} \|\rho\|_{L^{5/3}(\Gamma_R)} \|G_1\|_{L^{5/2}(\Gamma_1)},$$

and

$$\left| \iint \rho(x) G_R(x - y) \rho(y) \, dx \, dy \right| \leq R^{1/5} \int_{\Gamma_R} \rho \|G_1\|_{L^{5/2}(\Gamma_1)} \|\rho\|_{L^{5/3}(\Gamma_R)},$$

and finally

$$\int_{\Gamma_R} \rho \leq R^{6/5} \|\rho\|_{L^{5/3}(\Gamma_R)}.$$

Therefore, we have

$$E_{\text{per}, R}(\rho) \geq \frac{1}{R^2} (1 - C \cdot R^{1/5}),$$

for some positive constant C that is independent of R and for all ρ such that $\int_{\Gamma_R} \rho = 1$, whence

$$I_{\text{per}}(1, R) \xrightarrow{R \to 0} +\infty.$$

For the second assertion of (3.87), we proceed as follows. We first observe that we have

$$\left\| G_R(x) - \frac{1}{|x|} \right\|_{L^\infty(\Gamma_R)} = \left\| \frac{1}{R} \left(G_1 \left(\frac{x}{R} \right) - \frac{1}{|\frac{x}{R}|} \right) \right\|_{L^\infty(\Gamma_R)}$$

$$= \frac{1}{R} \left\| G_1(x) - \frac{1}{|x|} \right\|_{L^\infty(\Gamma_1)}$$

Hence

$$\left\| G_R(x) - \frac{1}{|x|} \right\|_{L^\infty(\Gamma_R)} \xrightarrow{R \to \infty} 0. \tag{3.88}$$

Let us now prove the following inequality:

$$I^{\text{TFW}}(1) \geq \limsup_{R \to +\infty} I_{\text{per}}(1, R). \tag{3.89}$$

Let us take some arbitrary $\rho \geq 0$ such that $\sqrt{\rho} \in \mathcal{D}(\mathbf{R}^3)$, $\int_{\mathbf{R}^3} \rho = 1$. For R large enough, ρ has its compact support contained in Γ_R; thus $\int_{\Gamma_R} \rho = 1$, and

$$E_{\text{per},R}(\rho) \geq I_{\text{per}}(1, R).$$

Now

$$\left| \int G_R \rho - \frac{1}{|x|} \rho \right| \leq \left\| G_R - \frac{1}{|x|} \right\|_{L^\infty(\Gamma_R)} = O\left(\frac{1}{R}\right),$$

and likewise, choosing R even larger in order to have $Supp(\rho) \subset \Gamma_{R/2}$,

$$\left| \int \rho(x) G_R(x - y) \rho(y) \, dx \, dy \right| = O\left(\frac{1}{R}\right).$$

It follows that

$$E_{\text{per},R}(\rho) - E^{\text{TFW}}(\rho) = O\left(\frac{1}{R}\right) \longrightarrow 0$$

as R goes to infinity, and thus

$$\limsup_{R \to +\infty} I_{\text{per}}(1, R) \leq \limsup_{R \to +\infty} E_{\text{per},R}(\rho) = E^{\text{TFW}}(\rho).$$

The above inequality holds for all $\rho \geq 0$ such that $\sqrt{\rho} \in \mathcal{D}(\mathbf{R}^3)$ and $\int_{\mathbf{R}^3} \rho = 1$. We next obtain (3.89) by density. We are next going to prove that

$$\liminf_{R \to +\infty} I_{\text{per}}(1, R) \geq I^{\text{TFW}}(1), \qquad (3.90)$$

which will conclude the proof of (3.87). We denote by ρ_R the minimizer of $I_{\text{per}}(1, R)$. In view of (3.89), we know that the sequence ρ_R satisfies

$$\begin{cases} \displaystyle\int_{\Gamma_R} |\nabla \sqrt{\rho_R}|^2 \leq C, \\[4mm] \displaystyle\int_{\Gamma_R} \rho_R = 1, \end{cases}$$

for some positive constant C that is independent of R. Extracting a subsequence if necessary, we may assume that $\sqrt{\rho_R}$ converges weakly in $H^1_{\text{loc}}(\mathbf{R}^3)$ to some $\sqrt{\rho_\infty}$.

Let us now fix R_0. For all $R \geq R_0$, we have

$$\int_{\Gamma_R} |\nabla \sqrt{\rho_R}|^2 \geq \int_{\Gamma_{R_0}} |\nabla \sqrt{\rho_R}|^2,$$

whence

$$\liminf_{R \to +\infty} \int_{\Gamma_R} |\nabla \sqrt{\rho_R}|^2 \geq \liminf_{R \to +\infty} \int_{\Gamma_{R_0}} |\nabla \sqrt{\rho_R}|^2 \geq \int_{\Gamma_{R_0}} |\nabla \sqrt{\rho_\infty}|^2.$$

Since the latter inequality holds for any R_0, we obtain

$$\liminf_{R\to+\infty} \int_{\Gamma_R} |\nabla\sqrt{\rho_R}|^2 \geq \int_{\mathbf{R}^3} |\nabla\sqrt{\rho_\infty}|^2.$$

Let us note that, the left-hand side being finite, it implies that $\sqrt{\rho_\infty} \in H^1(\mathbf{R}^3)$. Likewise, we find

$$\liminf_{R\to+\infty} \int_{\Gamma_R} \rho_R^{5/3} \geq \int_{\mathbf{R}^3} \rho_\infty^{5/3}.$$

Next, we use (3.88) to deal with the terms in G_R:

$$\left| \int_{\Gamma_R} \left(G_R - \frac{1}{|x|} \right) \rho_R \right| \leq \left\| G_R - \frac{1}{|x|} \right\|_{L^\infty(\Gamma_R)} = O\left(\frac{1}{R}\right),$$

while

$$\left| \int_{\Gamma_R} (\rho_R - \rho_\infty) \frac{1}{|x|} \right| \longrightarrow 0.$$

Finally, we have, in a similar way,

$$\left| \iint_{\Gamma_R \times \Gamma_R} \rho_R(x) \left(G_R(x-y) - \frac{1}{|x-y|} \right) \rho_R(y)\, \mathrm{d}x\, \mathrm{d}y \right| \longrightarrow 0.$$

and

$$\liminf_{R\to+\infty} \iint_{\Gamma_R \times \Gamma_R} \rho_R(x) \frac{1}{|x-y|} \rho_R(y)\, \mathrm{d}x\, \mathrm{d}y \geq D(\rho_\infty, \rho_\infty).$$

Collecting all these inequalities, we finally deduce that

$$\liminf_{R\to+\infty} I_{\mathrm{per}}(1, R) = \liminf_{R\to+\infty} E_{\mathrm{per}, R}(\rho_R) \geq E^{\mathrm{TFW}}(\rho_\infty).$$

In addition,

$$\int_{\mathbf{R}^3} \rho_\infty \leq \liminf_{R\to+\infty} \int_{\Gamma_R} \rho_R = 1.$$

Furthermore, recalling that the function $\lambda \mapsto I^{\mathrm{TFW}}(\lambda)$ is non-increasing with respect to λ, we have

$$E^{\mathrm{TFW}}(\rho_\infty) \geq I^{\mathrm{TFW}}(1).$$

Therefore, (3.90) follows.

(b) The function $R \longrightarrow I_{\mathrm{per}}(1, R)$ being continuous with respect to $R > 0$, the convergences (3.87) imply in particular that

$$\inf_{R>0} I_{\text{per}}(1, R) > -\infty.$$

Then the following question arises: Does there exist some optimal R_0 (the length of an 'optimal' cubic lattice) such that

$$I_{\text{per}}(1, R_0) = \inf_{R>0} I_{\text{per}}(1, R). \tag{3.91}$$

As far as we know, the existence of R_0 such that (3.91) holds is an open question. It is, of course, one of the simplest questions one may ask in the set of questions that are related to the far more general and fundamental questions of the existence of an optimized geometry (unconstrained to be a cube). On the way to tackling this very difficult question of the optimized unconstrained geometry, one may for instance investigate the basic question (3.91), i.e. the existence of a cube of optimized size (one degree of freedom only, the length of one side of the cube). Then, one can raise the following question: Does there exist a parallelepipedon that among the set of all parallelepipeda optimizes the energy (six degrees of freedom, three lengths and three angles). Indeed, with the proof of the convergence of the energy given in Chapter 6 (that gets rid of the symmetry assumptions that are crucial in Section 3.5), it is also a straightforward extension of our work (also made for the TF model in Lieb and Simon [40]) to prove that the thermodynamic limit exists for parallelepipeda. Note, of course, that once questions of this type have been answered, the fundamental problem is to prove that the sequence of optimized geometries for a fixed number $|\Lambda|$ of nuclei does converge to some optimized geometry, that may (or may not!) be one of these optimized periodic geometries. We shall come back on these issues in a future work. Let us also mention that Solovej has addressed some related questions for the reduced Hartree–Fock model in [54].

3.6.2   *Other shapes of nucleus*

In our work, we have so far considered the case when one nucleus is located at each point of the regular lattice $\mathbf{Z}^3$. In addition, when the nucleus is not a point, we have assumed that its shape (i.e. the compactly supported measure defining the nucleus) has cubic symmetry. A natural question is to extend our result to the case when there is more than one nucleus per cell, or the case when the shape of the nucleus does not share the symmetries of the cube. Of course, when there are many nuclei in the cell, but when the set of points (or measures) defining the location of these nuclei has the symmetries of the cube, then our arguments apply immediately. Less simple is the case when, for instance, the measure defining the nucleus (or the set of nuclei contained in the cell) does not share the symmetries of the cube. Indeed, we have strongly used in our arguments the fact that the periodic minimizer shares the symmetries of the cell in order to cancel the divergent terms in the expansion of the Coulomb potential (estimates of Section 3.5). When this symmetry is broken, a more detailed study should be made. We shall see in Chapter 6 that it is, however, possible to solve this problem by a different approach, based upon the convergence of the densities. We shall prove there both the convergence of the energy per unit volume and the convergence of the densities.

Whether or not we keep the symmetry inside the cell, there is another extension worth looking at, in particular thinking about applications: the case when we allow the function m defining the nucleus to take negative values. More precisely, suppose we take $m \in \mathcal{D}$, supported in Γ_0, of total mass 1 (think of a smeared ion of charge 1). In order to apply the arguments of this chapter, assume in addition that m has the symmetry of the unit cube, and keep the cell cubic of volume 1 (for the sake of simplicity, but other cases are tractable). The difficulty is that Solovej's inequality, giving in particular the bounds on θ_Λ, Φ_Λ, and ρ_Λ, does not hold. Therefore we cannot use the direct strategy of proof. However, we may conclude that the energy per unit volume still converges to the periodic energy, arguing as follows. The proof of the upper limit goes through without any modification. For the lower limit, we use the comparison with the Yukawa model. The convergence of the Yukawa energy holds because it does not make any use of the sign of the nucleus m. Finally, we let the Yukawa exponent a go to zero, and we conclude. It is of course clear that the same kind of argument that we have made in the point nuclei case applies if we take several Dirac masses (the set of the points in the cell sharing the symmetry of the cube) of total mass 1.

Nevertheless, we will see in Chapters 5 and 6 (see Remark 5.8) that we do not know how to extend to this case, when we allow m to take negative values, the argument that we put forward there to prove the convergence of the density. This has the following consequence: we do not know in this case how to get rid of the additional assumptions of symmetry to prove the convergence of the energy per unit volume.

In the same fashion, we are not able to extend our proof of the convergence of the energy per unit volume to the case when m is no longer a measure, nor a sum of Dirac masses, but is a distribution. Think for instance of the situation where there is in each cubic cell a dipole at the centre of the cell, and a smooth radially symmetric function m of mass 1. The singularity at zero is then too difficult to treat, as far as we now know.

3.6.3 The Thomas–Fermi–Dirac–von Weizsäcker model

We indicate in this section how the above arguments have to be modified in order to deal with the so-called Thomas–Fermi–Dirac–von Weizsäcker model (TFDW model for short). The TFDW functional is given by

$$E_\Lambda^{\mathrm{TFDW}}(\rho) = \int_{\mathbf{R}^3} |\nabla \sqrt{\rho}|^2 + c_1 \int_{\mathbf{R}^3} \rho^{5/3} - c_2 \int_{\mathbf{R}^3} \rho^{4/3}$$
$$- \int_{\mathbf{R}^3} V_\Lambda \, \rho + \tfrac{1}{2} D(\rho, \rho), \tag{3.92}$$

where the two constants c_1 and c_2 are positive ($c_1 = 1$, $c_2 = 0$ obviously corresponds to the case of the TFW model we have treated so far). We define the ground state energy I_Λ^{TFDW} by analogy with (3.2), by

$$I_\Lambda^{\text{TFDW}} = \inf \left\{ E_\Lambda^{\text{TFDW}}(\rho) + \tfrac{1}{2} \sum_{\substack{y,z \in \Lambda \\ y \neq z}} \frac{1}{|y - z|}; \right.$$

$$\left. \rho \geq 0, \sqrt{\rho} \in H^1(\mathbf{R}^3), \int_{\mathbf{R}^3} \rho = |\Lambda| \right\}. \qquad (3.93)$$

Let us recall that, from the physical viewpoint, the Dirac correction term $-c_2 \int_{\mathbf{R}^3} \rho^{4/3}$ may be interpreted as a model for the exchange energy of the electrons, which aims at 'weakening' the term $\frac{1}{2}D(\rho,\rho)$, which is an estimate from above of the inter-electronic repulsion.

Our purpose here is to show that, when c_2 is small (in a sense that will be made precise below; see (3.99)), the behaviour of the TFDW model is quite similar to the behaviour of the TFW model, i.e. the energy per unit volume $\frac{I_\Lambda^{\text{TFDW}}}{|\Lambda|}$ goes to some periodic energy. We shall not detail in the sequel whether we deal with the point nuclei model or the smeared nuclei model, because our arguments hold in both cases but for some obvious slight modifications. In addition, we shall not detail the proofs that basically follow the same patterns as in the TFW case, but only emphasize how they must be adapted to allow us to conclude in the TFDW case.

The main difference (and the main difficulty) between the TFW case and the TFDW case is the lack of convexity of the TFDW energy, because of the term $c_1 \rho^{5/3} - c_2 \rho^{4/3}$, which prevents us from making direct arguments as in the TFW case. However, we make the following simple observation: assume that we know that ρ is not too small, i.e. that it is essentially bounded from below, say, $\rho \geq \nu > 0$, and that the constant ν does not depend on c_2. Of course, this cannot be true at infinity as soon as ρ is an integrable function, but we may hope that it is true at finite distance of the nuclei. Then, we may choose c_2 small enough such that ν (and thus ρ) is in the 'good' zone of the function $t \longrightarrow c_1 t^{5/3} - c_2 t^{4/3}$, namely the zone where this function is both convex and positive. It follows that, with such a c_2, we may make use, with regard to the TFDW energy, of all the arguments we used in the TFW case, and that, consequently, if ρ_Λ satisfies the above requirement, the TFDW model has a thermodynamic limit like the TFW model.

At this point, we wish to make the following remark. The TFDW model is far from being one of the most commonly used in the modelling of molecules or crystals. However, from the mathematical viewpoint, it is interesting to treat, in this somewhat simple setting, the lack of convexity we have just pointed out. This will be particularly useful when we tackle more sophisticated models that are not convex either, like the Hartree–Fock model or models close to the Hartree–Fock model.

Our strategy mimics the formal argument we have just made. In a first step, we show that for any arbitrary c_2 we may choose a minimum ρ_Λ of the TFDW problem such that all the estimates we have established above in the TFW case also hold for the TFDW model. In particular, the Lagrange multiplier is bounded above uniformly with respect to Λ, ρ_Λ is bounded in L^∞ uniformly in Λ and c_2, and from this latter fact we deduce (step 2) by virtue of Harnack's inequality that, in 'most' of $\Gamma(\Lambda)$ (see a precise definition of this vague notion in (3.97) below), ρ_Λ is bounded away from 0 by a uniform positive constant (the ν above) that neither depends on Λ nor on c_2.

We next choose c_2 small enough such that $c_1\rho_\Lambda^{5/3} - c_2\rho_\Lambda^{4/3} > 0$ on most of $\Gamma(\Lambda)$. Since the contribution of the exterior of $\Gamma(\Lambda)$ to the energy is precisely irrelevant for the thermodynamic limit, it follows that the TFDW energy of ρ_Λ (i.e. $I_\Lambda^{\mathrm{TFDW}}$) roughly behaves like a convex energy, that it is not far from a TFW-type energy and has a thermodynamic limit. Our third step details this point.

Step 1. Let us now take an arbitrary c_2. The first fact we wish to point out is that, in the TFDW model, it is not known if the minimizing density of the problem with $|\Lambda|$ nuclei of unit charge and $|\Lambda|$ electrons is unique. We only know that such a minimizing density exists (see Lions [42]), and that, when c_2 is small enough for some given set of nuclei, it is unique (see Le Bris [28]). However, when Λ goes to infinity, such a constant c_2 below which the minimum is unique might possibly go to zero. Even if it is not very likely that such a situation occurs, we do not have any proof that it does not occur.

A consequence of this possible non-uniqueness is the following. Among 'all' the minimizing densities of $I_\Lambda^{\mathrm{TFDW}}$, we claim that we may choose one, that we hereafter denote by ρ_Λ, that satisfies the following property: the Lagrange multiplier θ_Λ associated to ρ_Λ is bounded above uniformly in Λ.

Let us prove our claim.

For this purpose, we remark that the inequality proved by Solovej in [53] for the TFW model, namely, for every $0 < \theta < 1$,

$$\tfrac{5}{3}\theta\, c_1\, \rho_\Lambda^{2/3} \le \Phi_\Lambda + [C_\theta - \theta_\Lambda]_+\,,$$

where C_θ is some fixed constant independent of Λ, may be extended to the TFDW model, with a constant C_θ that does not depend on the small coefficient c_2 in front of the Dirac term (provided that this coefficient is bounded from above). Indeed, if u_Λ is a solution to

$$-\Delta u_\Lambda + \tfrac{5}{3}c_1\, u_\Lambda^{7/3} - \tfrac{4}{3}c_2\, u_\Lambda^{5/3} - \Phi_\Lambda\, u_\Lambda + \theta_\Lambda u_\Lambda = 0,$$

following [53] we set

$$v = \lambda \rho_\Lambda^{2/3} - \Phi_\Lambda - [C - \theta_\Lambda]_+ \, ,$$

where C will be determined below.

We then have

$$\Delta v \geq \tfrac{4}{3}\lambda \, u_\Lambda^{1/3} \, \Delta u_\Lambda - 4\pi \, \rho_\Lambda$$

and thus

$$\begin{aligned}
\Delta v &\geq \tfrac{4}{3}\lambda \rho_\Lambda^{2/3} \left(\tfrac{5}{3}c_1\rho_\Lambda^{2/3} - \tfrac{4}{3}c_2\rho_\Lambda^{1/3} - \Phi_\Lambda + \theta_\Lambda \right) - 4\pi \, \rho_\Lambda \\
&= \tfrac{4}{3}\lambda \rho_\Lambda^{2/3} v + \tfrac{4}{3}\lambda(\tfrac{5}{3}c_1 - \lambda) \rho_\Lambda^{4/3} - (4\pi + \tfrac{16}{9}\lambda c_2) \rho_\Lambda \\
&\quad + \tfrac{4}{3}\lambda\rho_\Lambda^{2/3}(\theta_\Lambda + [C - \theta_\Lambda]_+).
\end{aligned}$$

If we now choose $\lambda = \tfrac{5}{6}c_1$ and remark that we have, for any $t \geq 0$,

$$t \leq \frac{\tfrac{25}{27}c_1^2}{4\pi + \tfrac{40}{27}c_1c_2} \, t^{4/3} + \frac{4\pi + \tfrac{40}{27}c_1c_2}{\tfrac{25}{27}c_1^2} \, t^{2/3}$$

we deduce that

$$\begin{aligned}
\Delta v &\geq \tfrac{10}{9}c_1 \, \rho_\Lambda^{2/3} \, v - \frac{27}{25\,c_1^2}(4\pi + \tfrac{40}{27}c_1c_2)^2 \, \rho_\Lambda^{2/3} \\
&\quad + \tfrac{10}{9}c_1\rho_\Lambda^{2/3}(\theta_\Lambda + [C - \theta_\Lambda]_+),
\end{aligned}$$

or, in other words,

$$\begin{aligned}
\Delta v &\geq \tfrac{10}{9}c_1 \, \rho_\Lambda^{2/3} \, v + \tfrac{10}{9}c_1\rho_\Lambda^{2/3}(\theta_\Lambda - C + [C - \theta_\Lambda]_+) \\
&\geq \tfrac{10}{9}c_1 \, \rho_\Lambda^{2/3} \, v,
\end{aligned}$$

where

$$C = \frac{9}{10\,c_1} \cdot \frac{27}{25\,c_1^2} \cdot (4\pi + \tfrac{40}{27}c_1c_2)^2.$$

We therefore obtain

$$\tfrac{5}{6}\,c_1\,\rho_\Lambda^{2/3} \leq \Phi_\Lambda + [C - \theta_\Lambda]_+ \, , \tag{3.94}$$

as in [53].

Remark 3.27 (On the 'Solovej inequality') The above argument shows that the inequality proved by Solovej in [53] does not only hold in the standard TFW case but also in more general situations. In particular, it may be of some interest to ask the following question: If the exponent $\tfrac{5}{3}$ is replaced by another arbitrary exponent p, for which range of p does this inequality hold?

Introducing the function $\lambda u^{2(p-1)} - \Phi$, we now extend the above argument. We first impose $p \geq \frac{3}{2}$ in order to use the convexity of the function $t \longrightarrow t^{2(p-1)}$, which yields

$$-\Delta(u^{2(p-1)}) \leq -2(p-1)u^{2p-3}\Delta u$$

and thus

$$-\Delta(\lambda u^{2(p-1)} - \Phi) \leq -2(p-1)(\lambda u^{2(p-1)} - \Phi) - 2(p-1)u^{4(p-1)}(p - \lambda)$$
$$-2\lambda(p-1)\theta u^{2(p-1)} + u^2.$$

Next, it remains to interpolate u^2 between $u^{2(p-1)}$ and $u^{4(p-1)}$. It is clearly possible if $2 \geq p > \frac{3}{2}$, and in the limit case $p = \frac{3}{2}$, provided that the term $\int |u|^{2p} = \int |u|^3$ in the energy is multiplied by a large enough constant. In both cases, the end of the argument is then straightforward and the so-called Solovej inequality then holds.

Therefore, for the TFW model with such an exponent p, we may prove that the bounds we have obtained in the standard TFW case and the arguments we have made also hold here. It follows that, in this setting, the thermodynamic limit for the energy per unit volume exists. We shall see in Chapter 5 that the densities also converge.

Let us now fix Λ, and take a minimum of the TFDW model for a total nuclear charge $|\Lambda|$ and an electronic charge $|\Lambda| + \varepsilon$. Recall that we know such a minimum exists for ε small enough (possibly depending on Λ, no matter what c_2 is; Le Bris [28]). We denote by a subscript $\Lambda + \varepsilon$ the quantities related to this minimum. In this case, $\Phi_{\Lambda+\varepsilon} < 0$ at infinity; thus by (3.94) we obtain $\theta_{\Lambda+\varepsilon} < C$, for all $\varepsilon > 0$ small. Hence, extracting a subsequence if necessary, we may assume that, when ε goes to zero, the sequence of minimizers $u_{\Lambda+\varepsilon}$ and its associated sequence of Lagrange multipliers $\theta_{\Lambda+\varepsilon}$ converge, respectively, to some minimum u_Λ of the neutral problem I_Λ^{TFDW} and its associated Lagrange multiplier θ_Λ, that therefore satisfies $\theta_\Lambda \leq C$. The desired minimum with uniformly bounded Lagrange multiplier is thus also obtained.

For this purpose, it suffices to remark that $v_{\Lambda+\varepsilon} = \sqrt{\frac{|\Lambda|}{|\Lambda|+\varepsilon}} \, u_{\Lambda+\varepsilon}$ is a minimizing sequence of I_Λ^{TFDW}, and thus is relatively compact in $H^1(\mathbf{R}^3)$. Then, arguing as in [42] or [28], it is easy to prove that it converges to a minimum u_Λ of I_Λ^{TFDW}. Next, we obtain that the associated Lagrange multiplier, say θ_Λ, is bounded above by passing to the limit in the Euler–Lagrange equation satisfied by $u_{\Lambda+\varepsilon}$ and recognizing θ_Λ as the limit of the $\theta_{\Lambda+\varepsilon}$. Hereafter, ρ_Λ denotes a minimum of the TFDW problem that is associated with a Lagrange multiplier θ_Λ that is uniformly bounded above.

Let us now make a few basic remarks. In view of Hölder's inequality, we have

$$\int_{\mathbf{R}^3} \rho^{4/3} \leq \left(\int_{\mathbf{R}^3} \rho\right)^{1/2} \left(\int_{\mathbf{R}^3} \rho^{5/3}\right)^{1/2}.$$

Thus, when $\int_{\mathbf{R}^3} \rho = |\Lambda|$,

$$c_1 \int_{\mathbf{R}^3} \rho^{5/3} - c_2 \int_{\mathbf{R}^3} \rho^{4/3} \geq c_1 \int_{\mathbf{R}^3} \rho^{5/3} - c_2 \sqrt{|\Lambda|} \left(\int_{\mathbf{R}^3} \rho^{5/3} \right)^{1/2},$$

$$\geq \frac{c_1}{2} \int_{\mathbf{R}^3} \rho^{5/3} - \frac{c_2^2}{2c_1} |\Lambda|.$$

It follows that, with obvious notation,

$$I_\Lambda^{\mathrm{TFW}}(c_1) \geq I_\Lambda^{\mathrm{TFDW}} \geq I_\Lambda^{\mathrm{TFW}}\left(\frac{c_1}{2}\right) - \frac{c_2^2}{2c_1} |\Lambda|.$$

Therefore, $\left| \frac{I_\Lambda^{\mathrm{TFDW}}}{|\Lambda|} \right|$ is bounded, whatever $c_1 > 0$ and $c_2 \geq 0$ are. In the same fashion, we deduce as in the proof of Proposition 3.8 by comparison with a TF-type model, that $\frac{1}{|\Lambda|} \int_{\mathbf{R}^3} |\nabla \sqrt{\rho_\Lambda}|^2$ and $\frac{1}{|\Lambda|^{1/p}} \|\rho_\Lambda\|_{L^p}$ are bounded. Using next the bound on θ_Λ, we obtain the analogue of Proposition 3.8 for the TFDW model. Likewise, the bounds established in Proposition 3.12 on Φ_Λ in $L^p_{\mathrm{unif}}(\mathbf{R}^3)$, $1 \leq p < 3$, hold

$$\exists C > 0, \quad \sup_{x \in \mathbf{R}^3} \|\Phi_\Lambda\|_{L^p(x + B_1)} \leq C. \tag{3.95}$$

In particular, the consequence is that we also have 'compactness' in the TFDW model, in the sense that

$$\frac{1}{|\Lambda|} \int_{\Gamma(\Lambda)} \rho_\Lambda \longrightarrow 1,$$

since it is based upon the estimates of Proposition 3.8.

Next, we notice that the proof of Proposition 3.10, which shows a uniform L^∞ bound on ρ_Λ, namely

$$\|\rho_\Lambda\|_{L^\infty} \leq C \tag{3.96}$$

for some $C > 0$ independent of Λ, still holds in the TFDW case, and that the constant C in (3.96) may be chosen independently of the coefficient c_2 of the Dirac correction term. Indeed, remark that the proof of (3.96) only makes use of the fact that

$$-\Delta u_\Lambda + (\tfrac{5}{3} c_1 u_\Lambda^{4/3} - \Phi_\Lambda + \theta_\Lambda) u_\Lambda \geq 0,$$

which is true in the TFDW model whatever c_2 is, since the left-hand side is $\tfrac{4}{3} c_2 u_\Lambda^{5/3}$.

Step 2. Now that we have such a L^∞ bound that is uniform in c_2 and in Λ, we deduce from Harnack's inequality a bound from below that is also uniform in c_2 and in Λ on a domain of $\mathbf{R}^3$ that is almost the whole domain $\Gamma(\Lambda)$. Let us detail this.

We then denote

$$A(\Lambda) = \{x \in \Gamma(\Lambda)/d(x, \partial\Gamma(\Lambda)) \geq |\Lambda|^{\alpha}\} \qquad (3.97)$$

for any constant $0 < \alpha < \frac{1}{3}$ (this is a particular choice of an 'interior domain' as defined in Chapter 1).

We wish to prove the existence of a constant $\nu > 0$ that is independent of c_2 and Λ such that

$$\inf_{A(\Lambda)} \rho_\Lambda \geq \nu > 0. \qquad (3.98)$$

For this purpose, we argue by contradiction and assume there exists a sequence $x_\Lambda \in A(\Lambda)$ such that $\rho_\Lambda(x_\Lambda) \longrightarrow 0$, where ρ_Λ is a minimum for the TFDW energy with a coefficient c_2 that may also depend on Λ. Let us recall then that $u_\Lambda = \sqrt{\rho_\Lambda}$ is a solution to

$$-\Delta u_\Lambda + \tfrac{5}{3}c_1 u_\Lambda^{7/3} - \tfrac{4}{3}c_2 u_\Lambda^{5/3} - \Phi_\Lambda u_\Lambda + \theta_\Lambda u_\Lambda = 0.$$

It follows that, for any fixed R_0, we have from Harnack's inequality (see [56] or [24]) that, for Λ large enough ($|\Lambda|^{\alpha} \geq R_0$),

$$\sup_{B(x_\Lambda, R_0)} \rho_\Lambda \leq C(R_0) \inf_{B(x_\Lambda, R_0)} \rho_\Lambda \leq C(R_0)\, \rho_\Lambda(x_\Lambda),$$

for some constant $C(R_0)$ that we may choose independently of Λ and c_2, since $\|u_\Lambda\|_{L^{\infty}}$, θ_Λ and $\sup_{x \in \mathbf{R}^3} \|\Phi_\Lambda\|_{L^p(B(x, R_0))}$ (for some $p > \frac{3}{2}$) are bounded independently of Λ and c_2.

Therefore, on $B(x_\Lambda, R_0)$,

$$-\Delta\Phi_\Lambda = 4\pi \left[\sum_{i \in \Lambda} \delta_i - \rho_\Lambda\right] \geq 4\pi \left[\sum_{i \in \Lambda \cap B(x_\Lambda, R_0)} \delta_i - \epsilon_\Lambda\right],$$

where ϵ_Λ goes to zero.

In addition, Solovej's inequality (3.94) holding also in our case, we have on $B(x_\Lambda, R_0)$

$$\Phi_\Lambda \geq \rho_\Lambda^{2/3} - C_0 \geq -C_0,$$

for some constant C_0 independent of Λ, c_2, and R_0. Besides,

$$\|\Phi_\Lambda\|_{L^p(B(x_\Lambda, 1))} \leq C_0'$$

for some constant C_0' that is independent of Λ, c_2, and R_0.

Next, we define Ψ_Λ by

$$\begin{cases} -\Delta \Psi_\Lambda = 4\pi \left[\sum_{i \in \Lambda \cap B(x_\Lambda, R_0)} \delta_i - \varepsilon_\Lambda \right] & \text{in } B(x_\Lambda, R_0), \\[2mm] \Psi_\Lambda = -C_0 & \text{on } \partial B(x_\Lambda, R_0). \end{cases}$$

Then, the maximum principle implies

$$\Phi_\Lambda \geq \Psi_\Lambda$$

on $B(x_\Lambda, R_0)$, and thus in particular on $B(x_\Lambda, 1)$, while

$$\inf_{B(x_\Lambda, 1)} \Psi_\Lambda \geq a\, R_0^{5/3} - b\, \varepsilon_\Lambda\, R_0^2 - C_0,$$

for some constants a, b that do not depend on Λ and c_2. Comparing with (3.95), we obtain a contradiction provided that we can make $a\, R_0^{5/3} - b\, \varepsilon_\Lambda\, R_0^2$ larger than some fixed positive constant C_0'' (which does not depend on R_0 nor on Λ). Then, we choose

$$R_0 > \left(\frac{C_0''}{a} \right)^{3/5},$$

and we reach the desired contradiction for Λ sufficiently large. Therefore, the inequality (3.98) holds.

We may now choose c_2 small enough in order to have

$$c_1 \left(\frac{\nu}{2} \right)^{5/3} - c_2 \left(\frac{\nu}{2} \right)^{4/3} \geq 0. \tag{3.99}$$

Step 3. At this stage, we have chosen a constant c_2 small enough to have

$$c_1 \rho_\Lambda^{5/3} - c_2 \rho_\Lambda^{4/3} \geq 0$$

on $A(\Lambda)$.

We therefore have

$$\frac{1}{|\Lambda|} I_\Lambda^{\mathrm{TFDW}} = \frac{1}{|\Lambda|} \left[\int_{\mathbf{R}^3} |\nabla \sqrt{\rho_\Lambda}|^2 - \int_{\mathbf{R}^3} V_\Lambda \rho_\Lambda + \frac{1}{2} D(\rho_\Lambda, \rho_\Lambda) + \frac{U_\Lambda}{2} \right.$$
$$\left. + \int_{\mathbf{R}^3} [c_1 \rho_\Lambda^{5/3} - c_2 \rho_\Lambda^{4/3}]_+ \right] + o(1), \tag{3.100}$$

because

$$\int_{\mathbf{R}^3} [c_1 \rho_\Lambda^{5/3} - c_2 \rho_\Lambda^{4/3}]_+ \leq (c_1 \|\rho_\Lambda\|_{L^\infty}^{5/3} + c_2 \|\rho_\Lambda\|_{L^\infty}^{4/3})\, vol(\Gamma(\Lambda) \backslash A(\Lambda))$$
$$+ (c_1 \|\rho_\Lambda\|_{L^\infty}^{2/3} + c_2 \|\rho_\Lambda\|_{L^\infty}^{1/3}) \int_{\Gamma(\Lambda)^c} \rho_\Lambda$$

$$= o(|\Lambda|).$$

Let us now define

$$I_{\Lambda,\varepsilon}^{\mathrm{TFDW}} = \inf \left\{ \int_{\mathbf{R}^3} |\nabla\sqrt{\rho}|^2 - \int_{\mathbf{R}^3} V_\Lambda \rho + \tfrac{1}{2} D(\rho,\rho) \right.$$
$$+\varepsilon \int_{\mathbf{R}^3} \rho^{5/3} + \int_{\mathbf{R}^3} [c_1\rho^{5/3} - c_2\rho^{4/3}]_+ + \frac{U_\Lambda}{2};$$
$$\left. \sqrt{\rho} \in H^1(\mathbf{R}^3), \int_{\mathbf{R}^3} \rho = |\Lambda| \right\}.$$

Clearly, since $[t]_+ \geq t$,

$$I_{\Lambda,\varepsilon}^{\mathrm{TFDW}} \geq I_\Lambda^{\mathrm{TFDW}}.$$

On the other hand, because of (3.100),

$$I_{\Lambda,\varepsilon}^{\mathrm{TFDW}} \leq E^{\mathrm{TFDW}}(\rho_\Lambda) + \varepsilon \int_{\mathbf{R}^3} \rho_\Lambda^{5/3} + \frac{U_\Lambda}{2} + o(|\Lambda|)$$
$$= I_\Lambda^{\mathrm{TFDW}} + \varepsilon \int_{\mathbf{R}^3} \rho_\Lambda^{5/3} + o(|\Lambda|).$$

It is straightforward to see that, since $\varepsilon\rho^{5/3} + [c_1\rho^{5/3} - c_2\rho^{4/3}]_+$ is a non-negative convex function that is greater than $\varepsilon\rho^{5/3}$ and smaller than $(c_1+\varepsilon)\rho^{5/3}$, $I_{\Lambda,\varepsilon}^{\mathrm{TFDW}}$ has a thermodynamic limit, i.e.

$$\frac{I_{\Lambda,\varepsilon}^{\mathrm{TFDW}}}{|\Lambda|} \longrightarrow I_{\mathrm{per},\varepsilon} + \frac{M}{2}$$

as Λ goes to infinity, where the periodic problem $I_{\mathrm{per},\varepsilon}$ is defined by

$$I_{\mathrm{per},\varepsilon} = \inf \left\{ \int_{\Gamma_0} |\nabla\sqrt{\rho}|^2 - \int_{\Gamma_0} G\rho + \tfrac{1}{2} \iint_{\Gamma_0 \times \Gamma_0} \rho(x)G(x-y)G(y)\,\mathrm{d}x\,\mathrm{d}y \right.$$
$$+\varepsilon \int_{\Gamma_0} \rho^{5/3} + \int_{\Gamma_0} [c_1\rho^{5/3} - c_2\rho^{4/3}]_+;$$
$$\left. \rho \geq 0, \sqrt{\rho} \in H_{\mathrm{per}}^1(\Gamma_0), \int_{\Gamma_0} \rho = 1 \right\}$$

Let us remark that, when ε goes to zero, $I_{\mathrm{per},\varepsilon}$ goes to the periodic problem

$$I_{\mathrm{per}}^{\mathrm{TFDW}} = \inf \left\{ \int_{\Gamma_0} |\nabla\sqrt{\rho}|^2 - \int_{\Gamma_0} G\rho + \tfrac{1}{2} \iint_{\Gamma_0 \times \Gamma_0} \rho(x)G(x-y)G(y)\,\mathrm{d}x\,\mathrm{d}y \right.$$

$$+ \int_{\Gamma_0} [c_1 \rho^{5/3} - c_2 \rho^{4/3}]_+;$$

$$\rho \geq 0, \sqrt{\rho} \in H^1_{\mathrm{per}}(\Gamma_0), \int_{\Gamma_0} \rho = 1 \bigg\}.$$

Indeed, if ρ_ε denotes the minimizer of $I_{\mathrm{per},\varepsilon}$, $\int_{\Gamma_0} \rho_\varepsilon^{5/3}$ is bounded independently of ε since, on the one hand,

$$\int_{\Gamma_0} |\nabla \sqrt{\rho_\varepsilon}|^2 - \int_{\Gamma_0} G \rho_\varepsilon \leq E_{\mathrm{per},\varepsilon}(\rho_\varepsilon) \leq E_{\mathrm{per},\varepsilon}\left(\frac{1}{|\Gamma_0|}\right) \leq E_{\mathrm{per},1}\left(\frac{1}{|\Gamma_0|}\right),$$

while, on the other hand,

$$\int_{\Gamma_0} \rho_\varepsilon^{5/3} \leq C \int_{\Gamma_0} \left(\rho_\varepsilon - \frac{1}{|\Gamma_0|} \int_{\Gamma_0} \rho_\varepsilon \right)^{5/3} + C$$

$$\leq C \left(\int_{\Gamma_0} |\nabla \sqrt{\rho_\varepsilon}|^2 \right)^{1/2} + C,$$

by the Poincaré–Wirtinger inequality, where C denotes various positive constants independent of ε.

Next, we have

$$\frac{I^{\mathrm{TFDW}}_{\Lambda,\varepsilon}}{|\Lambda|} - \varepsilon\, O(1) - o(1) \leq \frac{I^{\mathrm{TFDW}}_{\Lambda}}{|\Lambda|} \leq \frac{I^{\mathrm{TFDW}}_{\Lambda,\varepsilon}}{|\Lambda|}.$$

Thus, letting Λ go to infinity,

$$I_{\mathrm{per},\varepsilon} + \frac{M}{2} - \varepsilon\, O(1) \leq \liminf_{\Lambda \to \infty} \frac{I^{\mathrm{TFDW}}_{\Lambda}}{|\Lambda|}$$

$$\leq \limsup_{\Lambda \to \infty} \frac{I^{\mathrm{TFDW}}_{\Lambda}}{|\Lambda|}$$

$$\leq I_{\mathrm{per},\varepsilon} + \frac{M}{2}.$$

Now letting ε go to zero, we obtain

$$\frac{I^{\mathrm{TFDW}}_{\Lambda}}{|\Lambda|} \overset{\Lambda \to \infty}{\longrightarrow} I^{\mathrm{TFDW}}_{\mathrm{per}} + \frac{M}{2}.$$

Finally, we show that

$$I^{\mathrm{TFDW}}_{\mathrm{per}} = \inf \left\{ \int_{\Gamma_0} |\nabla \sqrt{\rho}|^2 - \int_{\Gamma_0} G\rho + \frac{1}{2} \iint_{\Gamma_0 \times \Gamma_0} \rho(x) G(x-y) G(y) \, dx \, dy \right.$$

$$+ \int_{\Gamma_0} c_1 \rho^{5/3} - c_2 \rho^{4/3};$$

$$\rho \geq 0, \sqrt{\rho} \in H^1_{\text{per}}(\Gamma_0), \int_{\Gamma_0} \rho = 1 \Bigg\}. \qquad (3.101)$$

(Let us emphasize the fact that the positive part has been removed.)

For this purpose, we take ρ as a minimum of the minimization problem appearing in the right-hand side. By the same kinds of arguments as the ones used above, we have

$$\inf_{\mathbf{R}^3} \rho \geq \nu > 0,$$

for a constant ν that does not depend on c_2 (small). Taking c_2 small enough we therefore have

$$c_1 \rho^{5/3} - c_2 \rho^{4/3} \geq 0$$

and thus $E(\rho) = I^{\text{TFDW}}_{\text{per}}$ (where E stands for the energy functional appearing in the right-hand side of (3.101)). For c_2 small, the minimizer is unique.

This concludes the proof of the existence of the thermodynamic limit for the TFDW model with c_2 small that we state now precisely through the following:

Theorem 3.28 (Convergence of the energy in the TFDW model) *Let us consider the TFDW model defined by (3.92) and (3.93).*

Then, there exists a positive constant $\bar{c}_2$ such that, for every parameter $0 < c_2 < \bar{c}_2$ arising in (3.92) and for any Van Hove sequence (Λ) satisfying

$$(c) \qquad\qquad \frac{|\Lambda^h|}{|\Lambda|} Log(|\Lambda^h|) \overset{\Lambda \to \infty}{\longrightarrow} 0,$$

in addition to the conditions (a) and (b) of Definition 1 of Chapter 1, we have

$$\lim_{\Lambda \to \infty} \frac{1}{|\Lambda|} I^{\text{TFDW}}_\Lambda = I^{\text{TFDW}}_{\text{per}} + \frac{M}{2},$$

where $I^{\text{TFDW}}_{\text{per}}$ is defined by (3.101) and $M = \lim_{x \to 0} G(x) - \dfrac{1}{|x|}$.

3.7   Appendix

We prove in this appendix further regularity results for the sequence of densities ρ_Λ. The following result is likely to be already known, but we prefer to redo the (rather simple) proof for the sake of completeness.

Lemma A   *Let $u_\Lambda = \sqrt{\rho_\Lambda}$, where ρ_Λ is the minimizer of I^{TFW}_Λ. Then, there exists a positive constant C that is independent of Λ such that*

$$\|\nabla u_\Lambda\|_{L^\infty(\mathbf{R}^3)} \leq C.$$

In particular, combining this bound with Proposition 3.10, we have shown that u_Λ is bounded in $W^{1,\infty}(\mathbf{R}^3)$ independently of Λ.

Proof of Lemma A We rewrite, for the reader's convenience, the Euler–Lagrange equation satisfied by u_Λ; namely,

$$-\Delta u_\Lambda = -\tfrac{5}{3}\, u_\Lambda^{7/3} + (\Phi_\Lambda - \theta_\Lambda)\, u_\Lambda \qquad \text{on } \mathbf{R}^3, \qquad (3.102)$$

for some $\theta_\Lambda > 0$.

Let us fix $0 < \delta < 1$. From the proof of Proposition 3.12, we know that $\Phi_\Lambda - \theta_\Lambda$ is bounded in $L^\infty(\mathbf{R}^3 \setminus \bigcup_{y \in \Lambda} B(y; \delta))$. Moreover, we know from Proposition 3.10 that u_Λ is bounded in $L^\infty(\mathbf{R}^3)$. Therefore, we deduce from (3.102) that Δu_Λ is bounded in $L^\infty(\mathbf{R}^3 \setminus \bigcup_{y \in \Lambda} B(y; \delta))$ and all the bounds mentioned above are independent of Λ. Thus, from standard elliptic regularity results (see [24], for example), ∇u_Λ is bounded in $L^\infty(\mathbf{R}^3 \setminus \bigcup_{y \in \Lambda} B(y; \delta))$, for all $0 < \delta < 1$ and independently of Λ.

Our next step now consists of obtaining bounds for ∇u_Λ in $L^\infty(B_\delta)$ (assuming that $0 \in \Lambda$), with the convention that $B_\delta = B(0, \delta)$. Next, we notice that the same bound holds in $L^\infty(B(y; \delta))$, for all $y \in \Lambda$, and is independent of y, since we recall from Propositions 3.10 and 3.12 that

$$\forall\, 1 \le p < 3, \qquad \sup_{x \in \mathbf{R}^3} \|\Phi_\Lambda\|_{L^p(x+B_1)} \le C, \qquad (3.103)$$

and

$$\|u_\Lambda\|_{L^\infty(\mathbf{R}^3)} \le C, \qquad (3.104)$$

where C denotes various positive constants that are independent of Λ.

For this purpose, we first decompose the term $\Phi_\Lambda\, u_\Lambda$ arising on the right-hand side of (3.102) in the following way:

$$\Phi_\Lambda\, u_\Lambda = \Phi_\Lambda\, (u_\Lambda - u_\Lambda(0)) + u_\Lambda(0)\left(\Phi_\Lambda(x) - \frac{1}{|x|}\right) + \frac{u_\Lambda(0)}{|x|}. \qquad (3.105)$$

Next, returning to the right-hand side of (3.102), we notice that, because of (3.104) and since θ_Λ is bounded independently of Λ, $-\tfrac{5}{3}\, u_\Lambda^{7/3} - \theta_\Lambda\, u_\Lambda$ is bounded in $L^\infty(\mathbf{R}^3)$ independently of Λ.

Besides, because of (3.103) and (3.104), we deduce from (3.102) and standard elliptic regularity results that u_Λ is bounded in $W^{2,p}_{\text{unif}}(\mathbf{R}^3)$, for all $1 \le p < 3$, and thus, from Sobolev's embeddings, u_Λ is bounded in $C^{0,\alpha}(\mathbf{R}^3)$ for all $0 < \alpha < 1$. In addition, from the proof of Proposition 3.12 (equation (3.33)), we obtain that

$$|\Phi_\Lambda(x)| \le \frac{C}{|x|} \qquad \text{a.e. in } B(0; \delta),$$

where, here and below, C denotes a positive constant that is independent of Λ. Thus, for every $0 < \alpha < 1$,

$$|\Phi_\Lambda(x)\,(u_\Lambda(x) - u_\Lambda(0))| \leq \frac{C}{|x|^{1-\alpha}};$$

hence $\Phi_\Lambda\,(u_\Lambda - u_\Lambda(0))$ is bounded in $L^q(B_\delta)$, for all $1 \leq q < +\infty$ (and independently of Λ).

Moreover, $\Phi_\Lambda - \frac{1}{|x|}$ is bounded in $L^\infty(B_\delta)$. Indeed, from Proposition 3.12, $\Phi_\Lambda - \frac{1}{|x|}$ is bounded in $L^p(B_\delta)$, for all $1 \leq p < 3$, while

$$-\Delta\left(\Phi_\Lambda - \frac{1}{|x|}\right) = -u_\Lambda^2 \qquad \text{in } B_\delta,$$

and thus $\Delta(\Phi_\Lambda - \frac{1}{|x|})$ is bounded in $L^\infty(B_\delta)$. So we conclude with standard elliptic regularity results.

At this stage, we rewrite u_Λ as the sum of two functions $u_\Lambda = v_\Lambda + w_\Lambda$, where v_Λ is the solution to

$$-\Delta v_\Lambda = -u_\Lambda^{7/3} - \theta_\Lambda\,u_\Lambda + \Phi_\Lambda\,(u_\Lambda - u_\Lambda(0)) + u_\Lambda(0)\left(\Phi_\Lambda(x) - \frac{1}{|x|}\right) \quad (3.106)$$

in B_δ, together with the boundary condition

$$v_\Lambda = u_\Lambda \qquad \text{on } \partial B_\delta,$$

and w_Λ is the solution to

$$\begin{cases} -\Delta w_\Lambda = \dfrac{u_\Lambda(0)}{|x|} & \text{in } B_\delta, \\ w_\Lambda = 0 & \text{on } \partial B_\delta. \end{cases} \qquad (3.107)$$

On the one hand, from the bounds gathered above, we know that the right-hand side of the equation (3.106) satisfied by v_Λ belongs to $L^q(B_\delta)$, for all $1 \leq q < +\infty$, with bounds independent of Λ. Therefore, from standard regularity results and by Sobolev's embeddings, v_Λ is bounded in $C^{1,\alpha}(B_\delta)$, for all $0 < \alpha < 1$, and thus, in particular, in $W^{1,\infty}(B_{\delta'})$ at least for any $0 < \delta' < \delta$ and this bound is independent of Λ.

On the other hand, we may compute explicitly the unique solution w_Λ of (3.107); that is,

$$w_\Lambda(x) = \tfrac{1}{2}\,u_\Lambda(0)\,[\delta - |x|].$$

Thus, since $u_\Lambda(0)$ is bounded independently of Λ, w_Λ is straightforwardly bounded in $W^{1,\infty}(B_\delta)$ independently of Λ.

Our claim follows, gathering all these results. $\diamondsuit$

Remark 3.29 It is worth noticing that the key point in the above argument is that the singularity at 0 of the right-hand side of the equation (3.102) behaves exactly like $\frac{1}{|x|}$. In particular, if one looks at the solution to

$$-\Delta u = f,$$

on a ball centered at 0, where $f(x) = \frac{a(\omega)}{|x|}$, $\omega = \frac{x}{|x|}$, and $a \in L^{2+\epsilon}(S^2)$, for any $\epsilon > 0$, we may still prove that u is in $W^{1,\infty}$ in the neighbourhood of 0. This choice is a particular choice for a function f in $L^{3,\infty}(\mathbf{R}^3)$ (which corresponds to the case when a is in $L^3(S^2)$). However, it is false in general that if $f \in L^{3,\infty}(\mathbf{R}^3)$, $\nabla u \in L^\infty_{\text{loc}}(\mathbf{R}^3)$.

CONVERGENCE OF THE DENSITY FOR THE THOMAS–FERMI–VON WEIZSÄCKER MODEL WITH YUKAWA POTENTIAL

4.1 Introduction

Whereas in Chapters 2 and 3 we were interested in the behaviour of the energy per unit volume when the molecular system asymptotically approached an infinite crystal, we begin with this chapter our investigation on the behaviour of the electronic density. Let us recall that the main questions on this density, that we introduced in Chapter 1, are: Does the electronic density go to a limit, and does this limit share the same periodicity as the infinite lattice of the crystal?

As in our study of the energy, we shall begin with the simpler case of a Yukawa potential to model the interactions between charged particles (nuclei and electrons). The next chapter will be devoted to the Coulomb case.

In the Yukawa case ($(Y-m)$ and $(Y-\delta)$ programs), we are going to prove that the answer to both questions above is positive. The limit density is indeed the minimizing density of the periodic minimization problem that we have obtained in Chapter 2 as the thermodynamic limit of the energy per unit volume.

Let us recall some notation before we state our main result.

When Λ is fixed, the minimization problem reads (in the $(Y - \delta)$ case; the definition of the $(Y - m)$ case is a straightforward modification of the formulae below; we refer the reader to Chapter 2)

$$E_\Lambda(\rho) = \int_{\mathbf{R}^3} |\nabla\sqrt{\rho}|^2 + \int_{\mathbf{R}^3} \rho^{5/3} - \int_{\mathbf{R}^3} \rho V_\Lambda$$
$$+ \frac{1}{2} \int\int_{\mathbf{R}^3 \times \mathbf{R}^3} \rho(x)\rho(y)V(x - y)\mathrm{d}x\,\mathrm{d}y,$$

with

$$V_\Lambda(x) = \sum_{k \in \Lambda} V(x - k),$$

$$V(x) = \frac{e^{-a|x|}}{|x|}, \tag{4.1}$$

$$I_\Lambda = \inf\left\{ E_\Lambda(\rho) + \frac{1}{2} \sum_{y \neq z \in \Lambda} V(y - z); \right.$$

$$\rho \geq 0, \sqrt{\rho} \in H^1(\mathbf{R}^3), \int_{\mathbf{R}^3} \rho = |\Lambda| \bigg\}.$$

The periodic problem that we have obtained in the thermodynamic limit is

$$I_{\text{per}}(\mu) = \inf\{E_{\text{per}}(\rho); \rho \geq 0, \sqrt{\rho} \in H^1_{\text{per}}(\Gamma_0), \int_{\Gamma_0} \rho = \mu\}, \tag{4.2}$$

for some $\mu > 0$, where the periodic energy is

$$E_{\text{per}}(\rho) = \int_{\Gamma_0} |\nabla\sqrt{\rho}|^2 + \int_{\Gamma_0} \rho^{5/3} - \int_{\Gamma_0} \rho V_\infty$$
$$+ \frac{1}{2} \iint_{\Gamma_0 \times \Gamma_0} \rho(x)\rho(y)V_\infty(x - y)\mathrm{d}x\,\mathrm{d}y$$
$$+ \frac{1}{2} \sum_{y \neq 0 \in \mathbf{Z}^3} V(y),$$

with

$$V_\infty(x) = \sum_{k \in \mathbf{Z}^3} V(x - k).$$

We define in (4.2) the 'charge' μ by

$$\mu = \min(\mu_0, 1) > 0, \tag{4.3}$$

with $\mu_0 = \int_{\Gamma_0} \rho_0$, where ρ_0 denotes the unique minimizing density that satisfies

$$E_{\text{per}}(\rho_0) = \inf\{E_{\text{per}}(\rho); \rho \geq 0, \sqrt{\rho} \in H^1_{\text{per}}(\Gamma_0)\}. \tag{4.4}$$

We recall that when the periodic cell and the nucleus inside the cell share the same symmetries, then the space $H^1_{\text{per}}(\Gamma_0)$ may be replaced by $H^1(\Gamma_0)$ in the definitions (4.2) and (4.4).

The main results that we are going to prove on the density may be regrouped in the following:

Theorem 4.1 ($(Y - m)$ and $(Y - \delta)$ programs, convergence of the electronic density to a periodic density) *The electronic density ρ_Λ converges to the periodic minimizing density ρ_{per} of (4.2)–(4.4) in the following senses:*

(i) Local convergence
$u_\Lambda = \sqrt{\rho_\Lambda}$ *converges to* $u_{\text{per}} = \sqrt{\rho_{\text{per}}}$ *strongly in* $H^1_{\text{loc}}(\mathbf{R}^3)$ *and in* $L^p_{\text{loc}}(\mathbf{R}^3)$ *for all* $1 \leq p < \infty$.

(ii) Uniform local convergence On any bounded domain Ω, ρ_Λ converges uniformly to ρ_{per}; that is,

$$\lim_{\Lambda \to \infty} \sup_{x \in \Omega} |\rho_\Lambda(x) - \rho_{\text{per}}(x)| = 0.$$

(iii) Uniform convergence on sequences of interior domains As Λ goes to infinity, ρ_Λ converges uniformly to ρ_{per} on any sequence of interior domains $\Gamma'(\Lambda)$ of $\Gamma(\Lambda)$ (see Definition 2 in Chapter 1); that is,

$$\lim_{\Lambda \to \infty} \sup_{x \in \Gamma'(\Lambda)} |\rho_\Lambda(x) - \rho_{\mathrm{per}}(x)| = 0.$$

Let us at this stage make the following observations.

In fact, we shall prove (iii) in Chapter 5, since the same result holds for the Coulomb case and its proof is quite similar, but with slight modifications. However, in order to be as complete as possible, we have regrouped in the above theorem the information we have obtained in our work on the density. In this chapter, we only prove (i) and (ii). Of course, as we shall see in Chapter 5, (iii) implies (i) and (ii).

It is worth noticing that the only (but striking) difference between the Yukawa case treated in this chapter and the Coulomb case treated in the next chapter is that in the latter, the number of electrons per cell obtained in the thermodynamic limit, namely μ, is necessarily 1, whereas in the Yukawa case there are some situations when $\mu < 1$ (see Chapter 2, Section 2.6).

In order to prove Theorem 4.1, we shall proceed as follows.

In Section 4.2, we shall prove that u_Λ converges locally to some function u_∞ which satisfies a partial differential equation that, apart from the convolution term, may be identified with the Euler–Lagrange equation of the periodic minimization problem (4.2), but that we cannot fully identify with it, because we do not yet know that u_∞ is periodic. This equation is

$$-\Delta u_\infty + \tfrac{5}{3} u_\infty^{7/3} - V_\infty u_\infty + (V \star u_\infty^2) u_\infty + \theta_{\mathrm{per}} u_\infty = 0, \qquad (4.5)$$

where θ_{per} is the Lagrange multiplier associated to the minimizing density of (4.2). Since we do not yet know that u_∞ is periodic, we cannot write that we have

$$V \star u_\infty^2(x) = \int_{\Gamma_0} V_\infty(x - y) u_\infty^2(y) \mathrm{d}y. \qquad (4.6)$$

However, we recall in this section that, in an average sense, that is to say using the $\sim$-transform, then the density converges locally to the periodic density. This fact has been proved in Chapter 2. The point here is of course that the periodicity of the limit, that is still lacking for the moment for the exact local convergence above, is automatically satisfied for the limit in the sense of the $\sim$-transform convergence.

In the same fashion, we give in Subsection 4.2.3 a proof of the periodicity of the limit density up to a translation. This means that we may find a sequence of unit cubes (possibly going to infinity) such that on this sequence of cubes the density converges to the periodic density. It is clear that this result is weaker than the full result of periodicity that we shall prove in the following section. However, we choose to give its proof because in some sense this result is, together with the $\sim$-transform convergence result, the optimal result one may hope to obtain by only using energetic estimates. Indeed, one may easily convince oneself that arguments based only upon estimates of the energy while Λ goes to infinity can only lead to what we may call average estimates at the order $|\Lambda|$ of periodicity (in the spirit of the result of Proposition 2.15 in Chapter 2). The exact local behaviour of the density cannot be deduced from such asymptotic energy estimates because these estimates cannot see the smallest local details of the density, and periodicity is a 'detail'.

The 'weak' results of Section 4.2 are summarized in the following:

Proposition 4.2 *(i) Convergence up to a translation*
   There exists a sequence $y_\Lambda \in \Lambda$ *such that* $u_\Lambda(\cdot + y_\Lambda) = \sqrt{\rho_\Lambda(\cdot + y_\Lambda)}$ *converges to* $u_{\text{per}} = \sqrt{\rho_{\text{per}}}$ *strongly in* $H^1(\Gamma_0)$, *and in* $L^p(\Gamma_0)$, *for all* $1 \leq p < +\infty$.
   (ii) Convergence in the sense of the $\sim$-transform
   $\sqrt{\tilde{\rho}_\Lambda}$ *converges to* $\sqrt{\rho_{\text{per}}}$ *strongly in* $H^1_{\text{loc}}(\mathbf{R}^3)$, *and in* $L^p_{\text{loc}}(\mathbf{R}^3)$, *for all* $1 \leq p < +\infty$.

To obtain exact results of periodicity, one must go deeper into the mathematical nature of the equation obtained in the limit, namely (4.5), and analyze this equation with the tools of the theory of non-linear elliptic equations. Basically, the question one has to solve is the uniqueness of the solution to such an equation. This is the purpose of Section 4.3.

The strategy to obtain uniqueness is as follows. We first prove the uniqueness for the equation without the convolution term, and then extend our result to (4.5) using the fact that the Yukawa potential is short-ranged, and thus that the convolution term is almost a local term. The mathematical tool that makes this formal argument rigorous is the implicit function theorem. Section 4.3 is therefore itself divided in four subsections, where we successively prove the existence of uniform bounds from above, then from below for the positive solutions to (4.5), next uniqueness for the equation without the convolution term, and eventually uniqueness for the full equation. Once uniqueness is obtained, we know that the solution is periodic and hence we may use (4.6) to transform (4.5) into

$$-\Delta u_\infty + \tfrac{5}{3} u_\infty^{7/3} - V_\infty u_\infty + \left(\int_{\Gamma_0} V_\infty(x-y) u_\infty^2(y) \mathrm{d}y \right) u_\infty + \theta_{\text{per}} u_\infty = 0, \quad (4.7)$$

and conclude that $u_\infty = \sqrt{\rho_{\text{per}}}$, where ρ_{per} is the minimizing density of (4.2).

In the course of Section 4.3, we shall give many existence and uniqueness lemmata concerning equations of the type (4.7). Our main result in this direction is Theorem 4.19.

Theorem 4.19 *Let* c *be a positive constant, let* $\lambda \geq 0$, *and let* Γ *be a periodic potential in* $L^p_{\text{loc}}(\mathbf{R}^3)$, *for some* $p > \frac{21}{8}$. *Assume that on the periodic cell of* Γ *(for the sake of simplicity, we assume this cell is the unit cube* Γ_0*), the first eigenvalue of the operator* $-\Delta - \Gamma$ *with periodic boundary conditions is negative. Let* $V \in L^1_{\text{loc}}(\mathbf{R}^3)$ *be a non-negative potential satisfying* $V(x) = O(\frac{1}{|x|^{3+\alpha}})$ *at infinity, for some* $\alpha > 0$ *(other short-ranged potential are tractable). Then, there exists a unique solution to*

$$\begin{cases} -\Delta u - \Gamma u + c u^{7/3} + \lambda (u^2 \star V) u = 0, \\ u \geq 0, \quad u \not\equiv 0. \end{cases}$$

This solution satisfies $u > 0$ on $\mathbf{R}^3$ and is periodic.

In addition, the assumption on $\lambda_1(-\Delta - \Gamma,\mathrm{per})$ is satisfied as soon as a solution to the above equation exists.

Let us finally say that the same study will be made in Chapter 5 for the Coulomb potential, and that we shall then see a proof, based upon different uniqueness arguments from those used here, that will also apply to the present Yukawa case.

4.2   Preliminary convergence results on the density

4.2.1   *Local convergence*

The information obtained on the sequence of densities ρ_Λ at the end of Chapter 2 is essentially contained in this subsection and the following one.

Proposition 4.3 *Let us define $u_\Lambda = \sqrt{\rho_\Lambda}$ and $u_{\mathrm{per}} = \sqrt{\rho_{\mathrm{per}}}$. Then, u_Λ is bounded in $L^\infty(\mathbf{R}^3) \cap H^1_{\mathrm{loc}}(\mathbf{R}^3)$, uniformly with respect to Λ. Thus, extracting a subsequence if necessary, we may assume that u_Λ converges to some nonnegative function $u_\infty \in L^\infty(\mathbf{R}^3) \cap H^1_{\mathrm{loc}}(\mathbf{R}^3)$ strongly in $H^1_{\mathrm{loc}}(\mathbf{R}^3) \cap L^p_{\mathrm{loc}}(\mathbf{R}^3)$, for all $1 \le p < +\infty$, and almost everywhere on $\mathbf{R}^3$.*

Moreover, u_∞ satisfies

$$-\Delta u_\infty + \tfrac{5}{3}u_\infty^{7/3} - V_\infty u_\infty + (V \star \rho_\infty)u_\infty + \theta_{\mathrm{per}}u_\infty = 0 \qquad (4.8)$$

at least in the sense of $\mathcal{D}'(\mathbf{R}^3)$, θ_{per} being the Lagrange multiplier associated with u_{per}.

Remark 4.4 (1) At this stage, we are not able to show that this local limit u_∞ does not depend on Λ. This will be a consequence of Section 4.3.

(2) Moreover, we shall have to work more to make sure that u_∞ may not vanish on $\mathbf{R}^3$ or, more precisely, that there exists a constant $\mu > 0$ such that $u_\infty \ge \mu$ a.e. on $\mathbf{R}^3$. We shall prove in Section 4.3 that this property, together with (4.8) and a L^∞ bound on u_∞, determines a unique solution to (4.8) and thus necessarily u_{per}.

(3) Unfortunately, we are not able to deduce immediately that

$$V \star \rho_\infty(x) = \int_{\Gamma_0} V_\infty(x - y)\rho_\infty(y)\, dy.$$

We need the periodicity first. Otherwise, we could of course say that, in view of (4.8), u_∞ is a critical point of the energy functional E_{per}, and then a minimizer of $I_{\mathrm{per}}(\lambda)$ for some $\lambda > 0$, since E_{per} is convex.

Proof of Proposition 4.3 The proof is essentially contained in the proof of Corollary 2.7 of Chapter 2. Only (4.8) remains to be shown. In addition, Corollary 2.26 in Chapter 2 gives that θ_Λ goes to θ_{per} as Λ goes to infinity.

We write down the Euler–Lagrange equation satisfied by u_Λ; namely,

$$-\Delta u_\Lambda + \tfrac{5}{3} u_\Lambda^{7/3} - V_\Lambda u_\Lambda + (V \star \rho_\Lambda) u_\Lambda + \theta_\Lambda u_\Lambda = 0. \qquad (4.9)$$

Now let φ be in $\mathcal{D}(\mathbf{R}^3)$ and let K be a compact subset of $\mathbf{R}^3$ such that $supp\, \varphi \subset K$.

We shall only prove that $\int_{\mathbf{R}^3} (V \star \rho_\Lambda) u_\Lambda\, \varphi\, dx$ goes to $\int_{\mathbf{R}^3} (V \star \rho_\infty) u_\infty\, \varphi\, dx$, as Λ goes to infinity, the treatment of the other terms in (4.9) being even easier.

The difference may be written as

$$\int_{\mathbf{R}^3} (V \star \rho_\Lambda)\, u_\Lambda\, \varphi\, dx - \int_{\mathbf{R}^3} (V \star \rho_\infty)\, u_\infty\, \varphi\, dx$$

$$= \int_{\mathbf{R}^3} V \star (\rho_\Lambda - \rho_\infty)\, u_\Lambda\, \varphi\, dx$$

$$+ \int_{\mathbf{R}^3} (V \star \rho_\infty)\, (u_\Lambda - u_\infty)\, \varphi\, dx$$

It is easily seen that the second term goes to zero as Λ goes to infinity, since $V \star \rho_\infty \in L^\infty(\mathbf{R}^3)$ ($V \in L^1(\mathbf{R}^3)$ and $\rho_\infty \in L^\infty(\mathbf{R}^3)$), and, for example, u_Λ converges to u_∞ in $L^2_{\text{loc}}(\mathbf{R}^3)$.

Concerning the first term, we use the uniform L^∞ bound on u_Λ; next we claim that $V \star (\rho_\Lambda - \rho_\infty)$ converges uniformly to zero on any compact subset K of $\mathbf{R}^3$. Indeed we have, for all x in K, for all $R > 0$,

$$|V \star (\rho_\Lambda - \rho_\infty)(x)| \leq \int_{\mathbf{R}^3} V(x+y)\, |\rho_\Lambda(y) - \rho_\infty(y)|\, dy$$

$$\leq \int_{B_R} V(x+y)\, |\rho_\Lambda(y) - \rho_\infty(y)|\, dy$$

$$+ 2\left(\int_{B_R^c} V(x+y) dy \right) \|\rho_\Lambda\|_{L^\infty(\mathbf{R}^3)}. \qquad (4.10)$$

For any fixed $R > 0$, the first term on the right-hand side goes to zero as Λ goes to infinity from the strong local convergence of ρ_Λ to ρ_∞, in, say, $L^2(B_R)$.

For the second term, using the monotonicity of the function $t \mapsto \frac{\exp(-at)}{t}$, we bound $V(x+y)$ by

$$\frac{\exp(-a|x-y|)}{|x-y|} \leq \frac{\exp(aM)}{d(K; B_R^c)} \exp(-a|y|), \qquad \text{for all } y \in B_R^c$$

with $M = \max_{x \in K} |x|$, R chosen large enough so that $K \cap B_R^c = \emptyset$, and where we used the fact that $|x - y| \geq |y| - |x|$.

Then, we conclude easily, letting Λ, and then R, go to infinity in (4.10). $\diamond$

In addition to Proposition 4.3, we may show that the convergence of u_Λ to u_∞ also holds for the uniform local topology. This is a straightforward consequence of elliptic regularity results. In view of (4.9), we obtain that u_Λ also converges in $W^{2,p}_{\mathrm{loc}}$, and thus in particular converges uniformly on any compact set. When we have identified u_∞ with u_{per} at the end of Section 4.3, this will yield (i) and (ii) of Theorem 4.1.

4.2.2 Convergence of the ~-transform

The first (weak) result towards the asymptotic periodicity of ρ_Λ concerns the convergence of the sequence $\tilde{\rho}_\Lambda = \dfrac{1}{|\Lambda|} \sum_{y \in \Lambda} \rho_\Lambda(. + y)$ to ρ_{per}, which we copy (without proof) from Corollary 2.26 of Chapter 2.

Proposition 4.5 *The sequence $\sqrt{\tilde{\rho}_\Lambda}$ converges to $\sqrt{\rho_{\mathrm{per}}}$ strongly in $H^1_{\mathrm{loc}}(\mathbf{R}^3)$, and in $L^p_{\mathrm{loc}}(\mathbf{R}^3)$, for all $1 \leq p < +\infty$.*

Remark 4.6 In Chapter 2, we announced a convergence in $H^1(\Gamma_0)$ (respectively, $L^p(\Gamma_0)$). However, it is easy to show that $\tilde{\rho}_\Lambda$ converges to a periodic function (see Chapter 2, Lemma 3.19). Then, we may recover a local limit on $\mathbf{R}^3$.

In the following subsection, we prove the strong convergence of u_Λ to u_{per} in $H^1_{\mathrm{loc}}(\mathbf{R}^3)$, up to a translation.

4.2.3 Periodicity up to a translation

The main result of this subsection is contained in the following:

Proposition 4.7 *For any Van Hove sequence (Λ), there exists a sequence y_Λ in Λ such that $u_\Lambda(\cdot + y_\Lambda) = \sqrt{\rho_\Lambda(\cdot + y_\Lambda)}$ converges to $u_{\mathrm{per}} = \sqrt{\rho_{\mathrm{per}}}$ strongly in $H^1(\Gamma_0)$, and in $L^p(\Gamma_0)$, for all $1 \leq p < +\infty$.*

Remark 4.8 (i) Unfortunately, we are not able to decide whether y_Λ remains bounded. Otherwise, we could recover the periodicity of u_Λ as well.

(ii) It is in fact possible to show the convergence in $H^1_{\mathrm{loc}}(\mathbf{R}^3)$ and $L^p_{\mathrm{loc}}(\mathbf{R}^3)$.

The main step towards the proof of Proposition 4.7 is contained in the following:

Lemma 4.9 *For any Van Hove sequence Λ, we have:*

(i) $\|u_\Lambda - u_{\mathrm{per}}\|_{L^p(\Gamma(\Lambda))} = o(|\Lambda|^{\frac{1}{p}})$, for all $1 \leq p < +\infty$;

(ii) $\|\nabla u_\Lambda - \nabla u_{\mathrm{per}}\|_{L^2(\Gamma(\Lambda))} = o(|\Lambda|^{\frac{1}{2}})$.

Proof of Lemma 4.9 In Chapter 2, we explained how to construct a sequence of functions $\bar{\rho}_\Lambda$ in $H^1(\mathbf{R}^3)$, such that

$$\begin{cases} \bar{\rho}_\Lambda = \rho_{\mathrm{per}} & \text{on } \Gamma(\Lambda), \\ supp(\bar{\rho}_\Lambda) \subset (\Gamma(\Lambda) \cup \Lambda^1), \\ \|\bar{\rho}_\Lambda\|_{L^p(\mathbf{R})^3} \leq C|\Lambda|^{1/p}, & \text{for all } 1 \leq p \leq +\infty, \\ \int_{\Gamma(\Lambda)^c} |\nabla \sqrt{\bar{\rho}_\Lambda}|^2 \leq C|\Lambda^1|, \end{cases} \qquad (4.11)$$

for some constants C that are independent of Λ, enjoying in addition the fundamental property that

$$\frac{1}{|\Lambda|}\left[E_\Lambda(\rho_\Lambda) - E_\Lambda(\overline{\rho}_\Lambda)\right] \overset{\Lambda\to\infty}{\longrightarrow} 0.$$

Therefore, we may copy, step by step, the argument of Chapter 2, based on the strong convexity of the functional, to obtain

$$\|\rho_\Lambda - \overline{\rho}_\Lambda\|_{L^p(\mathbf{R}^3)} = o(|\Lambda|^{\frac{1}{p}}) \qquad \text{for all } 1 \le p < +\infty. \tag{4.12}$$

Then, as an immediate consequence of (4.11) and (4.12), we deduce that we have

$$\|\rho_\Lambda - \rho_{\mathrm{per}}\|_{L^p(\Gamma(\Lambda))} = o(|\Lambda|^{\frac{1}{p}}) \qquad \text{for all } 1 < p < +\infty. \tag{4.13}$$

In fact, we may even include $p = 1$ in (4.13) using Hölder's inequality.

Now, noticing that

$$\forall\, a, b \in \mathbf{R}^+, (\sqrt{a} - \sqrt{b})^2 \le |a - b|,$$

and, using Hölder's inequalities once more, (4.13) implies that we have

$$\|u_\Lambda - u_{\mathrm{per}}\|_{L^p(\Gamma(\Lambda))} = o(|\Lambda|^{\frac{1}{p}}) \qquad \text{for all } 1 \le p < +\infty, \tag{4.14}$$

where

$$u_\Lambda = \sqrt{\rho_\Lambda} \qquad \text{and} \qquad u_{\mathrm{per}} = \sqrt{\rho_{\mathrm{per}}}.$$

It only remains to prove that

$$\|\nabla u_\Lambda - \nabla u_{\mathrm{per}}\|_{L^2(\Gamma(\Lambda))} = o(|\Lambda|^{1/2}). \tag{4.15}$$

For this purpose, we write down the Euler–Lagrange equation satisfied by u_{per}; namely,

$$-\Delta u_{\mathrm{per}} + \tfrac{5}{3}u_{\mathrm{per}}^{7/3} - \Phi_{\mathrm{per}}u_{\mathrm{per}} + \theta_{\mathrm{per}}u_{\mathrm{per}} = 0, \tag{4.16}$$

where

$$\Phi_{\mathrm{per}}(x) = V_\infty(x) - \int_{\Gamma_0} V_\infty(x - y)\rho_{\mathrm{per}}(y)\,\mathrm{d}y.$$

Then, if we multiply (4.16) by u_Λ, and then integrate over $\Gamma(\Lambda)$, we obtain (we recall that u_{per} is periodic—see Chapter 2)

$$\int_{\Gamma(\Lambda)} \nabla u_{\mathrm{per}} \cdot \nabla u_\Lambda\,\mathrm{d}x + \frac{5}{3}\int_{\Gamma(\Lambda)} u_{\mathrm{per}}^{7/3} u_\Lambda\,\mathrm{d}x - \int_{\Gamma(\Lambda)} \Phi_{\mathrm{per}}\, u_{\mathrm{per}}\, u_\Lambda\,\mathrm{d}x$$

$$+ \theta_{\mathrm{per}} \int_{\Gamma(\Lambda)} u_{\mathrm{per}}\, u_\Lambda\,\mathrm{d}x = 0.$$

The sum of the last three terms is equal to

$$|\Lambda|\left[\frac{5}{3}\int_{\Gamma_0} u_{\mathrm{per}}^{10/3}\,\mathrm{d}x - \int_{\Gamma_0} \Phi_{\mathrm{per}}\rho_{\mathrm{per}}\,\mathrm{d}x + \theta_{\mathrm{per}}\int_{\Gamma_0} \rho_{\mathrm{per}}\,\mathrm{d}x\right] + o(|\Lambda|)$$

$$= -|\Lambda| \int_{\Gamma_0} |\nabla u_{\text{per}}|^2 \, dx + o(|\Lambda|),$$

because of (4.16). Indeed, arguing for example on the second one, we have

$$\left| \int_{\Gamma(\Lambda)} \Phi_{\text{per}} u_{\text{per}} |u_\Lambda - u_{\text{per}}| \, dx \right| \leq \|u_{\text{per}}\|_{L^\infty} \|\Phi_{\text{per}}\|_{L^2(\Gamma(\Lambda))} \|u_\Lambda - u_{\text{per}}\|_{L^2(\Gamma(\Lambda))},$$

and we conclude with (4.14), together with

$$\|\Phi_{\text{per}}\|_{L^2(\Gamma(\Lambda))} \leq C \, |\Lambda|^{1/2}.$$

At this stage, we have just proved that the following equality holds:

$$\int_{\Gamma(\Lambda)} \nabla u_\Lambda \cdot \nabla u_{\text{per}} \, dx = |\Lambda| \int_{\Gamma_0} |\nabla u_{\text{per}}|^2 \, dx + o(|\Lambda|). \tag{4.17}$$

But, now, we recall that we know that $\lim_{\Lambda \to \infty} \dfrac{I_\Lambda}{|\Lambda|} = I_{\text{per}}(\mu)$, and that this convergence implies

$$\lim_{\Lambda \to \infty} \frac{1}{|\Lambda|} \int_{\Gamma(\Lambda)} |\nabla u_\Lambda|^2 \, dx = \int_{\Gamma_0} |\nabla u_{\text{per}}|^2 \, dx \tag{4.18}$$

(see Chapter 2).

Then, (4.17) and (4.18) yield (4.15), and our claim follows. $\diamond$

With the help of Lemma 4.9, we now go back to the proof of Proposition 4.7.

Proof of Proposition 4.7 We choose $y_\Lambda \in \Lambda$ among the $y \in \Lambda$ which satisfy

$$\|u_\Lambda - u_{\text{per}}\|_{H^1(\Gamma_y)} = \min \left\{ z \in \Lambda / \|u_\Lambda - u_{\text{per}}\|_{H^1(\Gamma_z)} \right\}.$$

Then,

$$o(|\Lambda|) = \|u_\Lambda - u_{\text{per}}\|^2_{H^1(\Gamma(\Lambda))}$$
$$= \sum_{y \in \Lambda} \|u_\Lambda - u_{\text{per}}\|^2_{H^1(\Gamma_y)}$$
$$\geq |\Lambda| \, \|u_\Lambda - u_{\text{per}}\|_{H^1(\Gamma_{y_\Lambda})},$$

by the definition of y_Λ.

This implies of course, in view of Lemma 4.9, that $u_\Lambda(\cdot + y_\Lambda)$ converges to u_{per} strongly in $H^1(\Gamma_0)$, and thus in $L^p(\Gamma_0)$ for all $p < 6$, because of Sobolev embeddings, and in fact in $L^p(\Gamma_0)$ for all $1 \leq p < +\infty$, because of Hölder's inequality and the L^∞ uniform bound on u_Λ. $\diamond$

Remark 4.10 In fact, it is even possible to show that $u_\Lambda(\cdot + y_\Lambda)$ converges to u_{per} in, say, $H^1(\frac{3}{2} \cdot \Gamma_0)$. For this purpose, we may assume, in order to simplify the presentation, that $0 \in \Lambda$, for all Λ, and we define $\dfrac{\Lambda}{2}$ as the subset of Λ consisting of the $\dfrac{|\Lambda|}{2}$ points of Λ which are the 'closest' to 0. Next, we define y_Λ as one of the y such that

$$\|u_\Lambda - u_{\mathrm{per}}\|_{H^1(y + \frac{3}{2}\cdot\Gamma_0)} = \min_{z \in \frac{\Lambda}{2}} \|u_\Lambda - u_{\mathrm{per}}\|_{H^1(z + \frac{3}{2}\Gamma_0)}.$$

Finally, the above argument goes through once we have noticed that

$$\sum_{y \in \frac{\Lambda}{2}} \|u_\Lambda - u_{\mathrm{per}}\|^2_{H^1(y + \frac{3}{2}\cdot\Gamma_0)} \leq 2 \sum_{y \in \Lambda} \|u_\Lambda - u_{\mathrm{per}}\|^2_{H^1(\Gamma_y)}$$
$$+ \|u_\Lambda - u_{\mathrm{per}}\|^2_{H^1(\partial\Lambda)},$$

where

$$\partial\Lambda = \{x \in \Gamma(\Lambda)^c / d(x, \Lambda) \leq 1\}.$$

The bound from above is once more of order $o(|\Lambda|)$, since by the definition of a Van Hove sequence we have

$$|\partial\Lambda| = |\Lambda^1| = o(|\Lambda|).$$

We are going to improve this result in the following section, since we are going to establish that the sequence u_Λ itself converges strongly to u_{per} in $H^1_{\mathrm{loc}}(\mathbf{R}^3)$.

4.3   Periodicity of the limit density

This section is devoted to the proof of the periodicity of the limit density that we have obtained by passing locally to the limit in the Euler–Lagrange equation corresponding to the TFW problem for Λ fixed (see Proposition 2.2 in Chapter 2).

For this purpose, we prove, under suitable conditions made precise below (see Theorem 4.19, also given in the introduction to this chapter), the uniqueness of the solution on $\mathbf{R}^3$ to

$$\begin{cases} -\Delta u - (\Gamma - \theta)u + cu^{7/3} + (u^2 \star V)u = 0, \\ u > 0, \end{cases} \tag{4.19}$$

where $c > 0$, $\theta \in \mathbf{R}$, V is a short-ranged potential, and Γ is a periodic function on $\mathbf{R}^3$. We prove that the unique solution to (4.19) is periodic, of the same period as Γ. Then taking $V = V_a$, $\Gamma = V_\infty$, $\theta = \theta_{\mathrm{per}}$, and $c = \frac{5}{3}$ yields the result of periodicity on the limit Yukawa density.

Our proof is divided into four steps. In the first one (Lemma 4.11), we show the existence of a uniform bound from above for all non-negative solutions. In the second one (Lemma 4.14), we deduce from this bound from above the existence of a bound from below by a positive constant ν for any solution to (4.19). In other words, $u \equiv 0$ is 'isolated' in the set of solutions.

The third step (Lemma 4.17) is the uniqueness result for the same problem as (4.19) where the non-local term $(u^2 \star V)u$ is turned off:

$$\begin{cases} -\Delta u - (\Gamma - \theta)u + cu^{7/3} = 0, \\ u > 0. \end{cases} \tag{4.20}$$

This step is as follows. We first prove the existence of a maximal solution to (4.20), that the maximal solution is periodic, and thus that the periodic solution is maximal. Next, using Lemma 4.14 for (4.20), we build a non-trivial minimal solution, that is therefore periodic, and thus also maximal. Uniqueness for (4.20) follows.

The fourth step explains how we obtain the uniqueness result for (4.19) from the uniqueness result for (4.20), by using the implicit function theorem. We embed (4.19) and (4.20) into the following family of equations:

$$\begin{cases} -\Delta u - (\Gamma - \theta)u + cu^{7/3} + \lambda(u^2 \star V)u = 0, \\ u > 0, \end{cases} \tag{4.21}$$

where λ is a parameter in $\mathbf{R}_+$. Roughly speaking, the initial problem being strictly convex, the linearization of (4.21) with respect to u is one-to-one at any $u \geq cst > 0$, and thus, in view of step 1, at any solution to (4.21) for $\lambda \in [0,1]$. Therefore, a unique curve of solutions to (4.21) links a solution at $\lambda = 1$ to a solution at $\lambda = 0$. The latter being unique by step 2, the uniqueness (and the periodicity) of the solution to (4.19) follows (Lemma 4.21). Eventually, we conclude for the Yukawa case we are interested in.

Let us also mention that in Chapter 5, when we deal with the periodicity of the limit density for the Coulomb potential, we shall give another (slightly more direct) proof of this periodicity in the $(Y - m)$ and the $(Y - \delta)$ cases.

Notation: In this section, we denote by $\lambda_1(L, \Omega)$ the first eigenvalue of the self-adjoint operator L on the domain Ω with Dirichlet boundary conditions. Likewise, $\lambda_1(L, \Omega, \text{per})$ (or $\lambda_1(L, \text{per})$ when there is no ambiguity) is the first eigenvalue of the self-adjoint operator L with some periodic potential, on the periodic cell Ω with periodic boundary conditions.

In addition, as long as we deal with some general operator, we replace $\Gamma - \theta$ appearing in (4.19)–(4.21) by some other periodic potential Γ. Of course, when we deal with the particular problem we are interested in (the Yukawa case), we shall return to the notation appearing in (4.19)–(4.21).

4.3.1   Bounds from above for solutions

We begin with the following:

Lemma 4.11 Let c be a positive constant and let Γ be a periodic potential in $L^p_{\text{loc}}(\mathbf{R}^3)$ for some $p > \frac{21}{8}$.

Then, there exists a constant C such that any solution to

$$\begin{cases} -\Delta u - \Gamma u + cu^{7/3} \le 0, \\ u > 0, \end{cases} \tag{4.22}$$

satisfies

$$\|u\|_{L^\infty} \le C. \tag{4.23}$$

Proof of Lemma 4.11 Set $c = 1$ for the sake of simplicity. The argument we make here has already been used in Chapter 2 (see Proposition 2.5). On the ball $B(0, r_0)$, any non-negative solution to

$$-\Delta u + u^{2q-1} - \Gamma u \le 0$$

is compared with $u_1 + u_2$, where u_1 is the solution to

$$\begin{cases} -\Delta u_1 = \Gamma^{q'} & \text{in } B(0, r_0), \\ u_1 = 0 & \text{on } \partial B(0, r_0), \end{cases}$$

where $q' = \frac{2q-1}{2q-2}$, and where u_2 is a solution to

$$\begin{cases} -\Delta u_2 + \frac{1}{2} u_2^{2q-1} \ge 0 & \text{in } B(0, r_0), \\ u_2 = +\infty & \text{on } \partial B(0, r_0). \end{cases}$$

The key point of the argument is the fact that u_1 is bounded. Now, the fact that it is bounded is ensured as soon as

$$-\Delta u_1 = \Gamma^{q'} \in L^r$$

for some $r > \frac{3}{2}$, which holds provided that Γ is in $L^p_{\text{loc}}(R^3)$ for some $p > \frac{3}{2} \frac{2q-1}{2q-2}$. The case $q = \frac{5}{3}$ corresponds to $p > \frac{21}{8}$. $\diamond$

As a consequence of the bound from above, we have the following result that will be useful in the next subsections.

Lemma 4.12 *Let c be a positive constant and let Γ be a periodic potential in $L^p_{\text{loc}}(\mathbf{R}^3)$ for some $p > \frac{21}{8}$. We assume that there exists some u satisfying*

$$\begin{cases} -\Delta u - \Gamma u + cu^{7/3} \le 0, \\ u > 0. \end{cases} \tag{4.24}$$

Then

$$\lambda_1(-\Delta - \Gamma, \text{per}) < 0.$$

Proof of Lemma 4.12 For the sake of simplicity, we assume that $c = 1$ and that the periodic cell of the periodic potential Γ is the unit cube Γ_0. Neither of these assumptions is of course necessary, and our proof may be extended through slight modifications to treat the other cases.

Let us first of all show that, for any $R > 0$, the first eigenvalue of $-\Delta - \Gamma$ on the 'large' cube $\Gamma(R) = [-2R, 2R]^3$ with periodic boundary conditions, denoted by $\lambda_1(-\Delta - \Gamma, \Gamma(R), \text{per})$, is equal to the first eigenvalue of this operator on the unit cube Γ_0 with periodic boundary condition, namely $\lambda_1(-\Delta - \Gamma, \text{per})$.

Let $u_1 > 0$ be the first eigenfunction of $-\Delta - \Gamma$ on $\Gamma(R)$. We recall that it is unique up to a sign.

Therefore, if e_1 is the unit vector in the x direction, we have $u_1 = u_1(\cdot + e_1)$. Likewise, $u_1 = u_1(\cdot + e_2) = u_1(\cdot + e_3)$, and u_1 is of the same period as Γ. It follows easily that

$$\lambda_1(-\Delta - \Gamma, \Gamma(R), \text{per}) = \lambda_1(-\Delta - \Gamma, \text{per}).$$

Let us now fix some cut-off function χ satisfying $\chi \equiv 1$ on the unit ball $B(0,1)$, $\chi \equiv 0$ on $B(0,2)^c$, and $|\nabla \chi| \leq C$, where C is a fixed constant. Let us denote $\chi_R = \chi(\frac{\cdot}{R})$ for any $R > 0$, and remark that we then have $|\nabla \chi_R| \leq \frac{C}{R}$. We multiply (4.22) by $u\chi_R^2$ and integrate over $\mathbf{R}^3$ to obtain

$$0 \geq \int (-\Delta u - \Gamma u + u^{7/3})u\chi_R^2$$
$$= -\int u^2 |\nabla \chi_R|^2 + \int |\nabla(u\chi_R)|^2 - \int \Gamma(u\chi_R)^2 + \int u^{10/3}\chi_R^2,$$

all integrals being taken on $\mathbf{R}^3$; that is to say,

$$\int |\nabla(u\chi_R)|^2 - \int \Gamma(u\chi_R)^2 + \int u^{10/3}\chi_R^2 \leq \int u^2 |\nabla \chi_R|^2$$
$$\leq \frac{C^2}{R^2}\int_{B(0,2R)} u^2.$$

We thus have, in view of the identification of the first eigenvalue on the large cube $\Gamma(R)$ with that on the small one Γ_0, and using that $\chi_R \equiv 1$ on $B(0,R)$,

$$\lambda_1(-\Delta - \Gamma, \text{per})\int u^2 \chi_R^2 + \int_{B(0,R)} u^{10/3} \leq \frac{C^2}{R^2}\int_{B(0,2R)} u^2. \qquad (4.25)$$

Let us now assume that $\lambda_1(-\Delta - \Gamma, \text{per}) \geq 0$.

Using Hölder's inequality to bound the right-hand side of (4.25) from above, we have

$$\int_{B(0,R)} u^{10/3} \leq C_0 R^{-4/5}\left(\int_{B(0,2R)} u^{10/3}\right)^{3/5}, \qquad (4.26)$$

where C_0 denotes a constant that is independent of u and R.

We next define the sequence

$$X_n = 4^n \int_{B(0,2^n)} u^{10/3}. \qquad (4.27)$$

Since $u \not\equiv 0$, it is clear that X_n goes to infinity as n goes to infinity, and therefore that, for p large enough, we have

$$X_p > \tfrac{1}{8}C_0^{5/2} = \alpha. \tag{4.28}$$

For some p such that (4.28) holds, we now write, using (4.26) with $R = 2^{n+p}$,

$$X_{n+p+1} \geq (4C_0^{-5/3})X_{n+p}^{5/3}. \tag{4.29}$$

Collecting the inequalities (4.29) for n, $n-1$, ..., 1,0, we obtain

$$X_{n+p} \geq (4C_0^{-5/3})^{\frac{3}{2}((\frac{5}{3})^n-1)}X_p^{(5/3)^n};$$

that is

$$X_{n+p} \geq \alpha(\tfrac{1}{\alpha}X_p)^{(5/3)^n}. \tag{4.30}$$

On the one hand, we next remark that, in view of Lemma 4.11, $u \in L^\infty$, and thus

$$X_{n+p} = 4^{n+p}\int_{B(0,2^{n+p})} u^{10/3} \leq C2^{5n+5p}\|u\|_{L^\infty}^{10/3}.$$

On the other hand, since we have chosen p such that (4.28) holds, we have $\beta = \tfrac{1}{\alpha}X_p > 1$.

Therefore we obtain, in (4.30),

$$\beta^{(5/3)^n} \leq C2^{5n},$$

for $\beta > 1$ and C that does not depend on n. We clearly reach a contradiction for n large enough. Therefore $\lambda_1(-\Delta - \Gamma,\text{per}) < 0$ and the proof is complete. $\diamond$

Remark 4.13 (The linear case) With a slight modification, the above argument shows that, for any periodic potential $\Gamma \in L_{\text{unif}}^{3/2,\infty}$ (for instance), if there exists some $u \in H^1 \cap L^\infty$ (or $u \in H^1 \cap L_{\text{unif}}^2$) such that $-\Delta u - \Gamma u \leq 0$ on $\mathbf{R}^3$, then $\lambda_1(-\Delta - \Gamma, \text{per}) \leq 0$ (possibly $\lambda_1 = -\infty$).

Indeed, arguing as above, we obtain here

$$\lambda_1(-\Delta - \Gamma, \text{per})\int u^2\chi_R^2 \leq \frac{C^2}{R^2}\int_{B(0,2R)} u^2$$

instead of (4.25). If we then assume that $\lambda_1 > 0$ and consider the sequence

$$X_n = \int_{B(0,2^n)} u^2,$$

we have

$$X_n \geq c2^{n^2-n},$$

for some constant $c > 0$. Next, using the fact that $u \in L^\infty$ by assumption, we remark that $X_n \leq C2^{3n}$ for some other constant C. Finally, we reach a contradiction, since 2^{4n-n^2} goes to zero as n goes to infinity.

It is important to note here that the point is that $u \in L^\infty$ (the same proof works if u grows at most like a power). This holds necessarily in the non-linear case, under a convenient hypothesis on the potential Γ, as shown by Lemma 4.11; but in the linear case this needs to be an assumption (think of the following case: $-\Delta u + u = 0$, $u(x,y,z) = e^x$).

4.3.2   *Bounds from below for solutions*

This subsection is devoted to the proof of the following:

Lemma 4.14 *Let c be a positive constant and let Γ be a periodic potential in $L^p_{loc}(\mathbf{R}^3)$ for some $p > \frac{21}{8}$. Assume that, on the periodic cell of Γ, the first eigenvalue of the operator $-\Delta - \Gamma$ with periodic boundary conditions is negative. Let V be a non-negative potential vanishing fast enough at infinity; more precisely satisfying, for instance, $V \in L^1(\mathbf{R}^3)$.*

We consider the equation, for some $\lambda \geq 0$,

$$\begin{cases} -\Delta u - \Gamma u + cu^{7/3} + \lambda(u^2 \star V)u = 0, \\ u > 0. \end{cases} \tag{4.31}$$

Then, there exists a constant $\nu > 0$ such that any solution u to (4.31) satisfies

$$u \geq \nu > 0. \tag{4.32}$$

Remark 4.15 (i) The assumption $\lambda_1(-\Delta - \Gamma, \Omega, \text{per}) < 0$ is satisfied, for instance, if

$$\int_\Omega \Gamma > 0. \tag{4.33}$$

Using the constant function $\frac{1}{|\Omega|}$ on Ω as a test function in the variational characterization of λ_1, we easily obtain

$$\lambda_1(-\Delta - \Gamma, \Omega, \text{per}) \leq -\frac{1}{|\Omega|} \int_\Omega \Gamma < 0.$$

Another case when this assumption is fulfilled is

$$\int_\Omega \Gamma = 0, \quad \Gamma \not\equiv 0.$$

The above argument yields $\lambda_1(-\Delta - \Gamma, \Omega, \text{per}) \leq 0$, and if $\lambda_1 = 0$ then we reach a contradiction, because $\frac{1}{|\Omega|}$ is a minimum but is not an eigenfunction.

(ii) The constant ν appearing in (4.32) may be chosen independently of $\lambda \in [0,1]$.

(iii) In view of the above subsection, we know that, as soon as a solution to (4.31) exists, then the assumption on the operator $-\Delta - \Gamma$, namely $\lambda_1(-\Delta - \Gamma, \text{per}) < 0$, is automatically fulfilled. (Note that this makes use of the fact that $V \geq 0$.)

(iv) If we replace the non-linear term $u^{\frac{7}{3}}$ in (4.31) by u^{2q-1} for some $q > \frac{3}{2}$, then we must assume $\Gamma \in L^p_{loc}(\mathbf{R}^3)$ for some $p > \frac{3}{2}\frac{2q-1}{2q-2}$.

Proof of Lemma 4.14 Set $c = 1$ for the sake of simplicity.

Step 1: $\lambda = 0$. We begin with the simpler case when $\lambda = 0$:

$$\begin{cases} -\Delta u - \Gamma u + cu^{7/3} = 0, \\ u > 0, \end{cases} \tag{4.34}$$

and next explain how our argument has to be modified in order to treat the case when $\lambda > 0$. We argue by contradiction and assume that there exists a sequence u_n of solutions to (4.34) such that

$$\inf_{\mathbf{R}^3} u_n \overset{n\to\infty}{\longrightarrow} 0.$$

Without loss of generality, translating u_n if necessary, we may assume that $u_n(0) \overset{n\to\infty}{\longrightarrow} 0$. Then, by Harnack's inequality, we have

$$u_n \overset{n\to\infty}{\longrightarrow} 0 \tag{4.35}$$

uniformly on any compact subset of $\mathbf{R}^3$. For the sake of simplicity, we assume hereafter that the cell of periodicity of the periodic potential Γ in (4.31) is a unit cube. Let $R \geq 0$ be a fixed integer. We introduce $K(R)$, which is the large cube centered at 0 and containing $(2R+1)^3$ unit cubes. On $K(R)$, we consider the minimization problem

$$I(n, R) = \inf \left\{ \int_{K(R)} |\nabla u|^2 + \frac{3}{5} \int_{K(R)} |u|^{10/3} - \int_{K(R)} \Gamma u^2; \right.$$

$$\left. u \in H^1(K(R)), u|_{\partial K(R)} = u_n \right\}. \tag{4.36}$$

The functional $\int_{K(R)} |\nabla u|^2 + \frac{3}{5} \int_{K(R)} |u|^{10/3} - \int_{K(R)} \Gamma u^2$ being convex with respect to u^2, any solution to the Euler–Lagrange equation associated with (4.36), namely

$$\begin{cases} -\Delta u - \Gamma u + c|u|^{4/3}u = 0, \\ u|_{\partial K(R)} = u_n, \end{cases}$$

is a minimizer of (4.36). Now, u_n obviously satisfies the above equation, and thus is a minimizer of (4.36). Hence

$$\int_{K(R)} |\nabla u_n|^2 + \frac{3}{5} \int_{K(R)} |u_n|^{10/3} - \int_{K(R)} \Gamma u_n^2 = I(n, R).$$

In fact, the energy functional being strictly convex, the minimizer is unique up to a sign, and u_n is therefore the positive minimizer of (4.36).

Therefore, since u_n converges uniformly to zero on $K(R)$ and is a solution to (4.34), we have

$$I(n, R) \overset{n\to\infty}{\longrightarrow} 0. \tag{4.37}$$

Indeed, in order to obtain (4.37), the point is to show that $\int_{K(R)} |\nabla u_n|^2$ goes to zero. One argues as follows. Choose some $\varphi \in \mathcal{D}(K(R+1))$ such that $|\nabla \varphi| \leq C$ and $\varphi \equiv 1$ on $K(R)$. Then, obviously,

$$\int_{\mathbf{R}^3}(-\Delta u_n)u_n\varphi^2 = \int_{\mathbf{R}^3}|\nabla u_n|^2\varphi^2 + 2\int_{\mathbf{R}^3}\varphi\nabla\varphi\cdot u_n\nabla u_n. \qquad (4.38)$$

The left-hand side goes to zero because of (4.34) and (4.35). The second term of the right-hand side may be bounded by the Cauchy–Schwarz inequality to obtain

$$\left|\int_{\mathbf{R}^3}\varphi\nabla\varphi\cdot u_n\nabla u_n\right| \leq \left(\int_{\mathbf{R}^3}\varphi^2|\nabla u_n|^2\right)^{1/2}\left(\int_{\mathbf{R}^3}u_n^2|\nabla\varphi|^2\right)^{1/2}$$

$$\leq C\sup_{K(R+1)}u_n\cdot\left(\int_{\mathbf{R}^3}\varphi^2|\nabla u_n|^2\right)^{1/2}, \qquad (4.39)$$

where C is a constant that does not depend on n. From (4.38) and (4.39), we deduce, for n large enough such that $\sup_{K(R+1)}u_n \leq 1$, that

$$\left(\int\varphi^2|\nabla u_n|^2\right) - 2C\left(\int\varphi^2|\nabla u_n|^2\right)^{1/2}$$

$$\leq \int\varphi^2|\nabla u_n|^2 - 2C\sup_{K(R+1)}u_n\left(\int\varphi^2|\nabla u_n|^2\right)^{1/2}$$

$$\leq \int\varphi^2|\nabla u_n|^2 + 2\int\varphi\nabla\varphi\cdot u_n\nabla u_n$$

$$= \int -\Delta u_n u_n\varphi^2.$$

The right-hand side goes to zero when n goes to infinity; in particular, it is bounded by a constant C'. Hence

$$\int\varphi^2|\nabla u_n|^2 - 2C\left(\int\varphi^2|\nabla u_n|^2\right)^{1/2} \leq C',$$

which shows that $\int\varphi^2|\nabla u_n|^2$ is bounded. Putting this information in (4.39), we obtain that $\int\varphi\nabla\varphi\cdot u_n\nabla u_n$ goes to zero when n goes to infinity. Next, (4.38) implies that $\int\varphi^2|\nabla u_n|^2$ also goes to zero. This yields $\int_{K(R)}|\nabla u_n|^2 \longrightarrow 0$, as n goes to infinity for any fixed R.

We now define

$$I_{\text{per}} = \inf\left\{\int_{K(0)}|\nabla u|^2 + \tfrac{3}{5}\int_{K(0)}|u|^{10/3} - \int_{K(0)}\Gamma u^2; u \in H^1_{\text{per}}(K(0))\right\}.$$
$$(4.40)$$

We claim that this infimum is negative. Indeed, let $\lambda_1 = \lambda_1(-\Delta-\Gamma, K(0), \text{per})$ be the first eigenvalue with periodic boundary conditions, and let v be the associated eigenfunction. We take $u = \varepsilon v$ and compute

$$\int_{K(0)}|\nabla u|^2 + \tfrac{3}{5}\int_{K(0)}|u|^{10/3} - \int_{K(0)}\Gamma u^2$$

$$= \varepsilon^2 \int_{K(0)} |\nabla v|^2 + \tfrac{3}{5}\varepsilon^{10/3} \int_{K(0)} |v|^{10/3} - \varepsilon^2 \int_{K(0)} \Gamma v^2.$$

For ε small enough, the right-hand side behaves like $\varepsilon^2 \lambda_1$ and thus is negative. Therefore

$$I_{\mathrm{per}} < 0.$$

We now denote by w the periodic minimizer of I_{per}, and we construct a function $w_{n,R}$ on the cube $K(R)$ that is equal to w on $K(R-1)$, and that is extended in a convenient way in order to satisfy the boundary condition $w_{n,R}|_{\partial K(R)} = u_n$. It is easy to see that

$$\int_{K(R)} |\nabla w_{n,R}|^2 + \tfrac{3}{5} \int_{K(R)} |w_{n,R}|^{10/3} - \int_{K(R)} \Gamma w_{n,R}^2 = (2R-1)^3 I_{\mathrm{per}} + O(R^2).$$

Hence, we may fix R (large enough) in such a way that, uniformly in n,

$$I(n,R) \le -1.$$

We then reach a contradiction with (4.37).

Step 2: $\lambda > 0$. We now turn to the case $\lambda > 0$. The proof follows the same pattern. Let us first check that, in addition to (4.35), we have

$$u_n^2 \star V \overset{n \to \infty}{\longrightarrow} 0$$

on any compact set K. This is a consequence of the short-range nature of the potential V. Let us take $x \in K$. We have

$$
\begin{aligned}
u_n^2 \star V(x) &= \int_{\mathbf{R}^3} V(y) u_n^2(x-y)\,\mathrm{d}y \\
&\le \|V\|_{L^1(K')} \sup_{x-K'} u_n^2 + C\|V\|_{L^1(\mathbf{R}^3 \setminus K')},
\end{aligned}
$$

for any compact set K', where C denotes a bound from above for u_n uniform in n. We recall that the existence of such a bound for any solution to (4.31) is based upon arguments developed by Véron in [57] (see the details in the above subsection). We now choose K' such that $\|V\|_{L^1(\mathbf{R}^3 \setminus K')}$ is small, and then n such that $\sup_{K-K'} u_n^2$ is small. We therefore obtain $u_n^2 \star V(x)$ as small as we wish, uniformly in $x \in K$.

Next, we consider, instead of (4.35),

$$
\begin{aligned}
I(n,R) = \inf \Big\{ &\int_{K(R)} |\nabla u|^2 + \tfrac{3}{5} \int_{K(R)} |u|^{10/3} - \int_{K(R)} \Gamma u^2 \\
&+ \tfrac{1}{2}\lambda \iint_{\mathbf{R}^3 \times \mathbf{R}^3} V(x-y) u^2(x) u^2(y)\,\mathrm{d}x\,\mathrm{d}y;
\end{aligned}
$$

$$u \in H^1_{\text{loc}}(\mathbf{R}^3), u|_{\mathbf{R}^3 \setminus K(R)} = u_n \Big\}. \tag{4.41}$$

Let us note that we need to extend u by u_n outside $K(R)$ in order to ensure that (4.31) is the Euler–Lagrange equation of (4.41) and that u_n is a critical point of the energy functional.

Arguing exactly as above, we have

$$I(n, R) \overset{n \to \infty}{\longrightarrow} 0. \tag{4.42}$$

Taking the same w as above (the periodic minimizing function of $I_{\text{per}} < 0$ given by (4.40)), we define $w_{n,R,\varepsilon}$ as follows. On $K(R-1)$, $w_{n,R,\varepsilon}$ is equal to εw for some $\varepsilon > 0$ small. We then extend $w_{n,R,\varepsilon}$ on $K(R) \setminus K(R-1)$ in order to satisfy the boundary condition $w_{n,R,\varepsilon}|_{\partial K(R)} = u_n$. We may assume that on $K(R) \setminus K(R-1)$, $w_{n,R,\varepsilon}$ and $|\nabla w_{n,R,\varepsilon}|$ are bounded by $Max(\varepsilon\|w\|_{L^\infty}, M(n, R))$, where $M(n, R) = \sup_{K(R)} u_n$.

Let us now compute the energy functional of $w_{n,R,\varepsilon}$. The local terms on $K(R-1)$ yield:

$$\int_{K(R-1)} \left[|\nabla w_{n,R,\varepsilon}|^2 + \tfrac{3}{5}|w_{n,R,\varepsilon}|^{10/3} - \Gamma w_{n,R,\varepsilon}^2 \right]$$
$$\leq \varepsilon^2 (2R-1)^3 I_{\text{per}}, \tag{4.43}$$

since $\varepsilon^{10/3} \leq \varepsilon^2$ as soon as $\varepsilon \leq 1$.

The local terms on $K(R) \setminus K(R-1)$ may be bounded as follows:

$$\left| \int_{K(R) \setminus K(R-1)} \left[|\nabla w_{n,R,\varepsilon}|^2 + \tfrac{3}{5}|w_{n,R,\varepsilon}|^{10/3} - \Gamma w_{n,R,\varepsilon}^2 \right] \right|$$
$$\leq C R^2 \, Max(\varepsilon\|w\|_{L^\infty}, M(n, R))^2, \tag{4.44}$$

where C denotes, here and below, various positive constants that are independent of R, ε, n.

Next, the convolution term on $K(R-1) \times K(R-1)$ contributes like

$$\int \int_{K(R-1)^2} V(x-y) w_{n,R,\varepsilon}^2(x) w_{n,R,\varepsilon}^2(y) \, dx \, dy = O(\varepsilon^4 R^6), \tag{4.45}$$

while for the terms on $(K(R) \setminus K(R-1)) \times K(R)$ we write

$$\left| \int \int_{(K(R) \setminus K(R-1)) \times K(R)} V(x-y) w_{n,R,\varepsilon}^2(x) w_{n,R,\varepsilon}^2(y) \, dx \, dy \right|$$
$$\leq C R^2 Max(\varepsilon\|w\|_{L^\infty}, M(n, R))^4 \|V\|_{L^1}. \tag{4.46}$$

With (4.43)–(4.46), we obtain

$$I(n, R) \leq \varepsilon^2 (2R-1)^3 I_{\text{per}}$$

$$+ c_1 \varepsilon^4 R^6$$
$$+ c_2 R^2 Max(\varepsilon \|w\|_{L^\infty}, M(n, R))^2$$
$$+ c_3 R^2 Max(\varepsilon \|w\|_{L^\infty}, M(n, R))^4, \qquad (4.47)$$

for $\varepsilon < 1$ and three constants c_i independent of n, R, and ε.

We now fix a constant $\nu > 0$ small enough such that

$$4\nu I_{\text{per}} + (c_1 + c_3 \|w\|_{L^\infty}^4)\nu^2 \le \nu I_{\text{per}}.$$

Next, we choose R large enough in order to have

$$8\nu I_{\text{per}} + c_2 \frac{\nu}{R} \|w\|_{L^\infty}^2 \le 4\nu I_{\text{per}},$$

and choose

$$\varepsilon^2 = \frac{\nu}{R^3}.$$

Finally, in view of (4.35), we choose N large enough such that

$$\forall n \ge N, \quad M(n, R) \le \varepsilon \|w\|_{L^\infty}.$$

Under these conditions, (4.47) yields successively

$$I(n, R) \le \varepsilon^2 (2R - 1)^3 I_{\text{per}} + c_1 \varepsilon^4 R^6$$
$$+ c_2 R^2 \varepsilon^2 \|w\|_{L^\infty}^2 + c_3 R^2 \varepsilon^4 \|w\|_{L^\infty}^4$$
$$= \frac{\nu}{R^3} (2R - 1)^3 I_{\text{per}} + c_1 \nu^2 + c_2 \frac{\nu}{R} \|w\|_{L^\infty}^2 + c_3 \frac{\nu^2}{R^4} \|w\|_{L^\infty}^4$$
$$\le 8\nu I_{\text{per}} + c_1 \nu^2 + c_2 \frac{\nu}{R} \|w\|_{L^\infty}^2 + c_3 \nu^2 \|w\|_{L^\infty}^4$$
$$\le \nu I_{\text{per}}.$$

Hence we have obtained, for some R and N,

$$\forall n \ge N, \quad I(n, R) \le \nu I_{\text{per}} < 0,$$

which contradicts the fact that $I(n, R) \longrightarrow 0$ while n goes to infinity. $\diamond$

Remark 4.16 Before turning to our first uniqueness result, we would like to emphasize that the arguments of the proof of Lemma 4.14 also show that the local limit u_∞ of the minimizing density u_Λ cannot identically vanish on $\mathbf{R}^3$.

Indeed, arguing as above by contradiction, one may prove that, for any compact set K, there exists some positive constant μ such that

$$u_\Lambda \ge \mu > 0,$$

for Λ large enough.

It follows that $u_\infty \not\equiv 0$. This allows us to apply Corollary 4.22 below to u_∞, and therefore to deduce $u_\infty = u_{\text{per}}$ and Theorem 4.1.

4.3.3   Uniqueness result without the convolution term

We now intend to prove uniqueness for (4.20).

Lemma 4.17 *Let c be a positive constant and Γ a periodic potential in $L^p_{\mathrm{loc}}(\mathbf{R}^3)$, for some $p > \frac{21}{8}$. Assume that, on the periodic cell of Γ, the first eigenvalue of the operator $-\Delta - \Gamma$ with periodic boundary conditions is negative. We consider the equation*

$$-\Delta u - \Gamma u + c u^{7/3} = 0.$$

Then, there exists a unique solution $u \geq 0$, $u \not\equiv 0 \in H^1_{\mathrm{loc}}(\mathbf{R}^3)$. In addition, $u > 0$ and u is periodic.

Remark 4.18 (i) As in Remark 4.15, it is worth noticing that, as soon as there exists a positive solution to the above equation, then the assumption $\lambda_1(-\Delta - \Gamma, \mathrm{per}) < 0$ is satisfied.

(ii) We refer the reader to Léon [30], [31] for some existence and uniqueness results on the same type of equations with different assumptions on the potential Γ.

Proof of Lemma 4.17

Step 0: uniqueness (up to a sign) of the periodic solution. First of all, let us notice that there exists a periodic solution (take $\lambda = 0$ in Lemma 4.20 below). Next, we remark that the periodic solution is unique, since the energy functional associated with the equation is strictly convex.

Indeed, since $\rho \longrightarrow \int_{\Omega} |\nabla \sqrt{\rho}|^2 - \Gamma\rho + \frac{3}{5}\rho^{5/3}$ is convex with respect to ρ, any periodic solution to $-\Delta u - \Gamma u + c u^{7/3} = 0$ is a minimizer of the minimization problem

$$\inf \left\{ \int_{\Omega} |\nabla u|^2 - \Gamma u^2 + \frac{3}{5} u^{10/3}; u \in H^1_{\mathrm{per}}(\Omega) \right\},$$

where Ω is the periodic cell of Γ. Now this problem is strictly convex with respect to $\rho = u^2$, and thus the minimum is unique up to a sign. Thus, there is one and only one non-negative periodic solution to $-\Delta u - \Gamma u + c u^{7/3} = 0$.

Step 1: existence of a maximal solution (a) We begin by considering the following equation:

$$\begin{cases} -\Delta u + c u^{7/3} - \Gamma u = 0 & \text{on } B_R, \\ u|_{\partial B_R} = A, \end{cases} \tag{4.48}$$

on a ball B_R, R being fixed for the moment, and with A denoting a positive constant (that will go to infinity later on).

There exists a solution $u_{R,A}$ to (4.48). Indeed, define the minimization problem

$$I(A, R) = \inf \left\{ \int_{B_R} |\nabla u|^2 + \frac{3}{5}c \int_{B_R} |u|^{10/3} - \int_{B_R} \Gamma u^2; \right.$$

$$u \in H^1(B_R), u|_{\partial B_R} = A \Big\}. \tag{4.49}$$

We have

$$\int_{B_R} |\Gamma u^2| \leq \|\Gamma\|_{L^p(B_R)} \|u\|^2_{L^q(B_R)},$$

for some $(p > \frac{3}{2}, q < 6), \frac{1}{p} + \frac{2}{q} = 1$.

Since $p \geq \frac{5}{2}$, we have $2 \leq q \leq \frac{10}{3}$ and

$$\left| \int_{B_R} \Gamma u^2 \right| \leq \Big| \Gamma \|_{L^p(B_R)} |B_R|^{1-3q/10} \Big(\int |u|^{10/3} \Big)^{3/5}.$$

Therefore

$$I(A, R) > -\infty,$$

which implies that, for a minimizing sequence u_n of (4.49), $\int_{B_R} |u_n|^{10/3}$ and $\int_{B_R} |\nabla u_n|^2$ are bounded. Thus, in view of Rellich's theorem, u_n converges strongly in $L^q(B_R)$ for $q < 6$.

The existence of a minimum for (4.49) follows. We may of course assume that this minimum, which we denote by $u_{R,A}$, is non-negative (possibly replacing u by $|u|$), and even positive by Harnack's inequality. This minimum satisfies (4.48).

(b) We now claim that the $u_{R,A}$ are increasing with respect to A.

Let us take $A > B$ and the respective solutions $u_{R,A}$ and $u_{R,B}$ to (4.49) on the same ball B_R. We first remark that $\lambda_1(-\Delta + cu_{R,A}^{4/3} - \Gamma, B_R)$, the first eigenvalue of $-\Delta + cu_{R,A}^{4/3} - \Gamma$ on B_R with Dirichlet boundary conditions, is non-negative. This is somewhat standard; we re-prove it for the reader's convenience. Let v be its first eigenfunction, i.e.

$$\begin{cases} (-\Delta + cu_{R,A}^{4/3} - \Gamma)v = \lambda_1 v, \\ v > 0, \qquad v|_{\partial B_R} = 0. \end{cases}$$

Comparing with (4.48), we obtain

$$\int_{\partial B_R} (\nabla v \cdot n) u_{R,A} = -\lambda_1 \int_{B_R} u_{R,A} v,$$

and thus $\lambda_1 \geq 0$.

We next compare $u_{R,A}$ to $u_{R,B}$ by defining $w = u_{R,A} - u_{R,B}$ and deducing from (4.49) and the convexity of $t \longrightarrow t^{7/3}$ the following inequality:

$$-\Delta w + \tfrac{7}{3} cu_{R,A}^{4/3} w - \Gamma w \geq 0.$$

Since

$$\lambda_1(-\Delta + \tfrac{7}{3} cu_{R,A}^{4/3} - \Gamma, B_R) > \lambda_1(-\Delta + cu_{R,A}^{4/3} - \Gamma, B_R) \geq 0,$$

because $\inf_{B_R} u_{R,A} > 0$ and $w \geq 0$ on ∂B_R, we deduce from the maximum principle that $w \geq 0$ on B_R, i.e. $u_{R,A} \geq u_{R,B}$.

We next use this monotonicity (together with the fact that any solution to (4.48) is bounded from above on any ball $B_{R'}$ with $R' < R$, which is true by the same argument as the one used to prove (4.23)) to define

$$u_R(x) = \lim_{A \to \infty} u_{R,A}(x),$$

for all $x \in B_R$, the limit being uniform on any ball $B_{R'}$ with $R' < R$. Clearly, u_R satisfies

$$\begin{cases} -\Delta u_R - \Gamma u_R + c u_R^{7/3} = 0 & \text{in } B_R, \\ u_R > 0 & \text{on } B_R, \\ u_R|_{\partial B_R} = +\infty. \end{cases} \qquad (4.50)$$

(c) The solutions u_R are decreasing with respect to R.

Indeed, let us fix $R < R'$. Because of (4.50), $\lambda_1(-\Delta + \frac{7}{3}cu_R^{4/3} - \Gamma, B_R)$, the first eigenvalue of $-\Delta + \frac{7}{3}cu_R^{4/3} - \Gamma$ on B_R with Dirichlet boundary condition, is positive.

Defining $w = u_R - u_{R'}$, we deduce from (4.50) and the convexity of $t \longrightarrow t^{7/3}$ that

$$-\Delta w + \tfrac{7}{3}cu_R^{4/3}w - \Gamma w \geq 0.$$

Besides, $w \geq 0$ on ∂B_R. Therefore, using the maximum principle, this implies that $w \geq 0$ on B_R, i.e. $u_R \geq u_{R'}$ on B_R.

(d) We may therefore define a solution $\bar{u}$ to

$$-\Delta \bar{u} - \Gamma \bar{u} + c\bar{u}^{7/3} = 0, \qquad (4.51)$$

on $\mathbf{R}^3$, which is the limit almost everywhere of the solutions u_R.

By the same comparison principle as above, we know that any solution u to (4.51) satisfies $u \leq u_R$ on B_R for any R. Thus $u \leq \bar{u}$. In other words, $\bar{u}$ is the maximal solution to (4.51).

Step 2: maximal is equivalent to periodic. Let $\bar{u}$ be the maximal solution. It is straightforward to see that $\bar{u}(\cdot + e_i)$, where e_i is one of the three unit vectors that define the periodic cell of the potential Γ, is also a solution. Therefore, we have

$$\bar{u}(\cdot + e_i) \leq \bar{u}.$$

Likewise, $\bar{u}(\cdot - e_i)$ is a solution, and thus

$$\bar{u}(\cdot - e_i) \leq \bar{u}.$$

It follows from these two observations that

$$\bar{u}(\cdot + e_i) = \bar{u},$$

i.e. that $\bar{u}$ is periodic.

In view of step 0, it follows that $\bar{u}$ is *the* periodic solution.

Step 3: bound from below. In this step, we only recall the result of Lemma 4.14, that we apply here in the case $\lambda = 0$. Any solution to

$$\begin{cases} -\Delta u - \Gamma u + cu^{7/3} = 0, \\ u \geq 0, \end{cases}$$

is positive by the Harnack inequality, and furthermore bounded below away from zero by a constant $\nu > 0$ independent of u.

Step 4: existence of a non-trivial minimal solution. We fix R and define on B_R the following minimization problem:

$$I(R) = \inf \left\{ \int_{B_R} |\nabla u|^2 + \tfrac{3}{5}c \int_{B_R} |u|^{10/3} - \int_{B_R} \Gamma u^2; \right.$$
$$\left. u \in H^1(B_R), u|_{\partial B_R} = \nu \right\}. \tag{4.52}$$

The energy functional being strictly convex, there exists a unique minimum of (4.52) that we denote by $\underline{u}_R$. It satisfies

$$\begin{cases} -\Delta \underline{u}_R - \Gamma \underline{u}_R + c\underline{u}_R^{7/3} = 0, \\ \underline{u}_R|_{B_R} = \nu. \end{cases}$$

The same comparison principle as above shows, in view of step 3, that any non-trivial solution u to

$$\begin{cases} -\Delta u - \Gamma u + cu^{7/3} = 0, \\ u \geq 0, \end{cases}$$

satisfies

$$u \geq \underline{u}_R$$

on B_R.

Next, we remark, by the same argument as the one used in the proof of Lemma 4.14, that for any compact set $K \subset \mathbf{R}^3$, there exists some constant $\mu > 0$ such that, for R large enough such that $K \subset B_R$,

$$\underline{u}_R \geq \mu > 0 \qquad \text{on } K. \tag{4.53}$$

Indeed, if the above inequality does not hold, then, extracting a subsequence if necessary, we have

$$\inf_K \underline{u}_R \overset{R\to\infty}{\longrightarrow} 0$$

on some compact set K, and thus

$$\inf_{B_R} \underline{u}_R \overset{R\to\infty}{\longrightarrow} 0,$$

and we reach a contradiction as in the proof of Lemma 4.14.

Since $\underline{u}_R$ is bounded in L^∞, then, by a standard argument, $\nabla \underline{u}_R$ is bounded in L^2_{loc}. It follows that $\underline{u}_R$ is bounded in H^1_{loc} and thus converges locally (weakly in H^1_{loc}, strongly in L^p_{loc} for $1 \le p < \infty$) to some $\underline{u}$ that is a solution on $\mathbf{R}^3$ to

$$\begin{cases} -\Delta \underline{u} + c\underline{u}^{7/3} - \Gamma \underline{u} = 0, \\ \underline{u} \ge 0. \end{cases}$$

Because of (4.53), $\underline{u} > 0$ and thus, in view of step 3,

$$\underline{u} \ge \nu.$$

Since any solution u satisfies $u \ge u_R$ on B_R for all R, we have $u \ge \underline{u}$, and thus $\underline{u}$ is the minimal solution.

Comparing $\underline{u}$ with $\underline{u}(\cdot \pm e_i)$, we deduce, as we did for the maximal solution, that $\underline{u}$ is periodic. It follows from step (b) that $\underline{u}$ is maximal. Thus we have obtained the uniqueness of the solution and the proof of Lemma 4.17 is complete. ◇

4.3.4   *Uniqueness result for the full equation: conclusion*

We conclude in this subsection the proof of the uniqueness of the solution to (4.19) and apply it to conclude the proof of the periodicity of the limit density in the Yukawa case. As announced in the introduction of this chapter, our main result here is the following:

Theorem 4.19 *Let c be a positive constant, let $\lambda \ge 0$, and let Γ be a periodic potential in $L^p_{\text{loc}}(\mathbf{R}^3)$, for some $p > \frac{21}{8}$. Assume that on the periodic cell of Γ (for the sake of simplicity, we assume this cell is the unit cube Γ_0), the first eigenvalue of the operator $-\Delta - \Gamma$ with periodic boundary conditions is negative. Let V be a non-negative potential $V \in L^1_{\text{loc}}(\mathbf{R}^3)$ satisfying $V(x) = O\left(\frac{1}{|x|^{3+\alpha}}\right)$ at infinity, for some $\alpha > 0$ (other short-ranged potentials are tractable). Then, there exists a unique solution to*

$$\begin{cases} -\Delta u - \Gamma u + cu^{7/3} + \lambda(u^2 \star V)u = 0, \\ u \ge 0, \quad u \not\equiv 0. \end{cases}$$

This solution satisfies $u > 0$ on $\mathbf{R}^3$ and is periodic.

In addition, the assumption on $\lambda_1(-\Delta - \Gamma, \text{per})$ is satisfied as soon as a solution $u > 0$ to the above equation exists.

In the above theorem, the two claims of existence and uniqueness are of unbalanced interest and difficulty. We first deal with the existence, which is simpler.

4.3.4.1   *Existence*   Clearly, the arguments we made in Chapter 2 show that there exists a solution to the equation; namely, the local thermodynamic limit of the subsequence of densities that converges. Let us recall that this subsequence is chosen arbitrarily, and thus that the limit depends (so far) on our choice.

However, even if we had not proved the existence by thermodynamic limit (which in fact relies upon the observation that $\lambda_1\left(-\Delta - \frac{e^{-a|x|}}{|x|}, \mathbf{R}^3\right) < 0$; see in Chapter 2 the proof of Proposition 2.2), we could still prove the existence, provided that we assume $\lambda_1(-\Delta - \Gamma, \text{per}) < 0$. This is the purpose of the following:

Lemma 4.20 *Let c be a positive constant, let $\lambda \geq 0$, and let Γ be a periodic potential in $L^p_{\text{loc}}(\mathbf{R}^3)$, for some $p > \frac{21}{8}$. Assume that on the periodic cell of Γ (for the sake of simplicity, we assume this cell is the unit cube Γ_0, but we may treat any other case), the first eigenvalue of the operator $-\Delta - \Gamma$ with periodic boundary conditions is negative. Let V be a short-ranged potential (say, $V(x) = O\left(\frac{1}{|x|^{3+\alpha}}\right)$ at infinity, for some $\alpha > 0$) such that $V_\infty = \sum_{y\in\mathbf{Z}^3} V(x - y)$ is a non-negative kernel. Then, there exists a (unique) periodic solution $u \geq 0$, $u \not\equiv 0$ in $H^1_{\text{loc}}(\mathbf{R}^3)$ to*

$$-\Delta u - \Gamma u + cu^{7/3} + \lambda(u^2 \star V)u = 0.$$

Proof of Lemma 4.20 We consider the following minimization problem

$$I = \inf\left\{ \int_{\Gamma_0} |\nabla u|^2 - \int_{\Gamma_0} \Gamma u^2 + \tfrac{3}{5}c \int_{\Gamma_0} u^{10/3} \right.$$
$$+ \tfrac{1}{2}\lambda \iint_{\Gamma_0\times\Gamma_0} u^2(x)V_\infty(x - y)u^2(y)\,\mathrm{d}x\,\mathrm{d}y;$$
$$\left. u \in H^1_{\text{per}}(\Gamma_0)\right\}, \tag{4.54}$$

where

$$V_\infty = \sum_{y\in\mathbf{Z}^3} V(x - y) = \sum_{y\in\mathbf{Z}^3} \frac{e^{-a|x-y|}}{|x - y|}.$$

We claim that the minimum I of (4.54) is achieved. Indeed, it is straightforward to see by standard arguments that any minimizing sequence of I is bounded in $H^1_{\text{per}}(\Gamma_0)$. It follows that, up to an extraction, a minimizing sequence converges weakly in $H^1_{\text{per}}(\Gamma_0)$, and strongly in L^p_{loc} for $1 \leq p < 6$. Its limit, say u, is a minimizer of I. We may always assume that u is non-negative (replace u by $|u|$ if necessary). Consequently, u is unique by the strict convexity of (4.54). It satisfies the Euler–Lagrange equation associated with (4.54), i.e.

$$-\Delta u - \Gamma u + cu^{7/3} + \lambda\left(\int_{\Gamma_0} u^2(y)V_\infty(x - y)\,\mathrm{d}y\right)u = 0.$$

Because u is periodic, this equation is also

$$-\Delta u - \Gamma u + cu^{7/3} + \lambda(u^2 \star V)u = 0.$$

Proving Lemma 4.20 amounts to proving that $u \not\equiv 0$.

For this purpose, let w be the first periodic eigenfunction of the operator $-\Delta - \Gamma$ on the cell Γ_0. It suffices to show that, for $\varepsilon > 0$ small enough, we have

$$\varepsilon^2 \int_{\Gamma_0} |\nabla w|^2 - \varepsilon^2 \int_{\Gamma_0} \Gamma w^2 + \tfrac{3}{5}\varepsilon^{10/3} c \int_{\Gamma_0} w^{10/3}$$
$$+ \tfrac{1}{2}\varepsilon^4 \lambda \iint_{\Gamma_0 \times \Gamma_0} u^2(x) V_\infty(x - y) u^2(y) \, dx \, dy < 0, \qquad (4.55)$$

which will imply that $I < 0$, and thus $u \not\equiv 0$. Now, (4.55) obviously holds for $\varepsilon > 0$ small enough because, by the definition of w, $\int_{\Gamma_0} |\nabla w|^2 - \int_{\Gamma_0} \Gamma w^2 < 0$. $\diamond$

4.3.4.2 *Uniqueness* We now turn to the uniqueness result contained in Theorem 4.19, which we isolate in the following lemma:

Lemma 4.21 *Let c be a positive constant, let $\lambda \geq 0$, and let Γ be a periodic potential in $L^p_{\text{loc}}(\mathbf{R}^3)$, for some $p > \frac{21}{8}$. Assume that, on the periodic cell of Γ, the first eigenvalue of the operator $-\Delta - \Gamma$ with periodic boundary conditions is negative. Let $V(x) = \frac{e^{-a|x|}}{|x|}$ for $a > 0$ (for instance). Finally, we consider $u \geq 0$, $u \not\equiv 0 \in H^1_{\text{loc}}(\mathbf{R}^3)$, a solution on $\mathbf{R}^3$ to*

$$-\Delta u - \Gamma u + cu^{7/3} + \lambda(u^2 \star V)u = 0.$$

Then, $u > 0$ on $\mathbf{R}^3$, u is unique and u is periodic.

Before we give the proof of the above lemma, we wish to make a few comments.

The first comment is about the short-range potential V. For the sake of simplicity, we have chosen $V(x) = \frac{e^{-a|x|}}{|x|}$ to state and prove the lemma. However, it will become clear for the reader in the course of our proof that other short-range potentials may be treated. We only make use of the three following facts: V blows up at most like $\frac{1}{|x|}$ at 0 (this fact may not be optimal; $\frac{1}{|x|^2}$ is likely to be the limit behaviour, but we do not want to enter such technical extensions here), V decays fast enough at infinity (e.g. $V \xrightarrow{|x| \to \infty} 0$, $|x|^{1/2}|V(x)|^{1/2} \in L^1(\mathbf{R}^3)$), and V is non-negative. All these properties together ensure that (4.69) and (4.71) hold. However, it is worth noticing that the fact that $V \geq 0$ is 'only' useful to obtain the uniform bound from below of Lemma 4.14, which in turn comes from the bound from above of Lemma 4.11. Once this bound is known, we only use in the proof of Lemma 4.21 the fact that V defines a non-negative kernel.

Our second comment concerns the assumption $\lambda_1(-\Delta - \Gamma, \text{per}) < 0$. As soon as the existence of a solution to

$$\begin{cases} -\Delta u - \Gamma u + cu^{7/3} + \lambda(u^2 \star V)u = 0, \\ u \geq 0, \qquad u \not\equiv 0 \end{cases}$$

is known, then the assumption is automatically satisfied (provided that we assume $V \geq 0$, which is of course the case for the Yukawa potential). Therefore, in the case we are interested in, we know that the assumption on λ_1 is satisfied for $\Gamma = V_\infty - \theta_a$, $V(x) = \frac{e^{-a|x|}}{|x|}$, since we have built such a solution by the thermodynamic limit process. In addition, in this case, there is another (more straightforward) way to check that $\lambda_1(-\Delta - \Gamma, \text{per}) < 0$, but this way requires the assumption that the exponent a defining the Yukawa potential is small enough.

For this purpose, we use Remark 4.15: it suffices to show (4.33), i.e.

$$\int_{\Gamma_0} V_\infty - \theta_a > 0.$$

Now,

$$\int_{\Gamma_0} V_\infty = \frac{4\pi}{a^2} \longrightarrow +\infty$$

as a goes to zero, while θ_a remains bounded (it is either zero or goes to the Lagrange multiplier of the Coulomb case). The condition (4.33) then follows for a small enough.

Therefore we have the following:

Corollary 4.22 *Let c be a positive constant, let $\lambda \geq 0$, let $V(x) = \frac{e^{-a|x|}}{|x|}$, and let $\Gamma = \sum_{y \in \mathbf{Z}^3} \frac{e^{-a|x-y|}}{|x-y|} - \theta_{\text{per}}$, where θ_{per} is the multiplier associated with (4.2). Then, there exists one and only one solution to*

$$\begin{cases} -\Delta u - \Gamma u + c u^{7/3} + \lambda(u^2 \star V)u = 0, \\ u \geq 0, \qquad u \not\equiv 0. \end{cases}$$

This solution satisfies $u > 0$ on $\mathbf{R}^3$ and is periodic.

By the remark we made at the end of Subsection 4.3.2, we know that $u_\infty \not\equiv 0$. Thus, Corollary 4.22 implies that $u_\infty = u_{\text{per}}$. This concludes the proof of Theorem 4.1.

We now turn to the proof of Lemma 4.21.

Proof of Lemma 4.21 As announced in Section 4.1, the proof of the uniqueness is based upon the implicit function theorem. The constant λ being a parameter in $\mathbf{R}_+$, we consider the equation

$$-\Delta u - \Gamma u + c|u|^{4/3}u + \lambda(u^2 \star V)u = 0. \tag{4.56}$$

We shall prove that any solution $u > 0$ is a function $u(\lambda)$ of λ, bounded away from 0, for all $\lambda \in [0, 1]$, and then deduce, by an argument detailed in step 3 below, that $u(\lambda = 1)$ is unique since $u(\lambda = 0)$ is unique by Lemma 4.17. In order to deduce such a result from the implicit function theorem, we first set the functional framework.

Step 1: the functional framework. We now define the uniform Sobolev spaces

$$L^p_{\text{unif}}(\mathbf{R}^3) = \left\{ u \in L^p_{\text{loc}}(\mathbf{R}^3); \sup_{x \in \mathbf{R}^3} \|u\|_{L^p(x+B_1)} < \infty \right\},$$

and

$$W^{2,p}_{\text{unif}}(\mathbf{R}^3) = \Big\{ u \in L^p_{\text{loc}}(\mathbf{R}^3);$$

$$\partial_i u \in L^p_{\text{loc}}(\mathbf{R}^3), \partial^2_{ij} u \in L^p_{\text{loc}}(\mathbf{R}^3), 1 \le i, j \le 3;$$

$$\sup_{x \in \mathbf{R}^3} \left(\int_{x+B_1} |u|^p + \sum_{i=1}^{3} |\partial_i u|^p + \sum_{i,j=1}^{3} |\partial^2_{ij} u|^p \right)^{1/p} < \infty \Big\}.$$

In the above formulae, B_1 denotes the unit ball centered at 0, and p is precisely the exponent such that $\Gamma \in L^p_{\text{loc}}(\mathbf{R}^3)$.

We recall that $W^{2,p}_{\text{unif}}$ is a Banach space, and that

$$W^{2,p}_{\text{unif}} \hookrightarrow C^{0,\alpha}_{\text{unif}}(\mathbf{R}^3), \tag{4.57}$$

where $C^{0,\alpha}_{\text{unif}}(\mathbf{R}^3)$ is the uniform Hölder space

$$C^{0,\alpha}_{\text{unif}}(\mathbf{R}^3) = \left\{ u \in C(\mathbf{R}^3); \sup_{\mathbf{R}^3} |u| < \infty, \sup_{x \ne y} \frac{|u(x) - u(y)|}{|x - y|^\alpha} < \infty \right\}$$

($C(\mathbf{R}^3)$ denotes the space of continuous functions on $\mathbf{R}^3$).

The exponent α in (4.57) is given by

$$\alpha = 2 - \frac{3}{p}.$$

Now consider the function F defined from $\mathbf{R}_+ \times W^{2,p}_{\text{unif}}(\mathbf{R}^3)$ into $L^p_{\text{unif}}(\mathbf{R}^3)$ by

$$F(\lambda, u) = -\Delta u - \Gamma u + c|u|^{4/3} u + \lambda(u^2 \star V) u.$$

The function F is differentiable with respect to u, and its first derivative $\frac{\partial F}{\partial u}$ at the point (λ, u) is the operator defined on $W^{2,p}_{\text{unif}}(\mathbf{R}^3)$ with values in $L^p_{\text{unif}}(\mathbf{R}^3)$ by

$$\frac{\partial F}{\partial u}(\lambda, u) \cdot \varphi = -\Delta \varphi - \Gamma \varphi + \tfrac{7}{3}c|u|^{1/3} u\varphi + \lambda(u^2 \star V)\varphi + 2\lambda(u\varphi \star V)u,$$

from $W^{2,p}_{\text{unif}}(\mathbf{R}^3)$ into $L^p_{\text{unif}}(\mathbf{R}^3)$.

In the next step, we show that this partial derivative is one-to-one at any point (λ, u) solution to (4.56) such that $u \ge \nu > 0$, where the constant ν is the constant appearing in Lemma 4.14.

Step 2: the linearized operator.

Consider a solution (λ, u) to (4.56) such that $u \geq \nu > 0$. We denote $G = \frac{\partial F}{\partial u}(\lambda, u)$; that is,

$$G\varphi = -\Delta\varphi - \Gamma\varphi + \tfrac{7}{3}cu^{4/3}\varphi + \lambda(u^2 \star V)\varphi + 2\lambda(u\varphi \star V)u.$$

This operator is defined on $W_{\text{unif}}^{2,p}(\mathbf{R}^3)$ and we intend to prove that it is one-to-one onto $L_{\text{unif}}^p(\mathbf{R}^3)$. We first prove some coercivity inequality for the operator G.

We start from $G\varphi = \psi$, for $\varphi \in W_{\text{unif}}^{2,p}(\mathbf{R}^3)$, $\psi \in L_{\text{unif}}^p(\mathbf{R}^3)$, i.e.

$$-\Delta\varphi - \Gamma\varphi + \tfrac{7}{3}c|u|^{4/3}\varphi + \lambda(V \star u^2)\varphi + 2\lambda(V \star u\varphi)u = \psi. \qquad (4.58)$$

Let $\omega_\varepsilon > 0$ be a cut-off function that we shall determine later (the subscript ε will also become clear below). Multiplying (4.58) by $\varphi\omega_\varepsilon^2$ and integrating over $\mathbf{R}^3$, we obtain

$$\int -\Delta\varphi(\varphi\omega_\varepsilon^2) - \int \Gamma\varphi^2\omega_\varepsilon^2 + \tfrac{7}{3}c\int |u|^{4/3}\varphi^2\omega_\varepsilon^2$$
$$+ \lambda\int (V \star u^2)\varphi^2\omega_\varepsilon^2 + 2\lambda\int (V \star u\varphi)\varphi u\omega_\varepsilon^2 = \int \varphi\psi\omega_\varepsilon^2,$$

where

$$\int -\Delta\varphi(\varphi\omega_\varepsilon^2) = \int \nabla\varphi \cdot \nabla(\varphi\omega_\varepsilon^2)$$
$$= \int |\nabla\varphi|^2\omega_\varepsilon^2 + 2\omega_\varepsilon\varphi\nabla\omega_\varepsilon \cdot \nabla\varphi$$
$$= \int |\nabla(\varphi\omega_\varepsilon)|^2 - \int \varphi^2|\nabla\omega_\varepsilon|^2.$$

Therefore, we deduce

$$\int |\nabla(\varphi\omega_\varepsilon)|^2 - \int \Gamma(\varphi\omega_\varepsilon)^2 + c\int |u|^{4/3}(\varphi\omega_\varepsilon)^2$$
$$+ \lambda\int (V \star u^2)(\varphi\omega_\varepsilon)^2 + \tfrac{4}{3}c\int |u|^{4/3}(\varphi\omega_\varepsilon)^2$$
$$= \int \varphi\psi\omega_\varepsilon + \int \varphi^2|\nabla\omega_\varepsilon|^2 - 2\lambda\int (V \star u\varphi)u\varphi\omega_\varepsilon^2. \qquad (4.59)$$

We now remark that the left-hand side may be written as

$$\langle L\varphi\omega_\varepsilon, \varphi\omega_\varepsilon\rangle + \tfrac{4}{3}\int |u|^{4/3}(\varphi\omega_\varepsilon)^2,$$

where

$$L = -\Delta - \Gamma + c|u|^{4/3} + \lambda(V \star u^2). \qquad (4.60)$$

Now, since we have on $\mathbf{R}^3$

$$\begin{cases} Lu = 0, \\ u > 0, \end{cases}$$

we deduce the following inequality:

$$\langle Lv, v \rangle \geq 0$$

for any function $v \in \mathcal{D}(\mathbf{R}^3)$, and also for any function v that vanishes fast enough at infinity—like, for instance, $v = O\left(\frac{1}{|x|^{3/2+\varepsilon}}\right)$ for some $\varepsilon > 0$.

We next remark, denoting

$$a = -\Gamma + c|u|^{4/3} + \lambda(V \star u^2), \qquad (4.61)$$

that, since $L = -\Delta + a$ is non-negative, we have in the sense of self-adjoint operators, for every $0 \leq \theta \leq 1$,

$$-\Delta + a = \theta(-\Delta + a) + (1 - \theta)(-\Delta) + (1 - \theta)a$$
$$\geq (1 - \theta)(-\Delta) - (1 - \theta)|a|.$$

Therefore,

$$\langle Lv, v \rangle \geq (1 - \theta) \int |\nabla v|^2 - (1 - \theta) \int |a|v^2,$$

for any function v as above.

Let us now focus on the term $\int |a|v^2$. We first remark that a given by (4.61) may be written $a = -\Gamma + b$, where b is L^∞ and Γ is periodic and belongs to L^p_{unif}.

Now let $\delta > 0$ be fixed. There exists some constant $A_\delta > 0$ such that

$$meas\{x \in \Omega / |\Gamma(x)| \geq A_\delta\} \leq \delta,$$

where we recall that Ω is the periodic cell of Γ. For the sake of simplicity, and without loss of generality, we may assume in this proof that this unit cell is the unit cube Γ_0.

For any function $v \in H^1_{\text{loc}}(\mathbf{R}^3)$, we may write

$$\int_{k+\Gamma_0} |\Gamma v^2| = \int_{x \in k+\Gamma_0, |\Gamma(x)| \geq A_\delta} |\Gamma|v^2 + \int_{x \in k+\Gamma_0, |\Gamma(x)| < A_\delta} |\Gamma|v^2$$

$$\leq \left(\int_{x \in k+\Gamma_0, |\Gamma(x)| \geq A_\delta} |\Gamma|^{3/2} \right)^{2/3} \|v\|^2_{H^1(k+\Gamma_0)} + A_\delta \|v\|^2_{L^2(k+\Gamma_0)}$$

$$\leq \delta^{1-\frac{3}{2p}} \|\Gamma\|_{L^p_{\text{unif}}} \|v\|^2_{H^1(k+\Gamma_0)} + A_\delta \|v\|^2_{L^2(k+\Gamma_0)}.$$

Hence, if we now consider a function v that vanishes fast enough at infinity, we have

$$\int_{\mathbf{R}^3} |a|v^2 = \sum_{k \in \mathbf{Z}^3} \int_{k+\Gamma_0} |a|v^2 \qquad (4.62)$$

$$\leq \sum_{k \in \mathbf{Z}^3} \int_{k+\Gamma_0} (|\Gamma| + |b|) v^2 \tag{4.63}$$

$$\leq \delta^{1-\frac{3}{2p}} \|\Gamma\|_{L^p_{\text{unif}}} \|v\|^2_{H^1(\mathbf{R}^3)} + (A_\delta + \|b\|_{L^\infty}) \|v\|^2_{L^2(\mathbf{R}^3)}. \tag{4.64}$$

From (4.64), we therefore deduce, for any $\delta > 0$,

$$\langle Lv, v \rangle \geq (1 - \theta)(1 - \delta^{1-\frac{3}{2p}} \|\Gamma\|_{L^p_{\text{unif}}}) \int |\nabla v|^2$$
$$- (1 - \theta)(A_\delta + \|b\|_{L^\infty} + \delta^{1-\frac{3}{2p}} \|\Gamma\|_{L^p_{\text{unif}}}) \int v^2.$$

Applying this inequality with $v = \varphi w_\varepsilon$, we obtain

$$\langle L\varphi w_\varepsilon, \varphi w_\varepsilon \rangle \geq (1 - \theta)(1 - \delta^{1-\frac{3}{2p}} \|\Gamma\|_{L^p_{\text{unif}}}) \int |\nabla(\varphi w_\varepsilon)|^2$$
$$- (1 - \theta)(A_\delta + \|b\|_{L^\infty} + \delta^{1-\frac{3}{2p}} \|\Gamma\|_{L^p_{\text{unif}}}) \int |\varphi w_\varepsilon|^2. \tag{4.65}$$

In addition, using the bound from below for u,

$$\int |u|^{4/3} (\varphi w_\varepsilon)^2 \geq c_0 \int \varphi^2 w_\varepsilon^2, \tag{4.66}$$

for some constant c_0. The inequalities (4.65) and (4.66) therefore give a lower bound to the left-hand side of (4.59).

We now turn to the right-hand side of (4.59).

In order to bound the term $\int \varphi \psi w_\varepsilon^2$ from above, we proceed in two different ways, depending upon whether or not $p \leq 2$. If $\frac{3}{2} < p \leq 2$, then q defined by $\frac{1}{p} + \frac{1}{q} = 1$ satisfies $2 \leq q < 3$. Hence $\varphi w \in L^q$ (since $\varphi \in L^\infty$ and $w \in L^q$ because $q \geq 2$). Therefore,

$$\int \varphi \psi w_\varepsilon^2 \leq \|\varphi w_\varepsilon\|_{L^q} \|\psi w_\varepsilon\|_{L^p}.$$

On the contrary, if $p > 2$, we use the Cauchy–Schwarz inequality:

$$\int \varphi \psi w_\varepsilon^2 \leq \|\varphi w_\varepsilon\|_{L^2} \|\psi w_\varepsilon\|_{L^2},$$

where $\|\psi w_\varepsilon\|_{L^2}$ is well defined because

$$\int \psi^2 w_\varepsilon^2 = \sum_{k \in \mathbf{Z}^3} \int_{k+\Gamma_0} \psi^2 w_\varepsilon^2$$
$$\leq \sum_{k \in \mathbf{Z}^3} \sup_{k+\Gamma_0} w_\varepsilon^2 \int_{k+\Gamma_0} \psi^2$$
$$\leq C\|\psi\|^2_{L^2_{\text{unif}}} \sum_{k \in \mathbf{Z}^3} \sup_{k+\Gamma_0} w_\varepsilon^2$$

$$\leq C'\|\psi\|^2_{L^2_{\text{unif}}} \sum_{k\in\mathbf{Z}^3} \int_{k+\Gamma_0} \omega^2_\varepsilon,$$

where C and C' are two constants, and where the last inequality holds as soon as there exists some constant C'' such that

$$|\nabla\omega_\varepsilon| \leq C''|\omega_\varepsilon|. \tag{4.67}$$

In both cases, $\frac{3}{2} < p \leq 2$ and $p \geq 2$, we obtain

$$\int \varphi\psi\omega^2_\varepsilon \leq \|\varphi\omega_\varepsilon\|_{H^1}\|\psi\omega_\varepsilon\|_{L^r}, \tag{4.68}$$

where $r = max(2,q)$, $\frac{1}{p} + \frac{1}{q} = 1$.

Next, for the last term of (4.59), we argue as follows:

$$-2\lambda \int (V \star u\varphi)u\varphi\omega^2_\varepsilon$$

$$= -2\lambda \int (V \star u\varphi\omega_\varepsilon)u\varphi\omega_\varepsilon$$

$$-2\lambda \int\int u(y)\varphi(y)V(x-y)u(x)\varphi(x)\omega_\varepsilon(x)(\omega_\varepsilon(x)-\omega_\varepsilon(y))\,dx\,dy.$$

The first term is non-positive, since V is a non-negative kernel. For the second term, we write

$$-2\lambda \int\int u(y)\varphi(y)V(x-y)u(x)\varphi(x)\omega_\varepsilon(x)(\omega_\varepsilon(x)-\omega_\varepsilon(y))\,dx\,dy$$

$$\leq 2\lambda\|u\|^2_{L^\infty} \int\int |\varphi(y)\omega_\varepsilon(y)||V(x-y)|\varphi(x)\omega_\varepsilon(x)|\frac{|\omega_\varepsilon(x)-\omega_\varepsilon(y)|}{|\omega_\varepsilon(y)|}\,dx\,dy$$

$$= 2\lambda\|u\|^2_{L^\infty} \int\int |\varphi(y)\omega_\varepsilon(y)|e^{-1/2a|x-y|}|\varphi(x)\omega_\varepsilon(x)|$$

$$\frac{e^{-1/2a|x-y|}|\omega_\varepsilon(x)-\omega_\varepsilon(y)|}{|x-y||\omega_\varepsilon(y)|}\,dx\,dy \tag{4.69}$$

Assume for the moment that, ε being fixed, we may choose a cut-off function ω_ε such that, for all y,

$$\sup_x \sup_{z\in]x,y[} |\nabla\omega_\varepsilon(z)|e^{-1/2a|x-y|} \leq \varepsilon\omega_\varepsilon(y), \tag{4.70}$$

and continue the argument.

Using (4.70) in (4.69), we obtain

$$-2\lambda \int\int u(y)\varphi(y)V(x-y)u(x)\varphi(x)\omega_\varepsilon(x)(\omega_\varepsilon(x)-\omega_\varepsilon(y)\,dx\,dy$$

$$\leq 2\lambda\|u\|^2_{L^\infty}\varepsilon \int\int |\varphi(y)\omega_\varepsilon(y)|e^{-1/2a|x-y|}|\varphi(x)\omega_\varepsilon(x)$$

$$\leq C\varepsilon\|\varphi\omega_\varepsilon\|^2_{L^2(\mathbf{R}^3)}, \tag{4.71}$$

for some constant C that is independent of ε, ω_ε and φ.

Now assume in addition to (4.70) that w_ε also satisfies

$$|\nabla w_\varepsilon| \leq \varepsilon w_\varepsilon, \tag{4.72}$$

which, in particular, implies (4.67).

This allows us to write

$$\int \varphi^2 |\nabla w_\varepsilon|^2 \leq \varepsilon^2 \int \varphi^2 w_\varepsilon^2. \tag{4.73}$$

Collecting the inequalities (4.65), (4.66), (4.68), (4.71) and (4.73), we obtain, with (4.59),

$$\|\psi w_\varepsilon\|_{L^r} \|\varphi w_\varepsilon\| \|_{H^1} \geq (1-\theta) \left(1 - \delta^{\frac{3}{2p}} \|\Gamma\|_{L^p_{\text{unif}}}\right) \int |\nabla(\varphi w_\varepsilon)|^2$$
$$+ \left[c_0 - \varepsilon^2 - C\varepsilon - (1-\theta)(A_\delta + \|b\|_{L^\infty}\right.$$
$$\left. + \delta^{1-\frac{3}{3p}} \|\Gamma\|_{L^p_{\text{unif}}})\right] \int |\varphi w_\varepsilon|^2.$$

We first choose ε such that

$$c_0 - \varepsilon^2 - C\varepsilon \geq \frac{c_0}{2}$$

and $\delta \leq 1$ such that

$$1 - \delta^{\frac{3}{2p}} \|\Gamma\|_{L^p_{\text{unif}}} \geq \tfrac{1}{2}.$$

Next, we choose θ in order to have

$$(1-\theta) \left(A_\delta + \|b\|_{L^\infty} + \delta^{1-\frac{3}{3p}} \|\Gamma\|_{L^p_{\text{unif}}}\right) \leq \frac{c_0}{4}.$$

We then obtain

$$\|\psi w_\varepsilon\|_{L^r} \|\varphi w_\varepsilon\| \|_{H^1} \geq \tfrac{1}{2}(1-\theta) \int |\nabla(\varphi w_\varepsilon)|^2 + \frac{c_0}{4} \int |\varphi w_\varepsilon|^2.$$

Finally, we fix some constant $c = min(\tfrac{1}{2}(1-\theta), \frac{c_0}{4})$, and we obtain

$$\|\psi w_\varepsilon\|_{L^r} \geq c \|\varphi w_\varepsilon\|_{H^1}, \tag{4.74}$$

where we recall that $r = max(2, q)$, $\frac{1}{p} + \frac{1}{q} = 1$, and that

$$\|\psi w_\varepsilon\|_{L^r} \leq C \|\psi\|_{L^p_{\text{unif}}}, \tag{4.75}$$

for some constant C that depends only on w_ε and p. The inequality (4.74) holds provided that we prove the existence of a convenient w_ε decaying fast enough at infinity and satisfying (4.70) and (4.72).

For this purpose, we introduce a function ω that satisfies

$$\begin{cases} |\nabla\omega| \leq C\omega, \\ \sup_x \sup_{z\in]x,y[} |\nabla\omega(z)|e^{-1/2a|x-y|} \leq C\omega(y). \end{cases} \qquad (4.76)$$

An example of such a function (the details of checking the above conditions are of course left to the reader) is provided by

$$\omega(x) = \frac{1}{(1+|x|^2)^{m/2}},$$

for m large enough.

We now consider

$$\omega_\varepsilon(x) = \omega(\varepsilon x),$$

and (4.70) and (4.72) follow easily from (4.76). Therefore, (4.74) holds.

The inequality (4.74) clearly shows that the operator G is injective from $W^{2,p}_{\text{unif}}(\mathbf{R}^3)$ into $L^p_{\text{unif}}(\mathbf{R}^3)$. It remains next to show that it is surjective. For the sake of simplicity, we forget the subscript ε in the sequel.

We now check that if an arbitrary ψ is given in $L^p_{\text{unif}}(\mathbf{R}^3)$, then there exists some $\varphi \in W^{2,p}_{\text{unif}}(\mathbf{R}^3)$ such that $G\varphi = \psi$. For this purpose, we consider a sequence $\psi_n \in \mathcal{D}(\mathbf{R}^3)$, supported in the ball B_n centered at 0 with radius n, such that ψ_n converges to ψ in $L^p_{\text{loc}}(\mathbf{R}^3)$ and

$$\sup_{x\in\mathbf{R}^3} \int_{x+B_1} |\psi_n|^p \leq \sup_{x\in\mathbf{R}^3} \int_{x+B_1} |\psi|^p + 1. \qquad (4.77)$$

For each n fixed, we may consider the following minimization problem:

$$\inf\left\{\tfrac{1}{2}\langle G\varphi, \varphi\rangle - \int_{B_n} \psi_n\varphi; \varphi \in H^1_0(B_n)\right\}. \qquad (4.78)$$

Let us observe that the fact that φ has compact support (in B_n) ensures that all the terms of $\frac{1}{2}\langle G\varphi, \varphi\rangle$ make sense.

We now prove that (4.78) has a minimum.

By a similar argument as the one used to prove (4.74), we have, for $\varphi \in H^1_0(B_n)$,

$$\langle G\varphi, \varphi\rangle \geq c\|\varphi\|^2_{H^1(B_n)}.$$

In addition,

$$\int |\psi_n\varphi| \leq C\|\psi_n\|_{L^p(B_n)}\|\varphi\|_{H^1(B_n)}.$$

Hence any minimizing sequence of (4.78) is bounded in $H^1_0(B_n)$, and thus converges weakly in $H^1(B_n)$ and strongly in $L^q(B_n)$ for any $2 \leq q < 6$, to some φ_n.

Since $G \geq L$, where L is given by (4.60), the limit φ_n is a minimum, and satisfies

$$G\varphi_n = \psi_n.$$

In addition, in view of the above inequality (4.74), we have, for all n,

$$\|\psi_n\omega\|_{L^r} \geq c\|\varphi_n\omega\|_{H^1}, \qquad (4.79)$$

where c does not depend on n, for $r = max(2, q)$, $\frac{1}{p} + \frac{1}{q} = 1$.

We now remark that since (4.77) and (4.75) hold, $\|\psi_n\omega\|_{L^r}$ is bounded independently of n.

It follows from (4.79) that $\varphi_n\omega$ is bounded in H^1, and thus weakly converges to some $\varphi\omega \in H^1$. Passing locally to the limit in the equation $G\varphi_n = \psi_n$, we obtain

$$G\varphi = \psi.$$

The same argument shows that

$$\sup_{x\in\mathbf{R}^3} \|\varphi\omega(\cdot + x)\|_{H^1} < \infty.$$

This implies first that φ belongs to $L^2_{\text{unif}}(\mathbf{R}^3)$, since $\omega(\cdot + x)$ is bounded from below on $x + B_1$ by a positive constant that does not depend on $x \in \mathbf{R}^3$. It follows that $\nabla\varphi$ also belongs to $L^2_{\text{unif}}(\mathbf{R}^3)$. Let us note that $\nabla(\varphi\omega) = \varphi\nabla\omega + \omega\nabla\varphi$ with $\varphi \in L^2_{\text{unif}}(\mathbf{R}^3)$ and $\nabla\omega \in \left(L^\infty(\mathbf{R}^3)\right)^3$.

Since $G\varphi = \psi$, where the potential Γ involved in G belongs to $L^p_{\text{unif}}(\mathbf{R}^3)$, we obtain by elliptic regularity that $\varphi \in W^{2,p}_{\text{unif}}(\mathbf{R}^3)$. This concludes the proof of surjectivity.

The fact that the linearized operator is one-to-one at any non-negative solution allows to apply the implicit function theorem. If (λ_0, u_0) is a solution to $F(\lambda, u) = 0$ such that $u_0 \geq 0$, $u_0 \not\equiv 0$, we first apply Lemma 4.14 to obtain $u_0 \geq \nu > 0$. Next, we remark that u_0 belongs to L^∞ by the proof of Chapter 2. Using the above argument, the linearized operator is thus one-to-one, and we obtain a neighbourhood of (λ_0, u_0) in which the solutions belong to a unique curve parameterized by λ around λ_0. We may, in addition, choose the neighbourhood such that it does not contain the solution $(\lambda, 0)$ since, in view of Lemma 4.14 again, zero is isolated. The curve is therefore contained in the set of functions u satisfying $u \geq \nu > 0$.

Step 3: conclusion. In this last step, we complete the proof of Lemma 4.21.

Let $u_1 \geq 0, \not\equiv 0$ be a solution to (4.56) for $\lambda = 1$. In view of Lemma 4.14, we have $u_1 \geq \nu > 0$. We now denote by Ω the set of $\lambda' \in [0, 1]$ satisfying the following property, which we refer to in the sequel as $(\mathcal{P})$:

There exists a neighbourhood $A_{\lambda'} \times U_{\lambda'}$ of $(1, u_1)$ in $\mathbf{R} \times V$ such that the set of solutions $(\lambda \in [\lambda', 1], u \geq 0, u \not\equiv 0)$ to $F(\lambda, u) = 0$ in $A_{\lambda'} \times U_{\lambda'}$ is a unique curve $(\lambda, u(\lambda))$ parameterized by $\lambda \in [\lambda', 1]$ that satisfies $u(\lambda) \geq \nu > 0$.

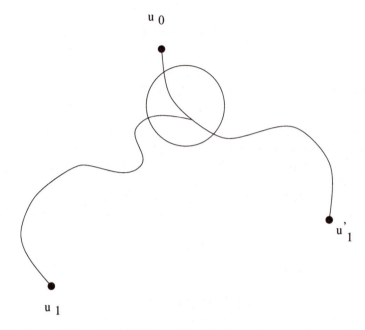

u 0

u '1

u 1

FIG. 4.1. **Argue by contradiction.** Two distinct solutions u_1 and u_1' for $\lambda = 1$ would be connected to the unique solution u_0 at $\lambda = 0$ by a unique curve, parameterized by λ. In the ball shown above, there is no local uniqueness of the solution, which contradicts the implicit function theorem.

Obviously, Ω is an interval. In view of the implicit function theorem applied in the neighbourhood of $(1, u_1)$, we have $1 \in \Omega$.

In addition, Ω is closed: if (λ_n, u_n) is a sequence of solutions to $F(\lambda, u) = 0$, and $\lambda_n \longrightarrow \lambda$, then u_n converges locally to some u that is a solution to $F(\lambda, u) = 0$ (use the same kind of arguments as in the proof of Corollary 2.7 in Chapter 2).

It follows that $\lambda_c = \inf \Omega$ belongs to Ω. Now assume that $\lambda_c > 0$. We know that $(\lambda_c, u(\lambda_c))$ is a solution and $u(\lambda_c) \geq \nu$. Applying the implicit function theorem at this point, we infer the existence of some neighbourhood A_{λ_c} of λ_c in $[0, 1]$ such that we may extend the curve of solutions beyond λ_c. We therefore reach a contradiction with the fact that λ_c is the smallest λ such that $(\mathcal{P})$ is satisfied. Hence $\lambda_c = 0$, i.e. there exists a neighbourhood $A_0 \times U_0$ of $(1, u_1)$ in $\mathbf{R} \times V$ such that the set of solutions $(\lambda \in [0, 1], u \geq 0, \not\equiv 0)$ of $F(\lambda, u) = 0$ in $A_0 \times U_0$ is a unique curve $(\lambda, u(\lambda))$ parameterized by $\lambda \in [0, 1]$ that satisfies $u(\lambda) \geq \nu > 0$.

Let us now consider another solution u_1' to (4.56) for $\lambda = 1$. We may argue as above and obtain some neighbourhood $A_0' \times U_0'$ of $(1, u_1')$. The point $(0, u_0)$ (recall that the solution at $\lambda = 0$ is unique by Lemma 4.17) belongs to both neighbourhoods $A_0 \times U_0$ and $A_0' \times U_0'$. By the local uniqueness (cf. $(\mathcal{P})$), the two curves coincide on the intersection of these neighbourhoods (see Figure 4.1 above), and therefore $u(\lambda = 1) = u'(\lambda = 1)$, i.e. $u_1 = u_1'$. Therefore the positive solution to (4.56) is unique (replace 1 in the above proof by any fixed $\lambda > 0$ to obtain the uniqueness with the coefficient λ). Since $u > 0$ is a solution, $u(\cdot + e_i)$ is also a solution, where e_i is one of the three unit vectors that define the cell of periodicity of the periodic potential Γ. Hence u is periodic. Lemma 4.21 is thus proved. $\Diamond$

As a conclusion of our study of the Yukawa case, we recall once more that we shall provide in Chapter 5 another proof of the uniqueness of the solution to (4.19).

In addition, we state in the framework of the Coulomb potential in Chapter 5 that u_Λ converges to u_{per} on a large part of $\Gamma(\Lambda)$, but it is easy to convince ourselves that the result and its proof go through *mutatis mutandis* to this case. For instance, if we denote

$$A(\Lambda) = \{x \in \Gamma(\Lambda) / \ d(x; \partial\Gamma(\Lambda)) \geq |\Lambda|^\alpha\},$$

for $\alpha > 0$ small enough ($\alpha < \frac{1}{3}$), then u_Λ converges to u_{per} uniformly on $A(\Lambda)$ (see Section 5.5 of Chapter 5 for more details).

CONVERGENCE OF THE DENSITY FOR THE THOMAS–FERMI–VON WEIZSÄCKER MODEL

5.1 Introduction

In this chapter, we are concerned with the convergence of the electronic density when the size of the neutral molecular system goes to infinity, thereby approaching an infinite crystal whose nuclei are located at the points of a regular lattice.

Let us recall that in Chapter 4 we proved that in the case when the interactions between charged particles are modelled by a short-range potential of the Yukawa type, the electronic density ρ_Λ converges to the periodic density ρ_{per}, minimizing the periodic problem $I_{\text{per}} = \lim_{\Lambda \longrightarrow \infty} \frac{I_\Lambda}{|\Lambda|}$ obtained by the thermodynamic limit (see Chapter 2).

Since in Chapter 3 we have proved that, in the Coulomb case, $\frac{I_\Lambda^{\text{TFW}}}{|\Lambda|}$ goes likewise to some periodic problem $I_{\text{per}}^{\text{TFW}}$ (up to some well identified additive constant, namely $\frac{M}{2}$), it is natural to ask the same question as in Chapter 4, but this time in the Coulomb case. More precisely, we recall that

$$I_\Lambda^{\text{TFW}} = \inf \left\{ E_\Lambda^{\text{TFW}}(\rho) + \tfrac{1}{2} \sum_{y \neq z \in \Lambda} \frac{1}{|y - z|}; \right.$$
$$\left. \rho \geq 0, \ \sqrt{\rho} \in H^1(\mathbf{R}^3), \ \int_{\mathbf{R}^3} \rho = |\Lambda| \right\},$$

where

$$E_\Lambda^{\text{TFW}}(\rho) = \int_{\mathbf{R}^3} |\nabla \sqrt{\rho}|^2 + \int_{\mathbf{R}^3} \rho^{5/3} - \int_{\mathbf{R}^3} \left(\sum_{k \in \Lambda} \frac{1}{|x - k|} \right) \rho(x) \mathrm{d}x$$
$$+ \tfrac{1}{2} \int_{\mathbf{R}^3 \times \mathbf{R}^3} \frac{\rho(x)\rho(y)}{|x - y|} \mathrm{d}x \, \mathrm{d}y.$$

Besides, the periodic minimization problem is

$$I_{\text{per}}^{\text{TFW}} = \inf \left\{ E_{\text{per}}^{\text{TFW}}(\rho); \rho \geq 0, \sqrt{\rho} \in H_{\text{per}}^1(\Gamma_0), \int_{\Gamma_0} \rho = 1 \right\},$$

where

$$E_{\text{per}}^{\text{TFW}}(\rho) = \int_{\Gamma_0} |\nabla\sqrt{\rho}|^2 + \int_{\Gamma_0} \rho^{5/3} - \int_{\Gamma_0} \rho(x)G(x)\mathrm{d}x$$
$$+ \tfrac{1}{2} \iint_{\Gamma_0 \times \Gamma_0} \rho(x)\rho(y)G(x-y)\mathrm{d}x\,\mathrm{d}y,$$

and $H_{\text{per}}^1(\Gamma_0)$ is the subset of the functions in $H^1(\Gamma_0)$ satisfying the periodic boundary conditions on the boundary of Γ_0. G is the periodic potential defined in Chapter 3 by (3.6) and (3.7) and

$$M = \lim_{x \to 0} G(x) - \frac{1}{|x|}.$$

We refer the reader to Chapters 1 and 3 for the necessary modifications when the nuclei are smeared out.

We proceed here as we did in the Yukawa case in Chapter 4.

We begin by proving in Section 5.2 preliminary convergence results that are based only on energetic arguments.

Then, in Section 5.3, we establish a uniqueness result for the partial differential equations system obtained in the thermodynamic limit. The proof also applies to the Yukawa case, therefore providing another proof than the one given in Chapter 4. More precisely, in the point nuclei case we prove the following:

Lemma 5.5 (Uniqueness in the Coulomb case) *Let* $(u; \Phi)$ *be a solution to*

$$\begin{cases} -\Delta u + u^{7/3} - \Phi\,u = 0, \\[2mm] -\Delta\Phi = 4\pi \left[\displaystyle\sum_{k \in \mathbf{Z}^3} \delta_k - u^2 \right], \end{cases}$$

where we assume that $(u; \Phi)$ *satisfies*

$$u \in L^\infty(\mathbf{R}^3),\ u \geq 0 \qquad and \qquad \Phi \in L_{\text{unif}}^1(\mathbf{R}^3),$$

and where we recall from Chapter 3 that $\Phi \in L_{\text{unif}}^1(\mathbf{R}^3)$ *means that there exists a positive constant* C *such that*

$$\sup_{x \in \mathbf{R}^3} \|\Phi\|_{L^1(x+B_1)} \leq C.$$

Then, $(u; \Phi)$ *is unique. In addition, we have*

$$\inf_{\mathbf{R}^3} u > 0,$$

$$\Phi \in L_{\text{unif}}^{3,\infty}(\mathbf{R}^3),$$

and u *and* Φ *are periodic.*

We first state and prove in Section 5.3.1 the uniqueness result in the smeared nuclei case; next, we give in Section 5.3.2 the proof of the above lemma. We want to emphasize at this stage that we shall establish in Chapter 6 a general existence and uniqueness theorem for solutions to coupled systems of partial differential equations of the Schrödinger–Poisson type that generalizes the uniqueness lemmata of this chapter.

In Section 5.4, we use this uniqueness result to prove the local convergence of the density in the Coulomb case.

Next, we establish in Section 5.5 a stronger result of convergence, the uniform convergence of the sequence ρ_Λ to ρ_{per} on a large part of the domain $\Gamma(\Lambda)$ (we shall make this statement precise in Section 5.5). This uniform convergence will be useful in Chapter 6 to give another (more general) proof of the convergence of the energy per unit volume.

5.2   Preliminary convergence results

As a consequence of the convergence of the energy proved in Chapter 3, we now deduce the convergence of the sequence of densities ρ_Λ to the density ρ_{per} minimizing the periodic minimization problem obtained by the thermodynamic limit both in the average sense of the $\sim$-transform and up to a translation. However, in Section 5.4, we shall have to work more with the specific structure of the Euler–Lagrange equations satisfied by the densities to understand why they become asymptotically periodic.

5.2.1   *Convergence of the $\sim$-transform*

The first preliminary result towards the asymptotic periodicity of ρ_Λ concerns the convergence of the sequence $\tilde{\rho}_\Lambda = \dfrac{1}{|\Lambda|} \displaystyle\sum_{y \in \Lambda} \rho_\Lambda(. + y)$ to ρ_{per}, whose proof may be found in Section 3.5 of Chapter 3 (Corollaries 3.25 and 3.26). The results that we state from now on hold both for the $(Cb - m)$ and the $(Cb - \delta)$ programs unless otherwise mentioned. We thus choose to keep the notation of the point nuclei setting for both of them.

Proposition 5.1 *Let ρ_{per} be the unique minimizer of $I_{\text{per}}^{\text{TFW}}$. Then, the sequence $\sqrt{\tilde{\rho}_\Lambda}$ converges to $\sqrt{\rho_{\text{per}}}$ strongly in $H_{\text{loc}}^1(\mathbf{R}^3)$ and in $L_{\text{loc}}^p(\mathbf{R}^3)$ for all $1 \leq p < +\infty$.*

Moreover, the sequence $\tilde{\Phi}_\Lambda - \int_{\Gamma_0} \tilde{\Phi}_\Lambda \, dx$ converges to Φ_{per} defined by $\int_{\Gamma_0} G(x - y)(m(y) - \rho_{\text{per}}(y)) \, dy$ (with the convention that m equals δ_0 in the $(Cb - \delta)$ case), strongly in $L_{\text{loc}}^p(\mathbf{R}^3)$ for all $1 \leq p < 3$ ($(Cb - \delta)$ case) or $1 \leq p < \infty$ ($(Cb - m)$ case).

This result is of course expected since $\tilde{\rho}_\Lambda$ is on purpose a kind of rearrangement of ρ_Λ that forces the limit to be periodic.

The following notion of convergence (up to a translation) is *per se* much more instructive.

5.2.2 *Convergence up to a translation*

As in the Yukawa case (see Chapter 4), we may show the following:

Proposition 5.2 *For any Van Hove sequence* (Λ), *there exists a sequence* y_Λ *in* Λ *such that:*

(i) $u_\Lambda(\cdot + y_\Lambda) = \sqrt{\rho_\Lambda(\cdot + y_\Lambda)}$ *converges to* $u_{\mathrm{per}} = \sqrt{\rho_{\mathrm{per}}}$ *strongly in* $H^1(\frac{3}{2}.\Gamma_0)$, *and in* $L^p(\frac{3}{2}.\Gamma_0)$, *for all* $1 \le p < +\infty$.

(ii) Moreover, in the $(Cb - m)$ *case,* $\Phi_\Lambda(\cdot + y_\Lambda) - \theta_\Lambda$ *converges to* $\Phi_{\mathrm{per}} - \theta_{\mathrm{per}}$ *strongly in* $H^1(\frac{3}{2}.\Gamma_0)$ *and in* $L^p(\frac{3}{2}.\Gamma_0)$ *for all* $1 \le p < +\infty$.

(iii) In the $(Cb - \delta)$ *case,* $\Phi_\Lambda(\cdot + y_\Lambda) - \theta_\Lambda$ *converges to* $\Phi_{\mathrm{per}} - \theta_{\mathrm{per}}$ *strongly in* $L^p(\frac{3}{2}.\Gamma_0)$, *for all* $1 \le p < 3$.

Proof of Proposition 5.2 The proof of Proposition 5.2 follows the same pattern as the proof of Proposition 4.7 that we proposed in Chapter 4 for the Yukawa case. Therefore, we shall not enter the details of the proof (that the reader will find in Subsection 4.2.3 of Chapter 4) but rather point out the modifications and adaptations to be made. The only new point concerns the claim on the effective potential Φ_Λ.

We first prove (ii). For this purpose, we define

$$a(\Lambda) = \|u_\Lambda - u_{\mathrm{per}}\|^2_{H^1(\Gamma(\Lambda))}.$$

The argument we made in the Yukawa case carries through *mutatis mutandis* so that

$$a(\Lambda) = o(|\Lambda|). \tag{5.1}$$

We next denote by y_Λ one point in Λ that achieves the minimum in

$$\min_{y \in \Lambda} \left\{ \|\Phi_\Lambda\|^2_{H^1(y + \frac{3}{2}.\Gamma_0)} + \frac{|\Lambda|}{a(\Lambda)} \|u_\Lambda - u_{\mathrm{per}}\|^2_{H^1(y + \frac{3}{2}.\Gamma_0)} \right\}.$$

Then, because of the uniform *a priori* bounds we obtained in Section 3.2 of Chapter 3, there exists a positive constant C that is independent of Λ such that

$$C\,|\Lambda| \ge \|\Phi_\Lambda\|^2_{H^1(\Gamma(\Lambda))} + \frac{|\Lambda|}{a(\Lambda)} \|u_\Lambda - u_{\mathrm{per}}\|^2_{H^1(\Gamma(\Lambda))}$$

$$\ge |\Lambda| \left\{ \|\Phi_\Lambda\|^2_{H^1(y_\Lambda + \frac{3}{2}.\Gamma_0)} + \frac{|\Lambda|}{a(\Lambda)} \|u_\Lambda - u_{\mathrm{per}}\|^2_{H^1(y_\Lambda + \frac{3}{2}.\Gamma_0)} \right\}.$$

Then, using (5.1), $u_\Lambda(\cdot + y_\Lambda)$ converges to u_{per} strongly in $H^1(\frac{3}{2}.\Gamma_0)$ and $\Phi_\Lambda(\cdot + y_\Lambda)$ is bounded in $H^1(\frac{3}{2}.\Gamma_0) \cap L^\infty(\mathbf{R}^3)$. In view of Rellich's theorem, we may extract a subsequence, if necessary, such that $\Phi_\Lambda(\cdot + y_\Lambda)$ converges to $\bar\Phi \in H^1(\frac{3}{2}.\Gamma_0) \cap L^\infty(\mathbf{R}^3)$ weakly in $H^1(\frac{3}{2}.\Gamma_0)$ and strongly in $L^p(\frac{3}{2}.\Gamma_0)$ for all $1 \le p < +\infty$.

With these ingredients, we pass to the limit in the partial differential equations satisfied by $u_\Lambda(\cdot + y_\Lambda)$ and $\Phi_\Lambda(\cdot + y_\Lambda)$, namely

$$\begin{cases} -\Delta u_\Lambda + \frac{5}{3} u_\Lambda^{7/3} - (\Phi_\Lambda - \theta_\Lambda) \, u_\Lambda = 0, \\ -\Delta \Phi_\Lambda = 4\pi \, (m_\Lambda - u_\Lambda^2), \end{cases} \tag{5.2}$$

and we obtain

$$-\Delta u_{\mathrm{per}} + \frac{5}{3} u_{\mathrm{per}}^{7/3} - \bar{\bar{\Phi}} \, u_{\mathrm{per}} + \theta_\infty \, u_{\mathrm{per}} = 0, \tag{5.3}$$

where θ_∞ is the limit of the sequence of Lagrange multipliers θ_Λ (see Chapter 3); of course, $\bar{\bar{\Phi}}$ solves

$$-\Delta \bar{\bar{\Phi}} = 4\pi \left(\sum_{y \in \mathbf{Z}^3} m(\cdot - y) - \rho_{\mathrm{per}} \right). \tag{5.4}$$

We have to compare this system of equations (5.3) and (5.4) with the Euler–Lagrange equation satisfied by u_{per}; that is,

$$-\Delta u_{\mathrm{per}} + \frac{5}{3} u_{\mathrm{per}}^{7/3} - \Phi_{\mathrm{per}} \, u_{\mathrm{per}} + \theta_{\mathrm{per}} \, u_{\mathrm{per}} = 0, \tag{5.5}$$

where $\Phi_{\mathrm{per}} = G_m(x) - \int_{\Gamma_0} G(x - y) \rho_{\mathrm{per}}(y) \, dy$ is the unique periodic solution to

$$-\Delta \Phi_{\mathrm{per}} = 4\pi \left(\sum_{y \in \mathbf{Z}^3} m(\cdot - y) - \rho_{\mathrm{per}} \right),$$

whose integral over Γ_0 vanishes.

The function u_{per} being periodic and positive (this is a consequence of Harnack's inequality which applies to the Euler–Lagrange equation satisfied by u_{per}), (5.3) implies that $\bar{\bar{\Phi}}$ is also periodic. Then, turning back to (5.4), we deduce that $\bar{\bar{\Phi}}$ and Φ_{per} are equal up to an additive constant; next, comparing (5.3) with the Euler–Lagrange equation satisfied by u_{per}, we identify $\bar{\bar{\Phi}} - \theta_\infty$ with $\Phi_{\mathrm{per}} - \theta_{\mathrm{per}}$. Our claim follows.

For the proof of (iii), we follow, step by step, the above proof, replacing $\|\Phi_\Lambda\|^2_{H^1(y + \frac{3}{2}.\Gamma_0)}$ by $\|\Phi_\Lambda\|^2_{L^2(y + \frac{3}{2}.\Gamma_0)}$ and using the uniform bounds on Φ_Λ in $L^2_{\mathrm{unif}}(\mathbf{R}^3)$ given in Proposition 3.12 of Chapter 3.

Finally, since the limit functions above do not depend on Λ, the convergence concerns the whole sequences $u_\Lambda(\cdot + y_\Lambda)$ and $\Phi_\Lambda(\cdot + y_\Lambda)$. The proof is then complete.   ◇

We now turn to the main result of this chapter, which is the proof of the local convergence of the sequence of densities ρ_Λ to the periodic minimizing density ρ_{per}.

5.3   A uniqueness result for a system of PDE's

To conclude our program about the thermodynamic limit of the TFW model, we wish to state and prove that the limit of the sequence of densities is periodic. For this purpose, we shall establish uniqueness results for a class of coupled Schrödinger–Poisson-type systems of partial differential equations that contains the periodic system we expect to obtain by passing to the limit as a particular case.

In order to give to the reader some insight into the strategy of proof, we first detail through Lemma 5.3 a simpler proof in the setting of the smeared nuclei and under restrictive assumptions on the regularity and the positivity of solutions. However, these assumptions are not necessary for our argument. In the important Coulomb case with point nuclei, we shall therefore state and prove the uniqueness result under less restrictive assumptions (that are still satisfied in our case).

5.3.1   *The smeared nuclei case*

Lemma 5.3 *Let m be a distribution on $\mathbf{R}^3$ and let us consider a solution $(u; \phi)$ of*

$$\begin{cases} -\Delta u + u^{7/3} - \phi u = 0, \\ -\Delta \phi = m - u^2, \end{cases} \tag{5.6}$$

where we assume that $u \in L^\infty(\mathbf{R}^3)$, $\phi \in L^\infty(\mathbf{R}^3)$ and $\inf_{\mathbf{R}^3} u > 0$.
   Then, such a solution is unique.

Remark 5.4 We emphasize once more the fact that the assumptions made on u and ϕ in the statement of Lemma 5.3 may be weakened. In particular, the condition $\inf_{\mathbf{R}^3} u > 0$ is fulfilled as soon as the distribution m is convenient (see the assumptions of Lemma 5.5 and Remark 5.6).

Proof of Lemma 5.3 Let us consider two solutions $(u; \phi)$, $(v; \psi)$ in $(L^\infty(\mathbf{R}^3))^2$ of the above system, and denote by $\mu > 0$ a constant such that both $u \geq \mu$ and $v \geq \mu$ hold. Recalling that we have

$$\begin{cases} -\Delta u + u^{7/3} - \phi u = 0, \\ -\Delta v + v^{7/3} - \psi v = 0, \end{cases} \tag{5.7}$$

we deduce, denoting $w = u - v$,

$$-\Delta w + u^{7/3} - v^{7/3} - (\phi u - \psi v) = 0.$$

We wish to prove that $w = 0$.

Let $\xi \in \mathcal{D}(\mathbf{R}^3)$. Multiplying the above equation by $w \xi^2$ and integrating over $\mathbf{R}^3$, we find that

$$\int -\Delta w\, w\, \xi^2 + \int (u^{7/3} - v^{7/3})w\, \xi^2 - \int (\phi u - \psi v)\, w\, \xi^2 = 0. \qquad (5.8)$$

We now remark that, on the one hand,

$$\int -\Delta w \cdot w\xi^2 = \int |\nabla(w\xi)|^2 - \int w^2|\nabla\xi|^2, \qquad (5.9)$$

while, on the other hand,

$$\phi u - \psi v = \frac{\phi + \psi}{2}w + \frac{\phi - \psi}{2}(u + v). \qquad (5.10)$$

Next, using the fact that u and v are bounded from below by $\mu > 0$, we know there exists a constant $\nu > 0$ such that

$$(u^{7/3} - v^{7/3})(u - v) \geq \tfrac{1}{2}(u^{4/3} + v^{4/3})(u - v)^2 + \nu(u - v)^2;$$

that is,

$$(u^{7/3} - v^{7/3})w \geq \tfrac{1}{2}(u^{4/3} + v^{4/3})w^2 + \nu w^2. \qquad (5.11)$$

Finally, denoting

$$L = -\Delta + \tfrac{1}{2}(u^{4/3} + v^{4/3}) - \frac{\phi + \psi}{2}, \qquad (5.12)$$

we remark that the operator L is non-negative over a set of functions that vanish fast enough at infinity. Indeed, since u and v are positive solutions to (5.7), we have

$$\begin{cases} \lambda_1(-\Delta + u^{4/3} - \phi, \Omega) > 0, \\ \lambda_1(-\Delta + v^{4/3} - \psi, \Omega) > 0, \end{cases} \qquad (5.13)$$

for any bounded open subset Ω of $\mathbf{R}^3$, where we denote as usual by $\lambda_1(H, \Omega)$ the first eigenvalue on Ω of the self-adjoint operator H with the Dirichlet boundary conditions. Then, (5.13) yields in a straightforward way

$$\lambda_1(L, \Omega) \geq 0. \qquad (5.14)$$

Collecting (5.9), (5.10), and (5.11), we deduce from (5.8)

$$\langle L(w\xi), (w\xi)\rangle + \nu \int w^2\xi^2 \leq \int w^2|\nabla\xi|^2 + \int \frac{\phi - \psi}{2}(u^2 - v^2)\xi^2. \qquad (5.15)$$

Now, since $-\Delta(\phi - \psi) = -(u^2 - v^2)$, we have

$$\int \frac{\phi - \psi}{2}(u^2 - v^2)\xi^2 = \frac{1}{2}\int (\phi - \psi)\Delta(\phi - \psi)\xi^2$$
$$= -\frac{1}{2}\int |\nabla((\phi - \psi)\xi)|^2 + \frac{1}{2}\int (\phi - \psi)^2|\nabla\xi|^2.$$

Therefore, (5.15) yields

$$\langle L(w\xi), (w\xi)\rangle + \nu \int w^2\xi^2 + \frac{1}{2}\int |\nabla((\phi - \psi)\xi)|^2$$
$$\leq \int w^2|\nabla\xi|^2 + \frac{1}{2}\int (\phi - \psi)^2|\nabla\xi|^2, \qquad (5.16)$$

hence, with (5.14),

$$\nu \int w^2\xi^2 + \frac{1}{2}\int |\nabla(\phi - \psi)\xi|^2 \leq \int w^2|\nabla\xi|^2 + \frac{1}{2}\int (\phi - \psi)^2|\nabla\xi|^2. \qquad (5.17)$$

Using

$$0 = \int div((\phi - \psi)^2\nabla(\xi^2)) = \int (\phi - \psi)^2\Delta(\xi^2) + 2\int (\phi - \psi)\nabla(\xi^2)\cdot\nabla(\phi - \psi),$$

(5.17) also implies

$$\frac{1}{2}\int |\nabla(\phi - \psi)|^2\xi^2 \leq \int w^2|\nabla\xi|^2 + \frac{1}{4}\int (\phi - \psi)^2|\Delta(\xi^2)|. \qquad (5.18)$$

If we apply the above inequalities (5.17) and (5.18) to a sequence of functions $\xi_n \in \mathcal{D}(\mathbf{R}^3)$ that converges to

$$\xi(x) = \frac{1}{(1 + |x|^2)^{m/2}}, \qquad (5.19)$$

for some exponent $m = \frac{1}{2} + \varepsilon$, $\varepsilon > 0$, we obtain the inequalities (5.17) and (5.18) for ξ given by (5.19). In addition, for this choice (5.19) of ξ we have

$$\int w^2|\nabla\xi|^2 \leq \|w\|_{L^\infty}^2 \int |\nabla\xi|^2 < \infty, \qquad (5.20)$$

$$\int (\phi - \psi)^2|\nabla\xi|^2 \leq \|\phi - \psi\|_{L^\infty}^2 \int |\nabla\xi|^2 < \infty,$$

and

$$\int (\phi - \psi)^2|\Delta(\xi^2)| \leq \|\phi - \psi\|_{L^\infty}^2 \int |\Delta(\xi^2)| < \infty.$$

Thus, we obtain, respectively, from (5.17) and (5.18)

$$\int w^2\xi^2 < \infty \tag{5.21}$$

and

$$\int |\nabla(\phi - \psi)|^2\xi^2 < \infty. \tag{5.22}$$

Our strategy of proof is now the following. Our next step will consist of proving that

$$\int (\phi - \psi)^2\xi^2 < \infty.$$

This information, together with (5.21), will allow us to apply a scaling argument to ξ in (5.17) which will imply that $w = 0$ because of the special choice we made for ξ in (5.19).

We now make the following observation. If a is a function in L^∞, and if $-\Delta + a \geq 0$ in the sense of self-adjoint operators, then, for any $0 < \theta < 1$, we have

$$-\Delta + a = \theta(-\Delta + a) + (1 - \theta)(-\Delta) + (1 - \theta)a$$
$$\geq (1 - \theta)(-\Delta) - (1 - \theta)\|a\|_{L^\infty}.$$

Therefore, with $a = \frac{1}{2}(u^{4/3} + v^{4/3}) - \frac{\phi+\psi}{2}$, we deduce, from (5.16),

$$(1 - \theta)\int |\nabla(w\xi)|^2 + (\nu - (1 - \theta)\|a\|_{L^\infty})\int w^2\xi^2$$
$$\leq \int w^2|\nabla\xi|^2 + \frac{1}{2}\int (\phi - \psi)^2|\nabla\xi|^2,$$

thus, for $1 - \theta$ small enough,

$$(1 - \theta)\int |\nabla(w\xi)|^2 \leq \int w^2|\nabla\xi|^2 + \frac{1}{2}\int (\phi - \psi)^2|\nabla\xi|^2,$$

which yields

$$\int |\nabla(w\xi)|^2 < \infty,$$

and thus

$$\int |\nabla w|^2\xi^2 < \infty. \tag{5.23}$$

We now return to (5.7) and use (5.10) to write

$$-\Delta w = \tfrac{1}{2}(\phi - \psi)(u + v) + \tfrac{1}{2}(\phi + \psi)(u - v) - (u^{7/3} - v^{7/3}),$$

from which we deduce that

$$\frac{-\Delta w}{u + v} = \tfrac{1}{2}(\phi - \psi) + \tfrac{1}{2}\frac{\phi + \psi}{u + v}w - \frac{u^{7/3} - v^{7/3}}{u + v}. \tag{5.24}$$

Each term of the right-hand side belongs to the set $\{f/ \int |\nabla f|^2 \xi^2 < \infty\}$. For the first one it is clear because of (5.22). For the second one and the third one, it is more involved. We remark first of all that u and v are L^∞. In addition, since we deal here with the smeared nuclei case, ϕ, ψ, $\nabla\phi$, and $\nabla\psi$ are also L^∞. In particular, this implies with (5.6) that Δu and Δv are L^∞, and thus, by elliptic regularity, u and v are in $W^{2,p}$ for all p, which ensures that ∇u and ∇v also are L^∞. We now write

$$\nabla(\frac{\phi + \psi}{u + v}w) = \frac{\nabla(\phi + \psi)}{u + v}w - \frac{(\phi + \psi)(\nabla u + \nabla v)}{(u + v)^2}w + \frac{\phi + \psi}{u + v}\nabla w.$$

Using that u and v are bounded from below by a positive constant, and that ϕ, ψ, ∇u, ∇v, $\nabla\phi$, and $\nabla\psi$ all are L^∞, we easily see that this expression is of the form $aw + b\nabla w$, where $a \in (L^\infty)^3$ and $b \in L^\infty$. It follows because of (5.21) and (5.23) that the second term of (5.24) is in the set $\{f/ \int |\nabla f|^2 \xi^2 < \infty\}$. For the third term of (5.24), we remark that

$$\nabla(\frac{u^{7/3} - v^{7/3}}{u + v}) = \frac{7}{3}\frac{u^{4/3} - v^{4/3}}{u + v}\nabla u + \frac{7}{3}\frac{v^{4/3}}{u + v}\nabla w - \frac{u^{7/3} - v^{7/3}}{(u + v)^2}(\nabla u + \nabla v).$$

Next, using

$$|u^p - v^p| \le C_p|u^{p-1} + v^{p-1}||w|,$$

(where C_p denotes a constant depending on $p \ge 1$ but not on u and v) both for $p = \tfrac{4}{3}$ and $p = \tfrac{7}{3}$, we obtain, arguing as above, that the third term of (5.24) is in the set $\{f/ \int |\nabla f|^2 \xi^2 < \infty\}$.

We therefore have

$$\int \left|\nabla(\frac{-\Delta w}{u + v})\right|^2 \xi^2 < \infty.$$

Next, we write

$$\int \left[\frac{1}{u + v}|\nabla(-\Delta w)| - |\Delta w||\nabla(\frac{1}{u + v})|\right]^2 \xi^2 \le \int \left|\nabla(\frac{-\Delta w}{u + v})\right|^2 \xi^2$$

and thus, since u and v are bounded below on $\mathbf{R}^3$,

$$\int \left[|\nabla(\Delta w)| - |\Delta w|^2 \left| \frac{\nabla(u+v)}{u+v} \right| \right]^2 \xi^2 \le C, \qquad (5.25)$$

where here and below C is a positive constant. Next, we develop the left-hand side of (5.25) above and use the Cauchy–Schwarz inequality and (5.25) to write

$$\int |\nabla(\Delta w)|^2 \xi^2 \le C + \left(\int |\Delta w|^2 \left| \frac{\nabla(u+v)}{u+v} \right|^2 \xi^2 \right)^{1/2} \left(\int |\nabla(\Delta w)|^2 \xi^2 \right)^{1/2}.$$

Thus, using the fact that ∇u and ∇v belong to L^∞ and the positive bound from below for u and v, we deduce from the above inequality that

$$\int |\nabla(\Delta w)|^2 \xi^2 \le C + C \left(\int |\Delta w|^2 \xi^2 \right)^{1/2} \left(\int |\nabla(\Delta w)|^2 \xi^2 \right)^{1/2}. \qquad (5.26)$$

On the other hand, integrating by parts and using the Cauchy–Schwarz inequality once more, we may write

$$\int |\Delta w|^2 \xi^2 = - \int \nabla w \cdot \nabla(\Delta w \, \xi^2)$$

$$= - \int \nabla w \cdot \nabla(\Delta w) \xi^2 + \int \Delta w \, \xi \, \nabla w \cdot \nabla \xi$$

$$\le \left(\int |\nabla(\Delta w)|^2 \xi^2 \right)^{1/2} \left(\int |\nabla w|^2 \xi^2 \right)^{1/2}$$

$$+ \left(\int |\Delta w|^2 \xi^2 \right)^{1/2} \left(\int w^2 |\nabla \xi|^2 \right)^{1/2}.$$

Finally, in view of (5.23), (5.21), and (5.20), we obtain

$$\int |\Delta w|^2 \xi^2 \le C + C \left(\int |\nabla(\Delta w)|^2 \xi^2 \right)^{1/2}. \qquad (5.27)$$

Comparing (5.26) and (5.27), we obtain

$$\int |\nabla(-\Delta w)|^2 \xi^2 < \infty,$$

and next, returning to (5.27),

$$\int |\Delta w|^2 \xi^2 < \infty. \qquad (5.28)$$

Using (5.7) and (5.10), we write

$$\phi - \psi = 2\frac{1}{u+v}\Delta w - 2\frac{u^{7/3} - v^{7/3}}{u+v} + \frac{\phi + \psi}{u+v}w. \tag{5.29}$$

By the same kinds of arguments as above, and using (5.28) to deal with the first term of the right-hand side, it is easy to see that (5.29) implies

$$\int |\phi - \psi|^2 \xi^2 < \infty. \tag{5.30}$$

At this stage, we therefore have obtained, for ξ given by (5.19),

$$\begin{cases} \int w^2\xi^2 < \infty, \\[2mm] \int |\phi - \psi|^2\xi^2 < \infty, \end{cases} \tag{5.31}$$

together with (5.17), that yields in particular

$$\nu \int w^2\xi^2 \le \int w^2|\nabla\xi|^2 + \tfrac{1}{2}\int (\phi - \psi)^2|\nabla\xi|^2. \tag{5.32}$$

We next replace ξ by $\xi_\varepsilon(x) = \xi(\varepsilon x)$. The whole analysis we made above with ξ goes through once again with ξ_ε and we therefore obtain an analogous estimate to (5.32), which now reads

$$\nu \int w^2\xi_\varepsilon^2 \le \int w^2|\nabla\xi_\varepsilon|^2 + \tfrac{1}{2}\int (\phi - \psi)^2|\nabla\xi_\varepsilon|^2. \tag{5.33}$$

Now,

$$\begin{aligned} |\nabla\xi_\varepsilon|^2 &\le C\varepsilon^2\xi_\varepsilon^2 \\ &= C\varepsilon^2\frac{1}{(1+\varepsilon^2|x|^2)^m} \\ &\le C\frac{\varepsilon^{2-2m}}{(1+|x|^2)^m} \\ &= C\varepsilon^{2-2m}\xi^2, \end{aligned} \tag{5.34}$$

for $\varepsilon \le 1$. Using (5.34) in (5.33), we obtain, for any radius R,

$$\frac{1}{(1+\varepsilon^2 R^2)^m}\int_{|x|\le R} w^2 \le \int_{|x|\le R} w^2\xi_\varepsilon^2 \le C\varepsilon^{2-2m}\int (w^2 + \tfrac{1}{2}(\phi - \psi)^2)\xi^2.$$

Therefore, letting ε go to 0 and using (5.31), we obtain $w = 0$ on $\{|x| \le R\}$, and thus $w = 0$ on $\mathbf{R}^3$, and finally $u = v$. Now, because (5.7) is satisfied by $u = v$, it is easily seen that $\phi = \psi$, since $u = v > 0$. Uniqueness is proved and the proof is complete.   ◇

We now turn to the point nuclei case.

5.3.2 *The point nuclei case*

The above Lemma 5.3 does not apply because the effective potential Φ to which the electrons are subjected does not belong to $W^{1,\infty}$: it behaves like $\frac{1}{|x-i|}$ in the neighbourhood of each point nuclei located at $i \in \mathbf{Z}^3$. We extend the previous result to this case in the following lemma.

Lemma 5.5 *Let $(u; \Phi)$ be a solution to*

$$
\begin{cases}
-\Delta u + u^{7/3} - \Phi u = 0, \\[2mm]
-\Delta \Phi = 4\pi \left[\sum_{k \in \mathbf{Z}^3} \delta_k - u^2 \right],
\end{cases}
\tag{5.35}
$$

where we assume that $(u; \Phi)$ satisfies

$$
u \in L^\infty(\mathbf{R}^3), \ u \geq 0 \qquad and \qquad \Phi \in L^1_{\text{unif}}(\mathbf{R}^3),
\tag{5.36}
$$

and where we recall from Chapter 3 that $\Phi \in L^1_{\text{unif}}(\mathbf{R}^3)$ means that there exists a positive constant C such that

$$
\sup_{x \in \mathbf{R}^3} \|\Phi\|_{L^1(x+B_1)} \leq C.
\tag{5.37}
$$

Then, $(u; \Phi)$ is unique. In addition,

$$
\inf_{\mathbf{R}^3} u > 0,
$$

$$
\Phi \in L^{3,\infty}_{\text{unif}}(\mathbf{R}^3),
$$

and u and Φ are periodic.

Let us make some comments:

Remark 5.6 (1) The same uniqueness result carries through with minor modifications (see Chapter 6) if we replace in (5.35) the measure $\sum_{k \in \mathbf{Z}^3} \delta_k$ by any non-negative measure m in

$$
\mathcal{M}^+_{\text{unif}}(\mathbf{R}^3) = \left\{ m \text{ non-negative measure on } \mathbf{R}^3, \ \sup_{x \in \mathbf{R}^3} m(x + B_1) < \infty \right\}
$$

enjoying the following additional property

$$
\lim_{R \to +\infty} \frac{1}{R} \inf_{x \in \mathbf{R}^3} m(x + B_R) = +\infty.
$$

Using this general statement (that includes both the smeared out and the point nuclei cases), we may extend our results, for example, to other shapes of periodic cells or we may allow several nuclei by periodic cell or even other situations (we shall make these extensions in Chapter 6).

(2) We may even weaken the assumptions (5.36) that we have imposed on u and Φ to ensure the uniqueness of a solution to (5.35). Indeed, under the assumptions on m given in point (1) of this remark, we shall prove in Chapter 6 a general existence and uniqueness result for the system (5.35) (which will recover the particular cases treated in Lemma 5.3 and Lemma 5.5) provided that u belongs to $L^2_{\text{unif}}(\mathbf{R}^3) \cap L^{7/3}_{\text{loc}}(\mathbf{R}^3)$ and Φ lies in $L^1_{\text{unif}}(\mathbf{R}^3)$.

Before giving the proof of this lemma, let us introduce some notation.

Notation We already have introduced the notation

$$L^p_{\text{unif}}(\mathbf{R}^3) = \left\{ \psi \in L^p_{\text{loc}}(\mathbf{R}^3) / \sup_{x \in \mathbf{R}^3} \|\psi\|_{L^p(x+B_1)} < \infty \right\},$$

for every $1 \leq p$, where B_1 stands for the ball of radius 1 centered at zero in $\mathbf{R}^3$. We define in the same manner

$$W^{p,q}_{\text{unif}}(\mathbf{R}^3) = \left\{ \psi \in W^{p,q}_{\text{loc}}(\mathbf{R}^3) / \sup_{x \in \mathbf{R}^3} \|\psi\|_{W^{p,q}(x+B_1)} < \infty \right\},$$

for every $1 \leq p, q$.

In addition, we recall the definition (together with some properties) of the Marcinkiewitz space $L^{3,\infty}(\Omega)$, for any open subset Ω of $\mathbf{R}^3$,

$$L^{3,\infty}(\Omega) = \left\{ \psi \in L^1_{\text{loc}}(\Omega) / \sup_{t>0} \left[t^3 \ meas\{y \in \Omega / |\psi(y)| \geq t\} \right] < \infty \right\}, \quad (5.38)$$

or, equivalently (see, for example, [8]),

$$L^{3,\infty}(\Omega) = \left\{ \psi \in L^1_{\text{loc}}(\Omega), \sup_K |K|^{-\frac{2}{3}} \int_K |\psi(x)| \, dx < \infty \right\}, \quad (5.39)$$

the supremum in the above definition being taken over all measurable subsets K of Ω of finite measure, denoted by $|K|$. With this second definition, $L^{3,\infty}$ is easily shown to be a Banach space, under the norm

$$\|\psi\|_{L^{3,\infty}(\Omega)} = min \left\{ C \in (0; +\infty) / \sup_K |K|^{-\frac{2}{3}} \int_K |\psi| \leq C \right\}.$$

Let us also define

$$L^{3,\infty}_{\text{unif}}(\mathbf{R}^3) = \left\{ \psi \in L^{3,\infty}_{\text{loc}}(\mathbf{R}^3) / \sup_{x \in \mathbf{R}^3} \|\psi\|_{L^{3,\infty}(x+B_1)} < \infty \right\}.$$

Moreover, we shall often need the following continuous embedding properties of $L^{3,\infty}_{\text{unif}}(\mathbf{R}^3)$:

$$L^q_{\text{unif}}(\mathbf{R}^3) \subset L^{3,\infty}_{\text{unif}}(\mathbf{R}^3) \subset L^p_{\text{unif}}(\mathbf{R}^3), \qquad \forall\, 1 \le p < 3,\ 3 \le q \le +\infty, \qquad (5.40)$$

(see [8] or [9] and the references therein for the proofs or more details on these spaces).

Proof of Lemma 5.5 The proof basically follows the same lines as the proof of Lemma 5.3. We therefore only indicate how the latter has to be modified. First of all, we begin with a few regularity results and estimates on any solution $(u; \Phi)$ to (5.35) satisfying the conditions (5.36).

Step 1: regularity results. We first remark that

$$\Phi \in L^{3,\infty}_{\text{unif}}(\mathbf{R}^3). \qquad (5.41)$$

Indeed, let us denote by m the measure

$$m = \sum_{k \in \mathbf{Z}^3} \delta_k - u^2.$$

Then, since, in particular, u is in $L^2_{\text{unif}}(\mathbf{R}^3)$, it is clear that m is uniformly locally bounded on $\mathbf{R}^3$, in the following sense:

$$\sup_{x \in \mathbf{R}^3} |m(x + B_1)| < +\infty.$$

Since Φ belongs to $L^1_{\text{unif}}(\mathbf{R}^3)$ and $\Delta\Phi$ belongs to $\mathcal{M}_{\text{unif}}(\mathbf{R}^3)$, it is a standard fact that $\Phi \in L^{3,\infty}_{\text{unif}}(\mathbf{R}^3)$ and $\nabla\Phi \in L^{3/2,\infty}_{\text{unif}}(\mathbf{R}^3)^3$. However, for the sake of completeness, we provide a proof of (5.41).

Let us fix x_0 in $\mathbf{R}^3$ and let us define

$$\Phi_{x_0} = (m\, \chi_{x_0 + B_2}) \star \frac{1}{|x|}.$$

Then, since $m\, \chi_{x_0 + B_2}$ is a bounded measure and since $|m(x_0 + B_2)|$ does not depend on x_0, Φ_{x_0} is in $L^{3,\infty}(\mathbf{R}^3)$ (and thus, in particular, in $L^{3,\infty}_{\text{unif}}(\mathbf{R}^3)$), and the norm of Φ_{x_0} in $L^{3,\infty}(\mathbf{R}^3)$ does not depend on x_0 (see, e.g., [8]). In particular, from (5.40), Φ_{x_0} is in $L^1_{\text{unif}}(\mathbf{R}^3)$.

Now, we use the fact that, by definition, Φ_{x_0} solves

$$-\Delta\Phi_{x_0} = 4\pi\, m\, \chi_{x_0 + B_2} \qquad \text{in } \mathbf{R}^3;$$

therefore, because (5.35) is satisfied by Φ, $\Phi - \Phi_{x_0}$ is harmonic in $x_0 + B_2$ and belongs to $L^1_{\text{unif}}(\mathbf{R}^3)$. Thus, using the mean value formula, we may write, for all $y \in x_0 + B_1$,

$$|(\Phi - \Phi_{x_0})(y)| \leq C \int_{y+B_1} |\Phi - \Phi_{x_0}|$$

$$\leq C \int_{x_0+B_2} |\Phi - \Phi_{x_0}|$$

$$\leq C,$$

for some positive constant C that is independent of y and x_0. Hence $\Phi - \Phi_{x_0}$ is bounded by a constant, say in $x_0 + B_1$, and this constant may be chosen independently of x_0. Since, from (5.40), $L^\infty(\mathbf{R}^3)$ is continuously embedded in $L^{3,\infty}_{\mathrm{unif}}(\mathbf{R}^3)$, we conclude.

From (5.41) and the embeddings recalled through (5.40), we now obtain that Φ is in $L^p_{\mathrm{unif}}(\mathbf{R}^3)$, for all $1 \leq p < 3$; hence, we obtain, from standard elliptic regularity results, that

$$u \in W^{2,p}_{\mathrm{unif}}(\mathbf{R}^3) \quad \text{for all } 1 \leq p < 3.$$

Indeed, from (5.35) and (5.36), $u \in L^\infty(\mathbf{R}^3)$ and $-\triangle u \in L^p_{\mathrm{unif}}(\mathbf{R}^3)$ for all $1 \leq p < 3$. In particular, from Sobolev's embeddings,

$$\nabla u \in L^p_{\mathrm{unif}}(\mathbf{R}^3) \quad \text{for all } 1 \leq p < +\infty. \tag{5.42}$$

In fact, following the proof of the Appendix of Chapter 3, we know that

$$\nabla u \in L^\infty(\mathbf{R}^3).$$

However, we shall not use this latter bound in the course of our proof, since we want to extend our result to more general situation (see Remark 5.6).

The second step of the proof consists of verifying that the restrictions (5.36) imposed on the solutions $(u; \Phi)$ to (5.35) ensure that u is bounded below away from 0 on $\mathbf{R}^3$.

Step 2. We now show that

$$\inf_{\mathbf{R}^3} u > 0.$$

We argue by contradiction. Then, there exists a sequence of points $x_n \in \mathbf{R}^3$ such that

$$u(x_n) \longrightarrow 0 \quad \text{as } n \to \infty.$$

Since $u^{4/3} - \Phi$ is in $L^p_{\mathrm{unif}}(\mathbf{R}^3)$, for some $p > \frac{3}{2}$, we deduce, from the equation satisfied by u in (5.35) and Harnack's inequality (see [56] or [24]) that the sequence $u(\cdot + x_n)$ converges to zero uniformly on the compact subsets of $\mathbf{R}^3$. We set $u_n = u(\cdot + x_n)$ and $\Phi_n = \Phi(\cdot + x_n)$. Then, from (5.36) and step 1, the sequences u_n and Φ_n are bounded, respectively, in $L^\infty(\mathbf{R}^3) \cap H^1_{\mathrm{unif}}(\mathbf{R}^3)$ and in $L^{3,\infty}_{\mathrm{unif}}(\mathbf{R}^3)$.

Thus, extracting subsequences if necessary, we may assume that they converge, respectively, to $\bar{u} \in L^\infty(\mathbf{R}^3) \cap H^1_{\mathrm{unif}}(\mathbf{R}^3)$ and $\bar{\Phi} \in L^{3,\infty}_{\mathrm{unif}}(\mathbf{R}^3)$.

Besides, we denote by $[x_n]$ the point of $\mathbf{Z}^3$ such that $x_n \in [x_n] + \Gamma_0$ and by $k_n = x_n - [x_n] \in \Gamma_0$. Then, up to a subsequence, k_n converges to some $\bar{k} \in \bar{\Gamma}_0$ and $\sum_{k \in \mathbf{Z}^3} \delta_k(\cdot + k_n)$ converges to $\sum_{k \in \mathbf{Z}^3} \delta_k(\cdot + \bar{k})$ in the sense of distributions. Thus, passing to the limit in (5.35), as n goes to infinity, we obtain that $\Phi_0 = \bar{\Phi}(\cdot - \bar{k})$ is a solution to

$$-\triangle \Phi_0 = 4\pi \sum_{k \in \mathbf{Z}^3} \delta_k \quad \text{in } \mathcal{D}'(\mathbf{R}^3), \tag{5.43}$$

belonging to $L^{3,\infty}_{\mathrm{unif}}(\mathbf{R}^3)$, and thus, in particular, to $L^1_{\mathrm{unif}}(\mathbf{R}^3)$. We now reach a contradiction.

Indeed, let $\xi \in \mathcal{D}(\mathbf{R}^3)$ be a cut-off function such that $0 \le \xi \le 1$, $supp\, \xi \subset B_2$, $\xi \equiv 1$ on B_1 and $|\triangle \xi| \le C$, for some positive constant C. Let $\xi_R = \xi\left(\frac{\cdot}{R}\right)$. We apply (5.43) to ξ_R to write

$$\langle -\triangle \Phi_0; \xi_R \rangle_{\mathcal{D}'(\mathbf{R}^3) \times \mathcal{D}(\mathbf{R}^3)} = 4\pi \left\langle \sum_{k \in \mathbf{Z}^3} \delta_k; \xi_R \right\rangle_{\mathcal{D}'(\mathbf{R}^3) \times \mathcal{D}(\mathbf{R}^3)},$$

or, integrating by parts,

$$-\frac{1}{R^2} \int_{R \le |x| \le 2R} \Phi_0 \, \triangle \xi\left(\frac{\cdot}{R}\right) = \sum_{k \in \mathbf{Z}^3 \cap B_{2R}} \xi_R(k). \tag{5.44}$$

We first bound the left-hand side of the above equality as follows:

$$\frac{1}{R^2} \int_{R \le |x| \le 2R} |\Phi_0| \left|\triangle \xi\left(\frac{\cdot}{R}\right)\right| \le \frac{C}{R^2} \int_{R \le |x| \le 2R} |\Phi_0|$$

$$\le \frac{C R^3}{R^2} \sup_{x \in \mathbf{R}^3} \int_{x + B_1} |\Phi_0|$$

$$\le C R.$$

On the other hand,

$$\frac{1}{R^3} \sum_{k \in \mathbf{Z}^3 \cap B_{2R}} \xi_R(k) = \frac{1}{R^3} \sum_{k \in \mathbf{Z}^3 \cap B_{2R}} \xi\left(\frac{k}{R}\right)$$

converges to $\int_{\mathbf{R}^3} \xi$, as R goes to infinity. Thus, the right-hand side of (5.44) is of the order of R^3 as R goes to infinity. Therefore, it cannot be of order R. We reach the desired contradiction.

We now turn to the proof of the uniqueness part of Lemma 5.5.

Step 3: uniqueness. The beginning of the argument is the same as in the proof of Lemma 5.3. We take two solutions u and v to (5.35) in the set (5.36), and if we define $w = u - v$, we obtain, as in (5.15),

$$\langle L(w\xi), (w\xi)\rangle + \nu \int_{\mathbf{R}^3} w^2 \xi^2$$
$$\leq \int_{\mathbf{R}^3} w^2 |\nabla \xi|^2 + \int_{\mathbf{R}^3} \frac{\phi - \psi}{2}(u^2 - v^2)\xi^2,$$

hence, and as in (5.16),

$$\langle L(w\xi), (w\xi)\rangle + \nu \int_{\mathbf{R}^3} w^2 \xi^2 + \tfrac{1}{2} \int_{\mathbf{R}^3} |\nabla((\phi - \psi)\xi)|^2$$
$$\leq \int_{\mathbf{R}^3} w^2 |\nabla \xi|^2 + \tfrac{1}{2} \int_{\mathbf{R}^3} (\phi - \psi)^2 |\nabla \xi|^2. \qquad (5.45)$$

At this point, with the same choice for ξ as given by (5.19), namely

$$\xi(x) = \frac{1}{(1 + |x|^2)^{m/2}},$$

for some exponent $m = \tfrac{1}{2} + \varepsilon$, $\varepsilon > 0$, we are going to show that

$$\int_{\mathbf{R}^3} w^2 |\nabla \xi|^2 < \infty, \qquad (5.46)$$

$$\int_{\mathbf{R}^3} (\phi - \psi)^2 |\nabla \xi|^2 < \infty, \qquad (5.47)$$

$$\int_{\mathbf{R}^3} (\phi - \psi)^2 |\Delta(\xi^2)| < \infty, \qquad (5.48)$$

and

$$\int_{\mathbf{R}^3} |\nabla(\phi - \psi)|^2 \xi^2 < \infty. \qquad (5.49)$$

We first prove (5.47). For this purpose, we remark that, from the expression (5.19) of ξ,

$$|\nabla \xi|^2 \leq \frac{C}{(1 + |x|^2)^{m+1}} \qquad \text{on } \mathbf{R}^3$$
$$\leq \frac{C}{(1 + |k|^2)^{m+1}} \qquad \text{on } k + \Gamma_0,$$

where C denotes various positive constants that are independent of x and $k \in \mathbf{Z}^3$. Hence, we may write, because of (5.36),

$$\int_{\mathbf{R}^3} (\phi - \psi)^2 |\nabla \xi|^2 \le C \sum_{k \in \mathbf{Z}^3} \frac{1}{(1 + |k|^2)^{m+1}} \int_{k+\Gamma_0} (\phi - \psi)^2$$

$$\le C \sup_{k \in \mathbf{Z}^3} (\|\phi\|_{L^2(k+\Gamma_0)}^2 + \|\psi\|_{L^2(k+\Gamma_0)}^2) \sum_{k \in \mathbf{Z}^3} \frac{1}{(1 + |k|^2)^{m+1}}.$$

The right-hand side of the above inequality is bounded since $2m + 2 > 3$ by assumption. A similar argument yields (5.48), so that (5.49) follows as in the proof of Lemma 5.3 and we compare with (5.45) to deduce

$$\int_{\mathbf{R}^3} w^2 \xi^2 < \infty. \tag{5.50}$$

We now want to prove that

$$\int_{\mathbf{R}^3} |\nabla w|^2 \xi^2 < \infty. \tag{5.51}$$

For this purpose, we set $a = \frac{1}{2}(u^{4/3} + v^{4/3}) - \frac{\phi + \psi}{2}$ and we bound from below $\langle L(w\xi), (w\xi) \rangle$ in (5.45) by

$$\int_{\mathbf{R}^3} |\nabla(w\xi)|^2 - \int_{\mathbf{R}^3} |a| w^2 \xi^2.$$

The same estimate (5.23) in the proof of Lemma 5.3 was based upon the assumption that $a \in L^\infty$, which no longer holds here. However, because of (5.36) and (5.41), a belongs to $L_{\mathrm{unif}}^{3,\infty}(\mathbf{R}^3)$, and thus, in particular, satisfies

$$\sup_{k \in \mathbf{Z}^3} \|a\|_{L^p(k+\Gamma_0)} < \infty \qquad \text{for all } 1 \le p < 3. \tag{5.52}$$

Then, by virtue of (5.46) and (5.47), there exists a positive constant C such that

$$\int_{\mathbf{R}^3} |\nabla(w\xi)|^2 \le C + \int_{\mathbf{R}^3} |a| w^2 \xi^2. \tag{5.53}$$

Now let K be a positive constant (that we shall send to $+\infty$ later on). We may write that

$$\int_{\mathbf{R}^3} |a| w^2 \xi^2 \le \int_{|a| \le K} |a| w^2 \xi^2 + \int_{|a| \ge K} |a| w^2 \xi^2$$

$$\le K \int_{\mathbf{R}^3} w^2 \xi^2 + \int_{|a| \ge K} |a| w^2 \xi^2$$

$$\le C K + \int_{|a| \ge K} |a| w^2 \xi^2, \tag{5.54}$$

because of (5.50). Besides, using Hölder's inequalities, we have

$$\int_{|a|\geq K} |a|w^2\xi^2 = \sum_{k\in \mathbf{Z}^3} \int_{(k+\Gamma_0)\cap(|a|\geq K)} |a|w^2\xi^2$$

$$\leq \sum_{k\in \mathbf{Z}^3} \left(\int_{(k+\Gamma_0)\cap(|a|\geq K)} |a|^{3/2} \right)^{2/3} \left(\int_{k+\Gamma_0} w^6\xi^6 \right)^{1/3} \quad (5.55)$$

On the one hand, by Sobolev's embeddings, we obtain

$$\left(\int_{k+\Gamma_0} w^6\xi^6 \right)^{1/3} \leq C \left(\int_{k+\Gamma_0} w^2\xi^2 + |\nabla(w\,\xi)|^2 \right), \quad (5.56)$$

where the constant C in the above inequality is independent of k.

On the other hand, since a is in $L^2_{\text{unif}}(\mathbf{R}^3) \cap L^{3,\infty}_{\text{unif}}(\mathbf{R}^3)$, it is clear, by using (5.38), that

$$\int_{(k+\Gamma_0)\cap(|a|\geq K)} |a|^{3/2} \leq \left(\int_{k+\Gamma_0} |a|^2 \right)^{3/4} meas\{x \in k + \Gamma_0/|a| \geq K\}^{1/4}$$

$$\leq C\, K^{-\frac{3}{4}}, \quad (5.57)$$

for some positive constant C that is independent of k. Next, collecting (5.56) and (5.57), we obtain from (5.55) that

$$\int_{|a|\geq K} |a|w^2\xi^2 \leq C\, K^{-2} \left(1 + \int_{\mathbf{R}^3} |\nabla(w\,\xi)|^2 \right), \quad (5.58)$$

because of (5.50). Then, we use (5.58) in (5.54), and then in (5.53), and we choose K large enough to ensure that

$$\int_{\mathbf{R}^3} |\nabla(w\xi)|^2 < \infty. \quad (5.59)$$

Correspondingly, (5.51) follows.

Next, we write

$$\frac{-\Delta w}{u+v} = F_1 + F_2, \quad (5.60)$$

with

$$F_1 = \frac{\phi+\psi}{2(u+v)}w \quad (5.61)$$

and

$$F_2 = \frac{\phi-\psi}{2} - \frac{u^{7/3} - v^{7/3}}{u+v}. \quad (5.62)$$

Bearing in mind that our objective is to prove that

$$\int_{\mathbf{R}^3} (\phi - \psi)^2 \xi^2 < \infty, \tag{5.63}$$

we now successively show that

$$\int_{\mathbf{R}^3} F_1^2 \xi^2 < \infty \tag{5.64}$$

and

$$\int_{\mathbf{R}^3} \left| \frac{\Delta w}{u + v} \right|^2 \xi^2 < \infty, \tag{5.65}$$

which will yield

$$\int_{\mathbf{R}^3} F_2^2 \xi^2 < \infty. \tag{5.66}$$

Then, (5.63) follows in a straightforward way, since $\inf_{\mathbf{R}^3}(u + v) > 0$ and $u, v \in L^\infty$.

First, (5.64) follows from the following observations:

$$\begin{aligned}
\int_{\mathbf{R}^3} F_1^2 \xi^2 &= \int_{\mathbf{R}^3} \left(\frac{\phi + \psi}{2(u + v)} w \right)^2 \xi^2 \\
&\leq C \int_{\mathbf{R}^3} \left(|\phi|^2 + |\psi|^2 \right) w^2 \xi^2 \qquad \text{since } \inf_{\mathbf{R}^3}(u + v) > 0, \\
&\leq C \sum_{k \in \mathbf{Z}^3} \int_{k + \Gamma_0} \left(|\phi|^2 + |\psi|^2 \right) w^2 \xi^2.
\end{aligned} \tag{5.67}$$

Next, we argue as follows. Since ϕ and ψ are in $L_{\text{unif}}^{3, \infty}(\mathbf{R}^3)$, $|\phi|^2$ and $|\psi|^2$ belong to $L_{\text{unif}}^{\frac{3}{2}, \infty}(\mathbf{R}^3)$. Next, we recall that, for every $k \in \mathbf{Z}^3$, the dual space of $L^{\frac{3}{2}, \infty}(k + \Gamma_0)$ is the Lorentz space $L^{3, 1}(k + \Gamma_0)$ (see [9], where the Lorentz spaces are introduced together with some properties). Thus, using interpolation inequalities, we may write

$$\begin{aligned}
\int_{k + \Gamma_0} \left(|\phi|^2 + |\psi|^2 \right) w^2 \xi^2 &\leq \left(\|\phi\|_{L^{3, \infty}(k + \Gamma_0)}^2 + \|\psi\|_{L^{3, \infty}(k + \Gamma_0)}^2 \right) \|w\xi\|_{L^{6, 2}(k + \Gamma_0)}^2 \\
&\leq C \|w\xi\|_{H^1(k + \Gamma_0)}^2,
\end{aligned}$$

for some positive constant C that is independent of k, since $L^{6, 2}(k + \Gamma_0)$ is continuously embedded in $H^1(k + \Gamma_0)$ (see [9]). We insert this estimate in (5.67), and then use (5.50) and (5.51) in order to obtain (5.64).

In order to prove (5.65), we proceed as follows:

$$\int_{\mathbf{R}^3} \left| \frac{\Delta w}{u+v} \right|^2 \xi^2 \leq C \int_{\mathbf{R}^3} \frac{|\Delta w|^2}{u+v} \xi^2$$

$$= C \int_{\mathbf{R}^3} \frac{\Delta w}{u+v} \Delta w \, \xi^2$$

$$= -C \int_{\mathbf{R}^3} (F_1 + F_2) \, \Delta w \, \xi^2.$$

Next, we have

$$\left| \int_{\mathbf{R}^3} F_1 \, \Delta w \, \xi^2 \right| \leq \left(\int_{\mathbf{R}^3} F_1^2 \, \xi^2 \right)^{1/2} \left(\int_{\mathbf{R}^3} |\Delta w|^2 \xi^2 \right)^{1/2} \tag{5.68}$$

and

$$\int_{\mathbf{R}^3} F_2 \, \Delta w \, \xi^2 = -\int_{\mathbf{R}^3} \nabla F_2 . \nabla w \, \xi^2 - 2 \int_{\mathbf{R}^3} \xi F_2 \, \nabla w . \nabla \xi$$

$$= -\int_{\mathbf{R}^3} \nabla F_2 . \nabla w \, \xi^2 + 2 \int_{\mathbf{R}^3} \nabla F_2 . \nabla \xi \, \xi \, w$$

$$+ \int_{\mathbf{R}^3} F_2 \, w \, \Delta(\xi^2). \tag{5.69}$$

On the one hand, we bound from above the first two terms of the right-hand side of (5.69), using the Cauchy–Schwarz inequality, by

$$\left(\int_{\mathbf{R}^3} |\nabla F_2|^2 \xi^2 \right)^{1/2} \left[\left(\int_{\mathbf{R}^3} |\nabla w|^2 \xi^2 \right)^{1/2} + \left(\int_{\mathbf{R}^3} w^2 |\nabla \xi|^2 \right)^{1/2} \right]. \tag{5.70}$$

Then, we first remark that, from expression (5.62) for F_2,

$$\int_{\mathbf{R}^3} |\nabla F_2|^2 \xi^2 \leq C \left(\int_{\mathbf{R}^3} |\nabla(\phi - \psi)|^2 \xi^2 + \int_{\mathbf{R}^3} |\nabla w|^2 \xi^2 \right.$$

$$\left. + \int_{\mathbf{R}^3} w^2 (|\nabla u|^2 + |\nabla v|^2) \xi^2 \right)$$

$$\leq C + C \int_{\mathbf{R}^3} w^2 (|\nabla u|^2 + |\nabla v|^2) \xi^2,$$

in view of (5.49) and (5.51). Besides, we make use of the bounds we have obtained in (5.42) for ∇u in $L_{\text{unif}}^p(\mathbf{R}^3)$, for all $1 \leq p < +\infty$ and of Sobolev's and Hölder's inequalities to write

$$\int_{\mathbf{R}^3} (|\nabla u|^2 + |\nabla v|^2) \, w^2 \xi^2$$

$$\leq \sum_{k \in \mathbf{Z}^3} \int_{k+\Gamma_0} (|\nabla u|^2 + |\nabla v|^2) \, w^2 \xi^2$$

$$\leq \sum_{k \in \mathbf{Z}^3} \|w\,\xi\|^2_{L^6(k+\Gamma_0)} \, (\|\nabla u\|^2_{L^3(k+\Gamma_0)} + \|\nabla v\|^2_{L^3(k+\Gamma_0)})$$

$$\leq C \int_{\mathbf{R}^3} |\nabla w \xi|^2,$$

for some constant C that is independent of k. The right-hand side of the above inequality is finite because of (5.59). Finally, we have

$$\int_{\mathbf{R}^3} |\nabla F_2|^2 \xi^2 < \infty.$$

In conclusion, the first two terms of (5.69) are finite.

On the other hand,

$$\left| \int_{\mathbf{R}^3} F_2 \, w \, \Delta(\xi^2) \right| \leq \left(\int_{\mathbf{R}^3} w^2 \xi^2 \right)^{1/2} \left(\int_{\mathbf{R}^3} F_2^2 \, \frac{|\Delta(\xi^2)|^2}{\xi^2} \right)^{1/2} \tag{5.71}$$

where

$$\frac{|\Delta(\xi^2)|^2}{\xi^2} \leq C \, \frac{1}{(1+|x|^2)^{m+2}}.$$

Thus, using (5.62),

$$\int_{\mathbf{R}^3} F_2^2 \, \frac{|\Delta(\xi^2)|^2}{\xi^2}$$

$$\leq C \int_{\mathbf{R}^3} F_2^2(x) \frac{1}{(1+|x|^2)^{m+2}} \, dx$$

$$\leq C \left(\int_{\mathbf{R}^3} \frac{w^2(x)}{(1+|x|^2)^{m+2}} \, dx + \int_{\mathbf{R}^3} \frac{|\phi|^2 + |\psi|^2}{(1+|x|^2)^{m+2}} \, dx \right)$$

$$\leq C\|w\|^2_{L^\infty} \int_{\mathbf{R}^3} \frac{dx}{(1+|x|^2)^{m+2}}$$

$$+ C \sum_{k \in \mathbf{Z}^3} \frac{1}{(1+|k|^2)^{m+2}} \sup_{k \in \mathbf{Z}^3} \int_{k+\Gamma_0} (|\phi|^2 + |\psi|^2).$$

The right-hand side of the above inequality is finite, since $2(m+2) > 3$ and because of (5.37). Collecting with (5.50), we obtain from (5.71)

$$\left| \int_{\mathbf{R}^3} F_2 \, w \, \Delta(\xi^2) \right| < \infty.$$

Now gathering with (5.68) and using the L^∞ bound on $u + v$, we thus have

$$\int_{\mathbf{R}^3} |\Delta w|^2 \xi^2 \leq C \int_{\mathbf{R}^3} \left| \frac{\Delta w}{u+v} \right|^2 \xi^2$$

$$\leq C + C \left(\int_{\mathbf{R}^3} |\Delta w|^2 \xi^2 \right)^{1/2}.$$

From which we deduce that

$$\int_{\mathbf{R}^3} |\Delta w|^2 \xi^2 < \infty$$

and thus (5.65) holds.

Now that (5.63) is proved, we have

$$\begin{cases} \displaystyle\int_{\mathbf{R}^3} w^2 \xi^2 < \infty, \\[2ex] \displaystyle\int_{\mathbf{R}^3} |\phi - \psi|^2 \xi^2 < \infty, \end{cases} \tag{5.72}$$

as in (5.31), and the rest of the argument is the same. We therefore obtain the uniqueness and, thus, the periodicity of u and Φ. $\diamond$

5.3.3 Corollaries and remarks

5.3.3.1 *The Yukawa case* In the case of a Yukawa potential, we are able to show an analogous result to those of Lemmas 5.3 and 5.5. Indeed, let us consider $(u; \Phi)$, a solution to

$$\begin{cases} -\Delta u + u^{7/3} - (\Phi - \theta_\infty)u = 0, \\[1ex] (-\Delta + a^2)\Phi = 4\pi \, (m - u^2), \end{cases} \tag{5.73}$$

with $a > 0$ (the exponent involved in the Yukawa potential), $\theta_\infty \geq 0$, where $u \in L^\infty(\mathbf{R}^3)$, $\inf_{\mathbf{R}^3} u > 0$, $\Phi \in L^1_{\mathrm{unif}}(\mathbf{R}^3)$, and where m satisfies the same assumption as in Remark 5.6. Then, such a couple $(u; \Phi)$ is unique.

Before giving the proof of this claim, let us make the following comment. In the Yukawa case, we have to impose the condition that $\inf_{\mathbf{R}^3} u > 0$. Indeed, if we skip this assumption, we obtain a solution to (5.73) by taking $u = 0$ and $\Phi = \frac{\exp(-a \, |x|)}{|x|} \star m$, that indeed belongs to $L^1_{\mathrm{unif}}(\mathbf{R}^3)$ (and, in fact, to $L^{3,\infty}_{\mathrm{unif}}(\mathbf{R}^3)$).

Let us now turn to the proof of the uniqueness of such a solution. For this purpose, we argue as in the above proof, consider two solutions $(u; \Phi)$ and $(v; \Psi)$, and remark first that (5.15) still holds in this setting. Next, since

$$(-\Delta + a^2)(\Phi - \Psi) = -4\pi \, (u^2 - v^2),$$

we obtain, instead of (5.16),

$$\langle L(w\xi), (w\xi) \rangle + (\nu + \theta_\infty) \int w^2 \xi^2 + \tfrac{1}{2} \int |\nabla(\Phi - \Psi)\xi|^2 + a^2 \int (\Phi - \Psi)^2 \xi^2$$
$$\leq \int w^2 |\nabla \xi|^2 + \tfrac{1}{2} \int (\Phi - \Psi)^2 |\nabla \xi|^2. \tag{5.74}$$

Therefore, we obtain (5.21) as above, i.e.

$$\int w^2 \xi^2 < \infty,$$

while (5.30), namely

$$\int (\Phi - \Psi)^2 \xi^2 < \infty,$$

is obtained directly from (5.74), and thus in a much more straightforward way than in the Coulomb case. Hence, (5.31) holds and the rest of the argument is the same.

Therefore, this provides another proof of the periodicity of the solution in the $(Y - m)$ and the $(Y - \delta)$ programs. Let us recall that the proof we have presented in Chapter 4 is based upon the implicit function theorem and the linearization of the operator.

Another comment is the following.

5.3.3.2 *Comparison with the TF case*   It is worth noticing that, in the Thomas–Fermi case, the analogous result holds, its proof being far less complicated.

Indeed, in the Thomas–Fermi case (see Lieb [33]), the system to consider is, up to some irrelevant constants,

$$\begin{cases} u^2 = \Phi^{3/2}, \qquad \Phi \geq 0, \\ -\Delta\Phi = m - u^2. \end{cases} \tag{5.75}$$

This yields

$$\begin{cases} -\Delta\Phi + \Phi^{3/2} = m & \text{on } \mathbf{R}^3, \\ \\ \Phi \geq 0. \end{cases}$$

In [12] (and also [7]), Brézis proved that, for every measure m, there exists a unique solution Φ in $L^{3/2}_{\text{loc}}(\mathbf{R}^3)$ to

$$-\Delta\Phi + |\Phi|^{1/2}\Phi = m \qquad \text{in } \mathcal{D}'(\mathbf{R}^3).$$

In addition, if $m \geq 0$ in the sense of measures, then $\Phi \geq 0$ and, of course, if m is periodic, Φ shares the same periodicity.

Note that this argument holds at the same time in the smeared out and the point nuclei cases.

5.4 Convergence of the density

The purpose of this section is to prove, using the uniqueness result of Section 5.3, that the sequence of densities ρ_Λ converges in a strong sense, that will be made precise below, to the periodic density ρ_{per} minimizing the periodic energy obtained by the thermodynamic limit in Chapter 3.

First of all, we recall that, in view of the results of Section 3.2 of Chapter 3 and from standard elliptic regularity results, the electronic density ρ_Λ and the effective potential Φ_Λ are bounded uniformly with respect to Λ, respectively, in $L^\infty(\mathbf{R}^3) \cap H^1_{\text{unif}}(\mathbf{R}^3)$ and in $L^p_{\text{unif}}(\mathbf{R}^3)$ for all $1 \leq p < 3$ (see Propositions 3.10 and 3.12). In the case of the smeared out nuclei ($(Cb-m)$ program), Φ_Λ also is bounded in $L^\infty(\mathbf{R}^3)$ uniformly with respect to Λ (see Proposition 3.5).

It follows that, in both the $(Cb-m)$ and the $(Cb-\delta)$ programs, we may pass locally to the limit in the system

$$\begin{cases} -\Delta u_\Lambda + \frac{5}{3}u_\Lambda^{7/3} - (\Phi_\Lambda - \theta_\Lambda)u_\Lambda = 0, \\ \\ -\Delta\Phi_\Lambda = 4\pi(m_\Lambda - u_\Lambda^2), \end{cases}$$

where the measure m_Λ is either

$$m_\Lambda = \sum_{y \in \Lambda} \delta_y$$

or

$$m_\Lambda = \sum_{y \in \Lambda} m(\cdot - y),$$

according to the problem we are considering.

Denoting by u_∞ and Φ_∞ the limits of u_Λ and $\Phi_\Lambda - \theta_\Lambda$, we obtain

$$\begin{cases} -\Delta u_\infty + \frac{5}{3}u_\infty^{7/3} - \Phi_\infty u_\infty = 0, \\ \\ -\Delta\Phi_\infty = 4\pi(m_\infty - u_\infty^2), \end{cases} \tag{5.76}$$

where the periodic measure m_∞ is, respectively, $\sum_{y \in \mathbf{Z}^3} \delta_y$ or $\sum_{y \in \mathbf{Z}^3} m(\cdot - y)$.

From the uniform bounds we have for u_Λ and Φ_Λ (together with (5.42)), we know that $u_\infty \in H^1_{\text{unif}}(\mathbf{R}^3) \cap L^\infty(\mathbf{R}^3)$, $u_\infty \geq 0$ and Φ_∞ is in $L^p_{\text{unif}}(\mathbf{R}^3)$ for all $1 \leq p < 3$ ($(Cb - \delta)$ case) or in $L^\infty(\mathbf{R}^3)$ ($(Cb - m)$ case). Consequently, we may apply the uniqueness lemmata of Section 5.3 to claim that u_∞ and Φ_∞ are periodic functions, whose periodic cell is the cell Γ_0 (because it is the one of m_∞), since for each unit vector e_i, $1 \leq i \leq 3$, defining Γ_0, $(u_\infty(\cdot \pm e_i); \Phi_\infty(\cdot \pm e_i))$ remains a solution to (5.76).

Now, Φ_∞ being periodic and satisfying

$$-\triangle \Phi_\infty = 4\pi\,(m_\infty - u_\infty^2) \qquad \text{on } \mathbf{R}^3, \tag{5.77}$$

this implies that

$$\int_{\Gamma_0} (m_\infty - u_\infty^2) = 0,$$

and thus

$$\int_{\Gamma_0} u_\infty^2 = 1. \tag{5.78}$$

Note, indeed, that if one integrates (5.77) over Γ_0, one finds that

$$\int_{\Gamma_0} -\triangle \Phi_\infty = \int_{\partial \Gamma_0} \nabla \Phi_\infty . n = 0,$$

which comes from the periodicity of Φ_∞ together with its H^1 regularity (at least) in the vicinity of $\partial \Gamma_0$.

At this stage, we may claim that

$$\Phi_\infty(x) = G(x) - \int_{\Gamma_0} G(x - y) u_\infty^2(y)\,dy + \phi_0 \tag{5.79}$$

(in the $(Cb - \delta)$ case), or

$$\Phi_\infty(x) = \int_{\Gamma_0} G(x - y)(m_\infty - u_\infty^2)(y)\,dy + \phi_0 \tag{5.80}$$

(in the $(Cb - m)$ case), for some constant ϕ_0 that is independent of the sequence Λ since, actually, $\phi_0 = \int_{\Gamma_0} \Phi_\infty$ and Φ_∞ is unique. Indeed, it is easily seen that the function arising on the right-hand side of (5.79) or (5.80) is another periodic solution to (5.77).

Gathering (5.76), (5.78)–(5.80), and the periodicity of u_∞ and Φ_∞, we see that u_∞ is a critical point of the energy functional $\rho \mapsto E^{\text{TFW}}_{\text{per}}(\rho)$ under the constraint $\int_{\Gamma_0} \rho = 1$. However, since $E^{\text{TFW}}_{\text{per}}$ is strictly convex, $\rho_\infty = u_\infty^2$ has to be the unique minimizer of $I^{\text{TFW}}_{\text{per}}(1)$, say ρ_{per}. Thus, comparing finally with the Euler–Lagrange equation satisfied by u_{per} (5.5), we conclude that $u_\infty = u_{\text{per}}$

and $\Phi_\infty = \Phi_{per} - \theta_{per}$. Returning to (5.79) and (5.80), we deduce at last that $\phi_0 = -\theta_{per}$.

Arguing as in the Yukawa case (see Chapter 4), we thus have the following:

Theorem 5.7 (Thermodynamic limit of the density) *The sequence $u_\Lambda = \sqrt{\rho_\Lambda}$ converges to the unique minimizer u_{per} of I_{per}^{TFW} strongly in $H^1_{loc}(\mathbf{R}^3) \cap L^p_{loc}(\mathbf{R}^3)$ for all $1 \le p < +\infty$ and uniformly on the compact subsets of $\mathbf{R}^3$.*

Moreover, $\Phi_\Lambda - \theta_\Lambda$ converges to $\Phi_{per} - \theta_{per}$ strongly in $L^p_{loc}(\mathbf{R}^3)$, for all $1 \le p < 3$ ((Cb − δ) program) or $1 \le p < \infty$ ((Cb − m) program).

Remark 5.8 (1) We shall even prove in the following section that u_Λ converges to u_{per} uniformly on any sequence of interior domains $\Gamma'(\Lambda)$ defined in Chapter 1. And this strong result of convergence will allow us to give another proof of the existence of the thermodynamic limit for the energy per unit volume without using the $\sim$-transform trick (see Chapter 6).

(2) In the Yukawa case, we can make almost the same argument: the bounds on u_Λ, θ_Λ, and Φ_Λ allow us to pass to the limit, and we also obtain some solutions u_∞, Φ_∞, and θ_∞ of the 'periodic system':

$$\begin{cases} -\Delta u_\infty + \frac{5}{3} u_\infty^{7/3} - (\Phi_\infty - \theta_\infty)\, u_\infty = 0, \\ -\Delta \Phi_\infty + a^2\, \Phi_\infty = 4\pi\, (m_\infty - u_\infty^2). \end{cases}$$

Next, by an argument quite similar to the proof of Lemma 4.14 in Chapter 4, Section 4.3, we may prove that

$$\inf_{\mathbf{R}^3} u_\infty > 0$$

(using variational arguments for this purpose; see the details in Chapter 4). Thus, we may apply the uniqueness result of Section 5.3.2 above. Therefore, u_∞ and Φ_∞ are periodic. However, the key point is to check that u_∞ has the right mass on Γ_0, namely

$$\int_{\Gamma_0} u_\infty^2 = \mu(1),$$

using the notation of Chapter 2. This is, of course, somewhat delicate to assert without using variational arguments, since $\mu(1)$ is determined in Chapter 2 through a variational characterization!

(3) The result still holds if we replace the exponent $\frac{5}{3}$ in the energy functional by some other exponent $2 \ge p > \frac{3}{2}$.

(4) If the function defining the nucleus is not non-negative, we cannot extend this theorem, since we do not have any bounds on the densities in order to obtain the convergence.

5.5   Uniform convergence on interior domains

We first recall the definition of a sequence of interior domains given in Chapter 1, Section 1.5.

Definition 1   Let Λ be a Van Hove sequence. We call a sequence of interior domains $\Gamma'(\Lambda)$ and we use the notation $\Gamma'(\Lambda) \subset\subset \Gamma(\Lambda)$, any sequence of subsets $\Gamma'(\Lambda)$ of $\mathbf{R}^3$ going to infinity in the sense that, for any compact subset $K \subset \mathbf{R}^3$, we have, for Λ large enough, $K \subset \Gamma'(\Lambda)$, and satisfying the following three conditions:

$$(i) \ \ \Gamma'(\Lambda) \subset \Gamma(\Lambda);$$

$$(ii) \ \ \frac{vol(\Gamma'(\Lambda))}{|\Lambda|} \longrightarrow 1 \qquad \text{as} \ \ \Lambda \to \infty; \tag{5.81}$$

$$(iii) \ \ dist(\Gamma'(\Lambda); \partial\Gamma(\Lambda)) \longrightarrow +\infty \qquad \text{as} \ \ \Lambda \to \infty. \tag{5.82}$$

It will be convenient for the sequel to introduce the following related definition:

Definition 1'   Let Λ be a Van Hove sequence. We use the notation $\Lambda' \subset\subset \Lambda$ for any sequence of subsets Λ' of $\mathbf{Z}^3$ going to infinity in the sense that, for any finite subset $A \subset \mathbf{Z}^3$, we have for Λ large enough, $A \subset \Lambda'$, and satisfying the following three conditions:

$$(i) \ \ \Lambda' \subset \Lambda;$$

$$(ii) \ \ \frac{|\Lambda'|}{|\Lambda|} \longrightarrow 1 \qquad \text{as} \ \ \Lambda \to \infty;$$

$$(iii) \ \ dist(\Lambda'; \partial\Gamma(\Lambda)) \longrightarrow +\infty \qquad \text{as} \ \ \Lambda \to \infty.$$

In particular, for any $\Lambda' \subset\subset \Lambda$, $\Gamma(\Lambda') = \cup_{y\in\Lambda'} y + \Gamma_0$ is a sequence of interior domains, according to Definition 1 above.

It is easy to build a sequence of interior domains for cubes, spheres, and so on. For example, if $\Gamma(\Lambda) = |\Lambda|^{1/3} \cdot \Gamma_0$ is the cube of volume $|\Lambda|$ centered at 0, it is clear that any subset $A(\Lambda)$ defined by

$$A(\Lambda) = \{x \in \Gamma(\Lambda) / \ d(x; \partial\Gamma(\Lambda)) \geq |\Lambda|^\alpha\}$$

satisfies the properties (5.81) and (5.82), for every $0 < \alpha < \frac{1}{3}$. However, the existence of the interior domains for general Van Hove sequences is a little more delicate to prove. We have made the choice to postpone its proof until the end of this section (see Lemma 5.11), and we now turn to the main result of this section. We keep from now on the notations of the Coulomb case even if the theorem also holds true in the smeared nuclei case.

Theorem 5.9 (Uniform convergence of the densities) *For any Van Hove sequence* $\Lambda \subset \mathbf{Z}^3$ *and for any sequence of interior domains, we have*

$$\|\rho_\Lambda - \rho_{\mathrm{per}}\|_{L^\infty(\Gamma'(\Lambda))} \longrightarrow 0 \tag{5.83}$$

and

$$\|\Phi_\Lambda - \theta_\Lambda - \Phi_{\mathrm{per}} + \theta_{\mathrm{per}}\|_{L^\infty(\Gamma'(\Lambda))} \longrightarrow 0 \tag{5.84}$$

as Λ *goes to infinity.*

Before giving the proof of this result, let us make some comments. First of all, this convergence result is likely to be the optimal result one may hope to obtain in this direction. Indeed, since ρ_Λ goes to zero as $|x|$ goes to infinity while ρ_{per} is a positive periodic function on $\mathbf{R}^3$, it is obvious that the uniform convergence cannot hold on the whole space.

Moreover, as a consequence of (5.83), we may precise the localization of the 'electron cloud' around the nuclei or the volume occupied by the molecule in the space. Indeed, let $\Lambda' \subset\subset \Lambda$ according to Definition 1': then, because of (5.81), we have that

$$\frac{1}{|\Lambda|} \int_{\Gamma(\Lambda')} |\rho_\Lambda - \rho_{\mathrm{per}}| \longrightarrow 0 \qquad \text{as } \Lambda \to \infty.$$

Thus,

$$\int_{\Gamma(\Lambda')} \rho_\Lambda = |\Lambda| + o(|\Lambda|),$$

since $\int_{\Gamma_0} \rho_{\mathrm{per}} = 1$ and $|\Lambda'| = |\Lambda| + o(|\Lambda|)$. It means that the main part of the electrons remains at the vicinity of the 'core' nuclei. By the way, let us remark that this is another strategy of proof of the compactness which need not appeal to bounds on the energy per unit volume.

Remark 5.10 It is important to note here that the proof of the convergences contained in Theorem 5.9 only makes use of the bounds on Φ_Λ and ρ_Λ, that we have deduced from the Euler–Lagrange equation of the minimization problems, and of the uniqueness result of Section 5.3. In particular, it is not based upon the existence of the thermodynamic limit for the energy per unit volume. This will allow us to appeal to this theorem in the next chapter in order to give another proof of this latter convergence.

Proof of Theorem 5.9 We shall only prove (5.83), thereby skipping the proof of (5.84), which follows exactly the same pattern.

Let Λ be a Van Hove sequence and $\Gamma'(\Lambda)$ be a sequence of interior domains. We assume by contradiction that (5.83) does not hold. Then, there exists $\varepsilon > 0$ and a sequence of points x_Λ in $\Gamma'(\Lambda)$ such that

$$|\rho_\Lambda(x_\Lambda) - \rho_{\text{per}}(x_\Lambda)| > \varepsilon, \qquad (5.85)$$

at least for a subsequence of (Λ) (still denoted by (Λ)).

Since, in particular, x_Λ is in $\Gamma(\Lambda)$, we may write that

$$x_\Lambda = [x_\Lambda] + k_\Lambda,$$

where $[x_\Lambda]$ belongs to Λ and k_Λ belongs to Γ_0. In particular, extracting a subsequence of (Λ), if necessary, we may assume that k_Λ converges to some $\bar{k}$ in $\bar\Gamma_0$. Thus, using the periodicity of ρ_{per}, we obtain from (5.85) that

$$|\rho_\Lambda(x_\Lambda) - \rho_{\text{per}}(\bar{k})| > \frac{\varepsilon}{2}, \qquad (5.86)$$

at least for Λ large enough.

We now aim to reach a contradiction with (5.86).

For this purpose, we recall from Propositions 3.10 and 3.12 (in Section 3.2 of Chapter 3) that $u_\Lambda = \sqrt{\rho_\Lambda}$ and $\Phi_\Lambda - \theta_\Lambda$ are bounded uniformly in Λ, respectively, in $L^\infty(\mathbf{R}^3) \cap H^1_{\text{unif}}(\mathbf{R}^3)$ and in $L^p_{\text{unif}}(\mathbf{R}^3)$ for all $1 \le p < 3$ (or even in $L^\infty(\mathbf{R}^3)$), in the case of smeared out nuclei (see Proposition 3.5 in Chapter 3). Therefore, extracting a subsequence if necessary and using Rellich's theorem, we deduce that the sequences $\bar{u}_\Lambda \equiv u_\Lambda(\cdot + x_\Lambda)$ and $\bar\Phi_\Lambda \equiv \Phi_\Lambda(\cdot + x_\Lambda) - \theta_\Lambda$ converge, respectively, to some functions $\bar{u} \ge 0$ in $L^\infty(\mathbf{R}^3)$ and $\bar\Phi$ in $L^p_{\text{unif}}(\mathbf{R}^3)$ for all $1 \le p < 3$.

We now turn to the system of equations satisfied by $\bar{u}_\Lambda$ and $\bar\Phi_\Lambda$; namely

$$\begin{cases} -\Delta\bar{u}_\Lambda + \frac{5}{3}\,\bar{u}_\Lambda^{7/3} - \bar\Phi_\Lambda\,\bar{u}_\Lambda = 0, \\[2mm] -\Delta\bar\Phi_\Lambda = 4\pi\,(\bar{m}_\Lambda - \bar{u}_\Lambda^2), \end{cases} \qquad (5.87)$$

where the measure $\bar{m}_\Lambda$ is either

$$\bar{m}_\Lambda = \sum_{y\in\Lambda} \delta_y(\cdot + x_\Lambda)$$

or

$$\bar{m}_\Lambda = \sum_{y\in\Lambda} m(\cdot + x_\Lambda - y)$$

according to the problem we are considering.

We admit for a while that the sequence $\bar{m}_\Lambda$ converges to the periodic measure $m_\infty(\cdot + \bar{k})$, where $m_\infty = \sum_{y\in\mathbf{Z}^3} m(\cdot - y)$ (with the convention that $m = \delta_0$ in the case of point nuclei). Then, passing to the limit in (5.87) when Λ goes to infinity, we obtain that $\bar{u}$ and $\bar\Phi$ are solutions to

$$\begin{cases} -\Delta\bar{u} + \frac{5}{3}\,\bar{u}^{7/3} - \bar\Phi\,\bar{u} = 0, \\[2mm] -\Delta\bar\Phi = 4\pi\,(m_\infty(\cdot + \bar{k}) - \bar{u}^2). \end{cases} \qquad (5.88)$$

Then, pointing out that the assumptions of the uniqueness Lemma 5.5 are fulfilled by $\bar{u}$ and $\bar\Phi$, we may identify $\bar{u}$ with $u_{\text{per}}(\cdot + \bar{k})$ (and $\bar\Phi$ with $\Phi_{\text{per}}(\cdot + \bar{k}) - \theta_{\text{per}}$).

We thus reach the desired contradiction with (5.86) by letting Λ go to infinity in (5.86) and by recalling that the convergence of $u_\Lambda(\cdot + x_\Lambda)$ to $u_{\text{per}}(\cdot + \bar{k})$ holds at each point of $\mathbf{R}^3$ (in particular, at zero).

It still remains to prove now that the sequence $\bar{m}_\Lambda$ converges to the periodic measure $m_\infty(\cdot + \bar{k})$. This is where the assumption that the sequence x_Λ lies in $\Gamma'(\Lambda)$ plays a role, together with the definition of the interior domains.

Indeed, it is obviously sufficient to check that, for any fixed compact subset K of $\mathbf{R}^3$,

$$\int_K m_\infty(\cdot + x_\Lambda) - m_\Lambda(\cdot + x_\Lambda) = 0,$$

for Λ large enough. The quantity appearing on the left-hand side of the above equality may conveniently be rewritten as

$$\sum_{y \in \mathbf{Z}^3 \setminus \Lambda} m(\cdot + x_\Lambda - y).$$

On the one hand, since $supp\, m \subset \Gamma_0$, the above sum cannot converge to zero on K provided that

$$(x_\Lambda + K) \cap \Gamma(\Lambda)^c \neq \emptyset.$$

On the other hand, because of the property (5.82) satisfied by $\Gamma'(\Lambda)$, it is clear that

$$x_\Lambda + K \subset \Gamma(\Lambda),$$

for Λ large enough, and we conclude. $\diamond$

As announced at the beginning of this section, let us now give the proof of the existence of a sequence of interior domains for any Van Hove sequence. More precisely, we now establish the following:

Lemma 5.11 (Existence of the interior domains) *Let $(\Lambda_n)_{n \geq 1}$ be a Van Hove sequence. Then, there exists a subsequence of (Λ_n) (still denoted by (Λ_n)) and a sequence of subsets $(\Lambda'_n)_{n \geq 1}$ of Λ_n, such that, for all finite subsets $A \subset \mathbf{Z}^3$, we have, for n large enough, $A \subset \Lambda'_n$, and enjoying in addition the following properties:*

(i) $\displaystyle \lim_{n \to +\infty} \frac{|\Lambda'_n|}{|\Lambda_n|} = 1;$

(ii) $\displaystyle \lim_{n \to +\infty} dist(\Lambda'_n; \partial\Gamma(\Lambda_n)) = +\infty.$

Proof of Lemma 5.11 For any fixed $h > 0$ and any subset $\Lambda \subset \mathbf{Z}^3$, we define

$$\Lambda^h = \{y \in \Lambda / dist(y; \partial\Gamma(\Lambda)) \leq h\}. \tag{5.89}$$

Next, we recall that, by the definition of a Van Hove sequence,

$$\lim_{n \to +\infty} \frac{|\Lambda_n^h|}{|\Lambda_n|} = 0. \tag{5.90}$$

In particular, there exists an integer $n_1 \geq 1$ such that

$$\forall n \geq n_1, \qquad \frac{|\Lambda_n^1|}{|\Lambda_n|} \leq 1.$$

We call $\varphi(1)$ the smallest integer n_1 satisfying this property and we set

$$\Lambda'_{\varphi(1)} = \Lambda_{\varphi(1)} \setminus \Lambda^1_{\varphi(1)}.$$

Now, using (5.90) once more, we claim that there exists a smallest integer $n_2 > n_1$ such that

$$\forall n \geq n_2, \qquad \frac{|\Lambda_n^2|}{|\Lambda_n|} \leq \frac{1}{2},$$

and we set, in the same manner,

$$\varphi(2) = n_2$$

and

$$\Lambda'_{\varphi(2)} = \Lambda_{\varphi(2)} \setminus \Lambda^2_{\varphi(2)}.$$

Finally, repeating the same argument for every integer $n \geq 1$, we build a one-to-one function $\varphi : \mathbf{N} \to \mathbf{N}$ and a sequence of subsets $\Lambda'_{\varphi(n)}$ of $\Lambda_{\varphi(n)}$ such that

$$\Lambda'_{\varphi(n)} = \Lambda_{\varphi(n)} \setminus \Lambda^n_{\varphi(n)}, \tag{5.91}$$

and

$$\forall n \geq 1, \qquad \frac{|\Lambda^n_{\varphi(n)}|}{|\Lambda_{\varphi(n)}|} \leq \frac{1}{n}. \tag{5.92}$$

From (5.91) and (5.92) it is easily seen that

$$1 - \frac{1}{n} \leq \frac{|\Lambda'_{\varphi(n)}|}{|\Lambda_{\varphi(n)}|} \leq 1.$$

Hence, (i) follows.

Besides, from (5.91) and (5.89), we have that

$$dist(\Lambda'_{\varphi(n)}; \partial\Gamma(\Lambda_{\varphi(n)})) \geq n,$$

from which (ii) follows. $\diamond$

In the following chapter, we are going to use this convergence in a strong sense to build another proof of the existence of the thermodynamic limit of the energy, more general than the one we presented in Chapter 3. In addition, we shall see how to extend the uniqueness result of Section 5.3 (and consequently the property of uniform convergence shown in this Section) to deal with some 'difficult' situations where the approach we have chosen so far fails.

6

CONVERGENCE OF THE ENERGY VIA THE
CONVERGENCE OF THE DENSITY

6.1 Introduction

In Chapters 2–5, we have shown that, both for Yukawa interaction and Coulomb interaction, there exists a thermodynamic limit for the energy per unit volume and for the electronic density, when the nuclei are located on a periodic cubic lattice. In addition, the density obtained in this limit is periodic, of the same period as the lattice. However, in the case of Coulomb forces, we have used some additional assumptions on the shape of the cell and of the nuclei, and also on the behaviour of the Van Hove sequence (see the statement of Theorem 3.1). The first goal of this chapter is to show that using a different strategy of proof allows us to get rid of these assumptions (see Section 6.2).

This strategy, which we call 'Convergence of the energy via uniform convergence of the density', consists of proving *first* that the electronic density converges to the 'good' density, and *then* to deduce from the convergence of the density that the energy per unit volume also converges. This will be detailed in Section 6.2 below. Unlike the direct strategy that we used in Chapters 2–5, the strategy we shall develop in the present chapter turns out to be convenient in very general situations. At the present time, we even believe that it is likely to be the strategy to be used in the optimized geometry case (see Section 1.6, the comments at the end of the present section, and [17]).

To briefly sketch our strategy of proof, let us say the proof falls into three steps. The first step consists of obtaining *a priori* estimates on the energy I_Λ, on the density ρ_Λ, on the Lagrange multiplier θ_Λ, and on the effective potential Φ_Λ. In a second step, we locally pass to the limit in the Euler–Lagrange equations; in view of a uniqueness result holding on the limit equation, we next identify the solution $(u_\infty; \Phi_\infty)$ obtained in the limit with *the* solution to the equation. Some qualitative properties (such as periodicity, for instance) of $(u_\infty; \Phi_\infty)$ follow. Roughly speaking, let us say that in particular the density inherits the properties of invariance from the set of nuclei. Once more using the uniqueness result, we deduce that the convergence of the density does not only hold locally but also in a strong sense (soon introduced in Section 5.5). In the third step, we make use of the strong convergence of the density to show the convergence of the energy per unit volume.

It must be clear to the reader that the key point of this new approach is the result of uniqueness of solutions to the system of PDEs obtained by passing locally to the limit in the Euler–Lagrange equation. In the Coulomb case for a periodic lattice of point nuclei, this result for uniqueness is contained in Lemma 5.5 of Chapter 5, and we make use of it in Section 6.2 below. It is to be mentioned here that, since a result for uniqueness also holds in the setting of the Yukawa potential (see Chapter 4, or Section 5.3.3.1), the same argument as in Section 6.2 could be made in the setting of the Yukawa potential, the results being the same *mutatis mutandis*. However, since the purpose of this section is to get rid of the additional assumptions used so far in the Coulomb case, there is no particular point in doing the proof in the Yukawa case. Let us recall, indeed, that these assumptions are not necessary when the potential is short-range.

Before treating more general situations where, for instance, the nuclei are not located on a periodic lattice, we therefore need to establish first a general result for uniqueness, in the spirit of Lemma 5.5. It is the purpose of Section 6.3 below to present the proof of this result, which we have already stated in Chapter 1, and that we reproduce here for the reader's convenience.

Theorem 6.5 *Let $c > 0$, and let m be a non-negative measure on $\mathbf{R}^3$ satisfying*

$$(H1) \quad \sup_{x \in \mathbf{R}^3} m(x + B_1) < \infty,$$

$$(H2) \quad \lim_{R \to +\infty} \inf_{x \in \mathbf{R}^3} \frac{1}{R} \, m(x + B_R) = +\infty,$$

where B_R denotes the ball of radius R centered at 0.

Then, there exists one and only one solution $(u; \Phi)$ on $\mathbf{R}^3$ to the system

$$\begin{cases} -\Delta u + c\, u^{7/3} - \Phi\, u = 0, \\ u \geq 0, \\ -\Delta \Phi = 4\pi\, [m - u^2], \end{cases}$$

with $u \in L^{7/3}_{loc} \cap L^2_{unif}(\mathbf{R}^3)$ and $\Phi \in L^1_{unif}(\mathbf{R}^3)$.

In addition, $\inf_{\mathbf{R}^3} u > 0$, $u \in L^\infty(\mathbf{R}^3) \cap C^{0,\alpha}(\mathbf{R}^3) \cap W^{2,p}_{unif}(\mathbf{R}^3)$ for all $0 < \alpha < 1$, $1 \leq p < 3$ and $\Phi \in L^{3,\infty}_{unif}(\mathbf{R}^3)$.

Let us now make some comments on this result.

First, we wish to emphasize the role of the assumptions (H1) and (H2).

From the physical point of view, (H1) means that there are no clusters of nuclei somewhere in the space, while (H2) says there is no hole of arbitrary large size in the 'lattice'. Of course, in the case of a periodic lattice like the ones we have considered so far, these conditions are automatically fulfilled, but recall once more that we aim at studying more general situations. In this spirit, these assumptions are standard. Indeed, as mentioned in Senechal [50], Delaunay has introduced the notion of the so-called Delaunay sets Λ, which are point sets satisfying the following two conditions: the set is *discrete*, that is to say there exists a positive real constant r such that, for every x, $y \in \Lambda$, $|x - y| \geq 2r$; and the set is *relatively dense*, which means there is a positive number R such that every sphere of radius greater than R contains at least one point of Λ in its interior. If we now set $m = \sum_{x \in \Lambda} \delta(\cdot - x)$, the condition of discreteness clearly corresponds to (H1); it is actually a little stronger than (H1) (it is in some sense more uniform). Likewise, the condition of relative density is the analogue of (H2), and is also stronger (it would correspond to a law in $\frac{1}{R^3}$ instead of the $\frac{1}{R}$ law appearing in (H2)).

From the mathematical standpoint, the assumptions (H1) and (H2) are of different use. On the one hand, (H1) (or some hypothesis of the same family) is necessary for a uniform bound on the potential Φ to exist. Typically, this bound is in some L_{unif}^p. In other words, it is a consequence of the bound on Φ_Λ that is uniform with respect to Λ obtained among the *a priori* estimates. On the other hand, the assumption (H2) is related to the density u. It prevents any solution u from vanishing. Consequently, the linearized operator corresponding to the non-linear operator is *strictly* coercive, and the uniqueness of u follows. Of course, all this is a bit formal, since the phenomena are closely coupled.

In any situation where the necessary *a priori* estimates are available, and where both assumptions (H1) and (H2) are fulfilled, we may attack the thermodynamic limit problem by the strategy of 'Convergence of the energy via the uniform convergence of the density'. We first prove the convergence of the density, and then the convergence of the energy.

In Section 6.4 below, we shall see some examples of convenient situations; namely, situations where the geometry of the lattice is the difficulty.

First of all, in Subsection 6.4.1, we shall consider the periodic case when the unit cell is not cubic. We shall also deal in this subsection with local perturbations of a periodic lattice (in other words, with lattices that are periodic at infinity). Let us already note that in both cases the two assumptions (H1) and (H2) are necessarily fulfilled as soon as the measure is locally bounded. The main point in this subsection is that it is still possible, since the set of nuclei presents some convenient long-range behaviour, to define the energy of the infinite system consisting of these nuclei and the associated electronic density.

On the contrary, we shall reach in Subsection 6.4.2 the first cases of infinite systems for which the energy is not simple to define. In order to treat the modelling of quasicrystals, we shall introduce there the notion of almost periodic measures and explain how we may deal with them in the context of this book.

Of course, the whole of Section 6.4 has to be seen as a first step towards the geometry optimization problem, which clearly is one of our goals in the long term. The examples of this section, apart from their physical interest (we shall indeed see below that quasicrystals do exist), are designed for cases when the non-periodicity of the set of nuclei is not too 'bad'. We have chosen situations where (H1) and (H2) are satisfied. However, with a view to solving the optimized geometry case, we have to make the following remark. We do not know if the assumptions (H1) and (H2) are necessary for the result for uniqueness (actually we suspect they are), while we deeply believe that uniqueness is necessary to conclude on the convergence of the densities, (on this latter point see Subsection 6.3.2). Therefore, attacking the optimized geometry case leads to the following difficulty. Either we show that, in some weak sense at least, the conditions (H1) and (H2) are satisfied for the limit of the sequence of optimized geometries, or we succeed in getting rid of these conditions, enlarging thereby the scope of Theorem 6.10. If we fail in both strategies, the only way to proceed is, according to some physical intuition, to *assume* that (H1) and (H2) are satisfied for the limit geometry of nuclei.

The second type of comment we would like to make on Theorem 6.10, and more generally on the new approach we detail in this chapter, is about the charge neutrality.

The strategy described here also allows us to conclude when the difficulty does not come from the geometry of the lattice, but rather from the charge balance. So far, in the Coulomb setting, we have considered only the case of global charge neutrality. Note, however, that in the Yukawa case, where the charge neutrality is less important, we have treated the non-neutral cases in Subsection 2.7.3 of Chapter 2. In the Coulomb case, the balance of charge plays a crucial role for the cancellation of the terms modelling the electrostatic interaction between particles. Therefore the direct approach of Chapter 3 makes an extensive use of the neutrality. We shall detail in Section 6.5 below how to deal with systems with a little excess or default of charge (in a sense to be defined there). In particular, we shall explain the intimate link between our problem and the problem of the determination of the maximal negative charge a given number of positive charges may bound.

6.2   A new proof of the convergence of the energy

In this section, we give a different strategy of proof of the convergence of the energy per unit cell by the thermodynamic limit. Until now, we have proved first the convergence of the energy per unit cell (under suitable assumptions, see Theorem 3.1 in Chapter 3), and next the convergence of the densities. With the forthcoming Theorem 6.6, we show that, in fact, we may first prove the (uniform) convergence of the densities (as stated in Theorem 5.9) and then deduce the convergence of the energy per unit cell. The improvements brought about by this strategy are the following.

If we compare the statement of Theorem 6.6 with that of Theorem 3.1, we may observe that two assumptions have disappeared. First of all, the proof of the convergence of the energy per unit volume that we now give holds true for any Van Hove sequence, and not only for Van Hove sequences, satisfying, in addition,

$$\forall h > 0, \frac{|\Lambda^h| log(|\Lambda^h|)}{|\Lambda|} \longrightarrow 0 \qquad as \quad \Lambda \to \infty.$$

Moreover, this new strategy of proof is also valid, for example, for non-symmetric measures m in the case of smeared nuclei and of a cubic unit cell. In addition, the strategy of proof detailed below allows us to extend the result of the existence of the thermodynamic limit for the energy per unit cell to a non-cubic unit cell, or to the case of a cell containing several nuclei, or to even more general situations including non-periodic structures (see Sections 6.4.1 and 6.4.2).

Furthermore, the argument developed below makes no use of the $\sim$-transform idea which we have extensively used in the proofs given in Chapters 2 and 3. Therefore, we may hope to apply the same argument to the case of non-convex functionals for example, provided that we still have uniqueness results for the Euler–Lagrange equation obtained by passing to the limit (see Section 6.3.2 for this important point).

Finally, let us emphasize that the argument we develop here may also be applied to the standard TF case, as we shall explain in Subsection 6.2.2.

Theorem 6.6 (Convergence of the energy) *Let* (Λ) *be a Van Hove sequence. Then,*

$$\lim_{\Lambda \to \infty} \frac{1}{|\Lambda|} I_\Lambda = I_{\text{per}} + \frac{M}{2},$$

where the notations I_Λ, I_{per}, *and* M *stand for both cases of point nuclei and smeared nuclei.*

6.2.1 Proof of Theorem 6.6

For the sake of consistency, we begin by recalling a few *a priori* bounds that we have obtained in the preceding chapters and that will be essential for our argument. We shall only give the results in the setting of the point nuclei. We shall give further bounds in the case of the smeared nuclei when it will become necessary in the course of our argument.

From Proposition 3.12 in Chapter 3, we know that

$$\forall\, 1 \leq p < 3, \qquad \sup_{x \in \mathbf{R}^3} \|\Phi_\Lambda\|_{L^p(x+B_1)} \leq C, \tag{6.1}$$

where here and below C denotes various positive constants that are independent of Λ. Besides, from Proposition 3.10, we have

$$\|\rho_\Lambda\|_{L^\infty(\mathbf{R}^3)} \leq C. \tag{6.2}$$

For the reader's convenience, we rewrite the Euler–Lagrange equation satisfied by $u_\Lambda = \sqrt{\rho_\Lambda}$; namely,

$$-\Delta u_\Lambda + \tfrac{5}{3} u_\Lambda^{7/3} - (\Phi_\Lambda - \theta_\Lambda)\, u_\Lambda = 0. \tag{6.3}$$

In view of Proposition 3.8, θ_Λ is bounded independently of Λ. Thus, comparing (6.1) and (6.2) with (6.3), it is standard to check that u_Λ is bounded in $W^{2,p}_{\text{unif}}(\mathbf{R}^3)$, for all $1 \leq p < 3$. Thus, in particular

$$\forall\, 1 \leq p < +\infty, \qquad \sup_{x \in \mathbf{R}^3} \|\nabla u_\Lambda\|_{L^p(x+B_1)} \leq C. \tag{6.4}$$

It is worth emphasizing once more the fact that these bounds hold true as soon as we know that the sequence of Lagrange multipliers associated with ρ_Λ is bounded (see also Section 6.5, which is devoted to the case of non-neutral systems). Let us now detail the proof of Theorem 6.6. Once the bounds (6.1), (6.2), and (6.4) are obtained, we may pass locally to the limit in (6.3). Then, we apply the uniqueness result of Lemma 5.5 in Section 5.3 of Chapter 5, and identify the local limits of u_Λ and $\Phi_\Lambda - \theta_\Lambda$, respectively, with the periodic minimizer u_{per} and $\Phi_{\mathrm{per}} - \theta_{\mathrm{per}}$, where

$$\Phi_{\mathrm{per}}(x) = G(x) - \int_{\Gamma_0} G(x-y)\, \rho_{\mathrm{per}}(y)\, \mathrm{d}y. \tag{6.5}$$

Next, appealing to Theorem 5.9 in Section 5.5 of Chapter 5, we obtain

$$\|\rho_\Lambda - \rho_{\mathrm{per}}\|_{L^\infty(\Gamma(\Lambda'))} \longrightarrow 0 \qquad \text{as} \quad \Lambda \to \infty, \tag{6.6}$$

and

$$\|\Phi_\Lambda - \theta_\Lambda - (\Phi_{\mathrm{per}} - \theta_{\mathrm{per}})\|_{L^\infty(\Gamma(\Lambda'))} \longrightarrow 0 \qquad \text{as} \quad \Lambda \to \infty, \tag{6.7}$$

for any sequence of interior domains $\Lambda' \subset\subset \Lambda$ (according to Definition 1' in Section 5.5).

Let us emphasize the fact that the arguments that make these convergence results hold true are, on the one hand, the *a priori* bounds on u_Λ and Φ_Λ and, on the other hand, the uniqueness Lemma 5.5. In other words, at this stage, we do not need to prove bounds on the energy per unit cell.

Let us now conclude the proof of Theorem 6.6.

(a) *We first show that*

$$\lim_{\Lambda \to \infty} \frac{1}{|\Lambda|} \int_{\mathbf{R}^3} \rho_\Lambda^{5/3} = \int_{\Gamma_0} \rho_{\mathrm{per}}^{5/3}. \tag{6.8}$$

From Section 3.3 in Chapter 3 we recall that

$$\lim_{\Lambda \to \infty} \frac{1}{|\Lambda|} \int_{\Gamma(\Lambda)^c} \rho_\Lambda = 0.$$

In the terminology we have used until now, this property was referred to as 'compactness'. It is worth noticing that, while in Section 3.3 we have deduced this compactness from the bounds on the energy per unit volume, it is, however, possible to deduce it directly from the convergence (6.6), as already mentioned in Section 5.5 of Chapter 5.

Therefore, using Hölder's inequalities together with (6.2), it is already clear that

$$\lim_{\Lambda \to \infty} \frac{1}{|\Lambda|} \int_{\Gamma(\Lambda)^c} \rho_\Lambda^{5/3} = 0. \tag{6.9}$$

Now let $\Lambda' \subset\subset \Lambda$, according to Definition 1' in Section 5.5 of Chapter 5.

We prove first that

$$\frac{1}{|\Lambda|} \int_{\Gamma(\Lambda)\setminus\Gamma(\Lambda')} \rho_\Lambda^{5/3} \longrightarrow 0 \qquad \text{as } \Lambda \to \infty. \tag{6.10}$$

Indeed,

$$\frac{1}{|\Lambda|} \int_{\Gamma(\Lambda)\setminus\Gamma(\Lambda')} \rho_\Lambda^{5/3} \leq \|\rho_\Lambda\|_{L^\infty(\mathbf{R}^3)}^{5/3} \frac{1}{|\Lambda|} |\Gamma(\Lambda)\setminus\Gamma(\Lambda')|$$

$$\leq C \frac{|\Lambda| - |\Lambda'|}{|\Lambda|},$$

for some constant C that is independent of Λ. And, we conclude, since according to Definition 1', $\frac{|\Lambda'|}{|\Lambda|}$ goes to 1 as Λ goes to infinity.

Let us now prove that

$$\lim_{\Lambda \to \infty} \frac{1}{|\Lambda|} \int_{\Gamma(\Lambda')} \rho_\Lambda^{5/3} = \int_{\Gamma_0} \rho_{\text{per}}^{5/3}. \tag{6.11}$$

For this purpose, we first write that

$$\frac{1}{|\Lambda|} \int_{\Gamma(\Lambda')} \rho_\Lambda^{5/3} = \frac{1}{|\Lambda|} \int_{\Gamma(\Lambda')} \rho_{\text{per}}^{5/3} + \frac{1}{|\Lambda|} \int_{\Gamma(\Lambda')} (\rho_\Lambda^{5/3} - \rho_{\text{per}}^{5/3})$$

$$= \frac{|\Lambda'|}{|\Lambda|} \int_{\Gamma_0} \rho_{\text{per}}^{5/3} + \frac{1}{|\Lambda|} \int_{\Gamma(\Lambda')} (\rho_\Lambda^{5/3} - \rho_{\text{per}}^{5/3}).$$

On the one hand, the first term on the right-hand side of the above equality converges to $\int_{\Gamma_0} \rho_{\text{per}}^{5/3}$ as Λ goes to infinity because of Definition 1'.

On the other hand, by using the inequality

$$|a^p - b^p| \leq p|a - b|(a \wedge b)^{p-1},$$

which holds for all real numbers $a, b \geq 0$, $p > 1$, where $a \wedge b = \max(a; b)$, we may bound from above the second term in the following way:

$$\frac{1}{|\Lambda|} \int_{\Gamma'(\Lambda)} |\rho_\Lambda^{5/3} - \rho_{\text{per}}^{5/3}| \leq \frac{5}{3} C \frac{|\Lambda'|}{|\Lambda|} \|\rho_\Lambda - \rho_{\text{per}}\|_{L^\infty(\Gamma(\Lambda'))},$$

where $C \geq \max(\|\rho_{\text{per}}\|_{L^\infty(\mathbf{R}^3)}; \|\rho_\Lambda\|_{L^\infty(\mathbf{R}^3)})^{2/3}$ is a positive constant that is independent of Λ. Now, we conclude that this term goes to zero, because of (6.6) and since $\frac{|\Lambda'|}{|\Lambda|}$ goes to 1.

Collecting (6.9), (6.10), and (6.11), (6.8) follows.

By the way, the same argument shows that

$$\lim_{\Lambda \to \infty} \frac{1}{|\Lambda|} \int_{\mathbf{R}^3} \rho_\Lambda^p = \int_{\Gamma_0} \rho_{\text{per}}^p,$$

for all $1 \leq p < +\infty$.

(b) *Let us now prove that*

$$\frac{1}{|\Lambda|} \int_{\mathbf{R}^3} |\nabla u_\Lambda|^2 \to \int_{\Gamma_0} |\nabla u_{\text{per}}|^2 \qquad \text{as } \Lambda \to \infty. \tag{6.12}$$

We shall only detail the proof in the standard Coulomb case, since the proof in the case of smeared nuclei is an easy variant.

From now on, and in order to simplify the presentation, we denote by $\bar{\Phi}_\Lambda$ the term $\Phi_\Lambda - \theta_\Lambda$. We multiply the Euler–Lagrange equation (6.3) by u_Λ, and then integrate over $\Gamma(\Lambda)^c$ to obtain

$$-\int_{\Gamma(\Lambda)^c} \Delta u_\Lambda \cdot u_\Lambda + \frac{5}{3} \int_{\Gamma(\Lambda)^c} \rho_\Lambda^{5/3} - \int_{\Gamma(\Lambda)^c} \bar{\Phi}_\Lambda \rho_\Lambda = 0. \tag{6.13}$$

From (6.9), we already know that

$$\frac{1}{|\Lambda|} \int_{\Gamma(\Lambda)^c} \rho_\Lambda^{5/3} \longrightarrow 0 \qquad \text{as } \Lambda \to \infty.$$

Besides, from Proposition 3.12 and the bound on θ_Λ,

$$\|\bar{\Phi}_\Lambda\|_{L^\infty(\Gamma(\Lambda)^c)} \leq C,$$

for some positive constant C that is independent of Λ. Gathering this with the compactness result on ρ_Λ, we obtain that

$$\frac{1}{|\Lambda|} \int_{\Gamma(\Lambda)^c} \bar{\Phi}_\Lambda \rho_\Lambda \longrightarrow 0 \qquad \text{as } \Lambda \to \infty.$$

Comparing with (6.13), we deduce that

$$-\frac{1}{|\Lambda|} \int_{\Gamma(\Lambda)^c} \Delta u_\Lambda \cdot u_\Lambda \longrightarrow 0 \qquad \text{as } \Lambda \to \infty,$$

or, equivalently, by integrating by parts,

$$\frac{1}{|\Lambda|} \int_{\Gamma(\Lambda)^c} |\nabla u_\Lambda|^2 + \frac{1}{|\Lambda|} \int_{\partial\Gamma(\Lambda)} \frac{\partial u_\Lambda}{\partial n} \cdot u_\Lambda \longrightarrow 0 \qquad \text{as } \Lambda \to \infty. \tag{6.14}$$

Now, since u_Λ is bounded in $W_{\text{unif}}^{2,p}(\mathbf{R}^3)$ and by standard trace theorems, $\dfrac{\partial u_\Lambda}{\partial n}$ is bounded in $L_{\text{unif}}^1(\partial\Gamma(\Lambda))$. Thus, using (6.2), we may write that

$$\left| \frac{1}{|\Lambda|} \int_{\partial \Gamma(\Lambda)} \frac{\partial u_\Lambda}{\partial n} \cdot u_\Lambda \right| \leq \frac{C}{|\Lambda|} \cdot |\partial \Gamma(\Lambda)|.$$

Moreover, $|\partial \Gamma(\Lambda)| = o(|\Lambda|)$, since we may bound the area $|\partial \Gamma(\Lambda)|$ of the boundary of $\Gamma(\Lambda)$ very roughly by

$$6 \cdot \#\{y \in \Lambda, dist(y; \partial \Gamma(\Lambda)) \leq 1\},$$

and this quantity is negligible in front of $|\Lambda|$ by definition of a Van Hove sequence. Returning to (6.14), we thus obtain that

$$\frac{1}{|\Lambda|} \int_{\Gamma(\Lambda)^c} |\nabla u_\Lambda|^2 \longrightarrow 0 \qquad \text{as } \Lambda \to \infty. \tag{6.15}$$

On the other hand, we write

$$\frac{1}{|\Lambda|} \int_{\Gamma(\Lambda) \backslash \Gamma(\Lambda')} |\nabla u_\Lambda|^2 = \frac{1}{|\Lambda|} \sum_{y \in \Lambda \backslash \Lambda'} \int_{y+\Gamma_0} |\nabla u_\Lambda|^2$$

$$\leq \frac{|\Lambda| - |\Lambda'|}{|\Lambda|} \cdot \sup_{y \in \mathbf{R}^3} \|\nabla u_\Lambda\|_{L^2(y+\Gamma_0)}^2.$$

Thus,

$$\frac{1}{|\Lambda|} \int_{\Gamma(\Lambda) \backslash \Gamma(\Lambda')} |\nabla u_\Lambda|^2 \longrightarrow 0 \qquad \text{as } \Lambda \to \infty, \tag{6.16}$$

since $\frac{|\Lambda'|}{|\Lambda|} \to 1$ and by virtue of (6.4).

It now just remains to prove that

$$\frac{1}{|\Lambda|} \int_{\Gamma(\Lambda')} |\nabla u_\Lambda|^2 \longrightarrow \int_{\Gamma_0} |\nabla u_{\text{per}}|^2 \qquad \text{as } \Lambda \to \infty. \tag{6.17}$$

In order to do so, we subtract from the Euler–Lagrange equation (6.3) satisfied by u_Λ the one satisfied by u_{per} (5.5), to write

$$-\Delta(u_\Lambda - u_{\text{per}}) + \frac{5}{3}(u_\Lambda^{7/3} - u_{\text{per}}^{7/3}) - (\Phi_\Lambda - \theta_\Lambda - \Phi_{\text{per}} + \theta_{\text{per}})u_\Lambda$$
$$- (\Phi_{\text{per}} - \theta_{\text{per}})(u_\Lambda - u_{\text{per}}) = 0. \tag{6.18}$$

Then, because of (6.6), (6.2), and (6.7), together with the fact that $\Phi_{\text{per}} - \theta_{\text{per}}$ lies in $L_{\text{unif}}^p(\mathbf{R}^3)$, for all $1 \leq p < 3$, it is clear that

$$\sup_{y \in \Lambda'} \|\Delta(u_\Lambda - u_{\text{per}})\|_{L^p(y+\Gamma_0)} \longrightarrow 0 \qquad \text{as } \Lambda \to \infty, \tag{6.19}$$

for all $1 \leq p < 3$. To conclude, we multiply (6.18) by $u_\Lambda - u_{\text{per}}$ and integrate over $y + \Gamma_0$ for some $y \in \Lambda'$. Then, by using (6.6) and (6.19) together with the bounds on u_Λ in $W_{\text{unif}}^{2,p}(\mathbf{R}^3)$, we check easily that

$$\sup_{y \in \Lambda'} \| \nabla(u_\Lambda - u_{\text{per}}) \|_{L^2(y+\Gamma_0)} \longrightarrow 0 \qquad \text{as} \quad \Lambda \to \infty.$$

Correspondingly, (6.17) follows .

Collecting (6.15), (6.16) and (6.17), we finally obtain (6.12).

(c) The rest of the proof is devoted to the Coulomb case with point nuclei. We shall mention at the end of the argument (point (d)) what modifications we have to take into account in the case of smeared out nuclei.

In the following, we denote by θ_∞ the limit of the sequence θ_Λ as Λ goes to infinity. We first prove that

$$\lim_{\Lambda \to \infty} \frac{U_\Lambda^{Cb}}{2|\Lambda|} - \frac{1}{2|\Lambda|} \int_{\mathbf{R}^3} V_\Lambda \rho_\Lambda = \tfrac{1}{2} \left[M - \int_{\Gamma_0} G(x)\rho_{\text{per}}(x)dx + \theta_\infty - \theta_{\text{per}} \right]. \quad (6.20)$$

For this purpose, we rewrite this quantity as

$$\tfrac{1}{2} \lim_{\Lambda \to \infty} \frac{1}{|\Lambda|} \sum_{y \in \Lambda} \lim_{x \to y} \left[\Phi_\Lambda(x) - \frac{1}{|x-y|} \right],$$

and we distinguish in the above sum the contribution of the y in Λ' and those outside Λ'.

From (6.7), and then the periodicity of Φ_{per}, we obtain that

$$\lim_{\Lambda \to \infty} \frac{1}{|\Lambda|} \sum_{y \in \Lambda'} \lim_{x \to y} \left[\Phi_\Lambda(x) - \frac{1}{|x-y|} \right]$$

$$= \lim_{\Lambda \to \infty} \frac{1}{|\Lambda|} \sum_{y \in \Lambda'} \left\{ \lim_{x \to y} \left[\Phi_{\text{per}}(x) - \frac{1}{|x-y|} \right] \right\} + \theta_\Lambda - \theta_{\text{per}}$$

$$= \lim_{x \to 0} \left[\Phi_{\text{per}}(x) - \frac{1}{|x|} \right] + \theta_\infty - \theta_{\text{per}},$$

since $\frac{|\Lambda'|}{|\Lambda|} \to 1$ as $\Lambda \to \infty$. Besides, using the definition (6.5) of Φ_{per}, we obtain

$$\lim_{x \to 0} \left[\Phi_{\text{per}}(x) - \frac{1}{|x|} \right] = M - \int_{\Gamma_0} G(x)\rho_{\text{per}}(x) \, dx.$$

Thus,

$$\lim_{\Lambda \to \infty} \frac{1}{|\Lambda|} \sum_{y \in \Lambda'} \lim_{x \to y} \left[\Phi_\Lambda(x) - \frac{1}{|x-y|} \right] = \theta_\infty - \theta_{\text{per}} + M - \int_{\Gamma_0} G(x)\rho_{\text{per}}(x)dx.$$

$$(6.21)$$

We now intend to prove that

$$\lim_{\Lambda \to \infty} \frac{1}{|\Lambda|} \sum_{y \in \Lambda \setminus \Lambda'} \lim_{x \to y} \left[\Phi_\Lambda(x) - \frac{1}{|x-y|} \right] = 0. \tag{6.22}$$

Indeed, from (6.2), we have that, for all $0 < \delta < 1$,

$$\forall y \in \Lambda, \ |-\Delta\Phi_\Lambda - 4\pi\delta_y| \le C \qquad \text{in } B(y; \delta),$$

where C denotes a positive constant that is independent of Λ and $y \in \Lambda$, and thus, equivalently,

$$\left| -\Delta\left(\Phi_\Lambda - \frac{1}{|x-y|}\right) \right| \le C \qquad \text{in } B(y; \delta).$$

Since $\Phi_\Lambda - \frac{1}{|x-y|}$ is bounded in $L^p_{\text{unif}}(\mathbf{R}^3)$, for all $1 \le p < 3$, independently of Λ, we then have, by a straightforward application of the mean value inequality,

$$\left| \lim_{\substack{x \to y \\ x \ne y}} \left(\Phi_\Lambda(x) - \frac{1}{|x-y|} \right) \right| \le C,$$

and C is independent of Λ and y. Therefore,

$$\lim_{\Lambda \to \infty} \frac{1}{|\Lambda|} \sum_{y \in \Lambda \setminus \Lambda'} \lim_{x \to y} \left| \Phi_\Lambda(x) - \frac{1}{|x-y|} \right| \le \frac{C(|\Lambda| - |\Lambda'|)}{|\Lambda|},$$

and thus goes to zero as Λ goes to infinity.

So (6.20) follows, gathering together (6.21) and (6.22) .

Our last step consists of showing that

$$\lim_{\Lambda \to \infty} \left[-\frac{1}{2|\Lambda|} \int_{\mathbf{R}^3} V_\Lambda \rho_\Lambda + \frac{1}{2|\Lambda|} D^{Cb}(\rho_\Lambda, \rho_\Lambda) \right]$$
$$= -\frac{1}{2} \int_{\Gamma_0} G(x)\rho_{\text{per}}(x) \, \mathrm{d}x + \tfrac{1}{2}D_G(\rho_{\text{per}}, \rho_{\text{per}}) - \frac{\theta_{\text{per}}}{2} + \frac{\theta_\infty}{2}. \tag{6.23}$$

We first rewrite the quantity on the left-hand side of the above equality as $\frac{1}{2|\Lambda|} \int_{\mathbf{R}^3} \Phi_\Lambda \rho_\Lambda$. Next, we recall from Proposition 3.12 that

$$\|\Phi_\Lambda\|_{L^\infty(\Gamma(\Lambda)^c)} \le C,$$

while, from the compactness,

$$\frac{1}{|\Lambda|} \int_{\Gamma(\Lambda)^c} \rho_\Lambda \longrightarrow 0 \qquad \text{as } \Lambda \to \infty.$$

From these two facts, we infer that

$$\frac{1}{|\Lambda|} \int_{\Gamma(\Lambda)^c} \Phi_\Lambda \rho_\Lambda \longrightarrow 0 \qquad \text{as } \Lambda \to \infty. \tag{6.24}$$

Besides, on the one hand, we may bound

$$\left| \frac{1}{|\Lambda|} \int_{\Gamma(\Lambda) \setminus \Gamma(\Lambda')} \Phi_\Lambda \rho_\Lambda \right| \le \|\rho_\Lambda\|_{L^\infty} \frac{|\Lambda| - |\Lambda'|}{|\Lambda|} \sup_{y \in \mathbf{Z}^3} \|\Phi_\Lambda\|_{L^1(y + \Gamma_0)}. \tag{6.25}$$

Hence, the left-hand side of the above inequality goes to zero as Λ goes to infinity, by definition of Λ' and from (6.2) and (6.1).

On the other hand, using (6.7) and the periodicity of Φ_{per}, we obtain that

$$\frac{1}{|\Lambda|} \int_{\Gamma(\Lambda')} \Phi_\Lambda \rho_\Lambda = \frac{1}{|\Lambda|} \int_{\Gamma_0} \Phi_{\mathrm{per}} \rho_\Lambda - \theta_{\mathrm{per}} + \theta_\infty + o(1)$$

as Λ goes to infinity, since $\frac{|\Lambda'|}{|\Lambda|} \to 1$ and $\frac{1}{|\Lambda|} \int_{\Gamma'(\Lambda)} \rho_\Lambda \to 1$, while, from (6.6),

$$\frac{1}{|\Lambda|} \int_{\Gamma(\Lambda')} \Phi_{\mathrm{per}} \rho_\Lambda = \int_{\Gamma_0} \Phi_{\mathrm{per}} \rho_{\mathrm{per}} + o(1),$$

since $|\Lambda'| = |\Lambda| + o(|\Lambda|)$. Therefore,

$$\frac{1}{|\Lambda|} \int_{\Gamma(\Lambda')} \Phi_\Lambda \rho_\Lambda = \int_{\Gamma_0} \Phi_{\mathrm{per}} \rho_{\mathrm{per}} + o(1). \tag{6.26}$$

Therefore, (6.23) follows, collecting (6.26), (6.24) and (6.25).

Finally, Theorem 6.6 follows in the case of point nuclei by gathering together (6.8), (6.12), (6.20), and (6.23).

The rest of the proof is now devoted to the case of smeared out nuclei.

(d) In the smeared nuclei case, it is obvious that (6.8) and (6.12) are true by the same argument. Therefore, the only remaining point consists of showing that

$$\frac{1}{|\Lambda|} \int_{\mathbf{R}^3} |\nabla \Phi_\Lambda^m|^2 \longrightarrow \int_{\Gamma_0} |\nabla \Phi_{\mathrm{per}}^m|^2 \qquad \text{as } \Lambda \to \infty, \tag{6.27}$$

where $\int_{\Gamma_0} |\nabla \Phi_{\mathrm{per}}^m|^2$ may equivalently be expressed as

$$- \int_{\Gamma_0} G_m(x) \rho_{\mathrm{per}}(x) \, dx + \frac{1}{2} D_G(\rho_{\mathrm{per}}, \rho_{\mathrm{per}}) + \frac{M}{2},$$

with $G_m = G \star m$ and

$$M = \iint_{\Gamma_0 \times \Gamma_0} m(x) m(y) \left[G(x - y) - \frac{1}{|x - y|} \right] dx \, dy.$$

We first recall from Proposition 3.5 in Section 3.2 of Chapter 3, that Φ_Λ^m is bounded in $L^\infty(\mathbf{R}^3)$ (independently of Λ) and satisfies

$$-\Delta\Phi_\Lambda^m = 4\pi\left[m_\Lambda - \rho_\Lambda\right] \quad \text{in } \mathbf{R}^3.$$

Thus, because of (6.2), $-\Delta\Phi_\Lambda^m$ also is bounded in $L^\infty(\mathbf{R}^3)$. Therefore, $\nabla\Phi_\Lambda^m$ is bounded in $L^\infty(\mathbf{R}^3)$ independently of Λ. Then, it is easy to convince ourselves that the strategy of proof to establish (6.12) carries through to check (6.27), and that it is even simpler to reproduce.

This concludes the proof of Theorem 6.6. ◇

Remark 6.7 The above argument has been made either in the case of point nuclei (Dirac masses) or in the case of regular smeared nuclei (smooth functions m). It is important to note here that it can be extended to cover the case of other 'shapes' of nuclei. Indeed, if one takes a non-negative measure m to define the nucleus in each cell, and assume that this measure has compact support in the unit cell and lies in $L_{\text{loc}}^{6/5}$, then the argument that we have just made above in the setting of regular smeared nuclei can be straightforwardly adapted. On the contrary, if the measure defining the nuclei is not in $L_{\text{loc}}^{6/5}$, we have to apply the argument we have made in the case of the Dirac masses, through slight modifications.

6.2.2 Adaptation to the TF case

Let us make here some comments on how our methods adapt to the Thomas–Fermi case, and, by the way, improve the results obtained by Lieb and Simon in that setting (see [40]). More precisely, we now establish the following:

Proposition 6.8 (Thermodynamic limit in the TF case) Let (Λ) be a Van Hove sequence. Then,

$$\lim_{\Lambda\to\infty} \frac{I_\Lambda^{\text{TF}}}{|\Lambda|} = I_{\text{per}}^{\text{TF}} + \frac{M}{2}, \tag{6.28}$$

and, for any sequence of interior domains $\Gamma'(\Lambda) \subset\subset \Gamma(\Lambda)$,

$$\lim_{\Lambda\to\infty} \sup_{x\in\Gamma'(\Lambda)} |\rho_\Lambda^{\text{TF}}(x) - \rho_{\text{per}}^{\text{TF}}(x)| = 0, \tag{6.29}$$

I_Λ^{TF} and $I_{\text{per}}^{\text{TF}}$ being defined in Chapter 1 by the formulae (1.5)-(1.6) and (1.8)-(1.9), respectively, while ρ_Λ^{TF} and $\rho_{\text{per}}^{\text{TF}}$ are the corresponding minimizers.

In addition, we also have

$$\lim_{\Lambda\to\infty} \sup_{x\in\Gamma'(\Lambda)} |\Phi_\Lambda^{\text{TF}}(x) - \Phi_{\text{per}}^{\text{TF}}(x)| = 0, \tag{6.30}$$

where

$$\Phi_{\text{per}}^{\text{TF}}(x) = G(x) - \int_{\Gamma_0} G(x-y)\rho_{\text{per}}^{\text{TF}}(y)\,\mathrm{d}y + \psi_0, \qquad (6.31)$$

for some positive constant ψ_0 that is independent of the sequence Λ, such that the Euler–Lagrange equation satisfied by $\rho_{\text{per}}^{\text{TF}}$ reads

$$\tfrac{5}{3}\rho_{\text{per}}^{\text{TF}\ 2/3} = \Phi_{\text{per}}^{\text{TF}}. \qquad (6.32)$$

Remark 6.9 (1) If we compare with [40], we have improved the results of Lieb and Simon in two directions. First of all, in [40], the convergences of $\rho_{\Lambda}^{\text{TF}} - \rho_{\text{per}}^{\text{TF}}$ and $\Phi_{\Lambda}^{\text{TF}} - \Phi_{\text{per}}^{\text{TF}}$ to zero are shown to be uniform on the compact subsets of $\mathbf{R}^3$, while we establish the uniform convergence on a large part of $\mathbf{R}^3$. Moreover, our results hold true for any Van Hove sequence, while the proof of the convergence of the densities proposed by Lieb and Simon is based on the additional monotonicity assumption that $\Lambda_n \subset \Lambda_{n+1}$, at least for n large enough.

(2) We shall prove first (6.29) and (6.30), and next deduce (6.28). By the way, let us emphasize the fact that the extensions made in the setting of the TFW model in the following Section 6.4 carry through straightforwardly to the TF case.

Proof of Proposition 6.8 In order to simplify the notations, we skip the superscript TF from now on. The proof of Proposition 6.8 falls into three steps.

We first establish some *a priori* estimates for the sequences ρ_{Λ} and Φ_{Λ}, together with some local convergence results. Let us emphasize the fact that in [40], the pointwise convergence of Φ_{Λ} (and thus ρ_{Λ}) came from the monotonicity of these sequences with respect to Λ, and then from the dominated convergence theorem. We propose here an alternative strategy of proof which needs no monotonicity assumption on the Van Hove sequences (Λ) to be considered.

Once these bounds are obtained, we may pass to the limit in the Euler–Lagrange equations satisfied by ρ_{Λ} and Φ_{Λ}. Next, we use the uniqueness result for the limit Thomas–Fermi equation that we have already mentioned in Subsection 5.3.3.2 of Chapter 5, to identify the limits of ρ_{Λ} and Φ_{Λ} to, respectively, ρ_{per} the minimizer of $I_{\text{per}}^{\text{TF}}$ and Φ_{per} defined by (6.31).

Next, by virtue of the uniqueness result mentioned above, we argue as in Section 5.5 to deduce (6.29) and (6.30). Let us notice that, even if Φ_{Λ} and Φ_{per} (and thus ρ_{Λ} and ρ_{per}) do not belong to L^{∞}, their difference does, therefore making sense to (6.30) and (6.29).

Finally, we show how to adapt the proof of Theorem 6.6 to our case to obtain (6.28).

Step 1: a priori bounds. The Euler–Lagrange equation satisfied by ρ_{Λ} reads

$$\tfrac{5}{3}\rho_\Lambda{}^{2/3} = \Phi_\Lambda \qquad \text{on } \mathbf{R}^3, \tag{6.33}$$

while Φ_Λ is the solution to

$$-\triangle\Phi_\Lambda = 4\pi\left[\sum_{y\in\Lambda}\delta_y - \rho_\Lambda\right], \tag{6.34}$$

going to zero at infinity. Combining (6.33) and (6.34), we identify Φ_Λ with the unique solution to the so-called Thomas–Fermi equation (see [40] and [13]):

$$-\triangle\Phi_\Lambda + c\,\Phi_\Lambda{}^{3/2} = 4\pi\sum_{y\in\Lambda}\delta_y, \tag{6.35}$$

with $c = 4\pi\,3^{3/2}\,5^{-3/2}$. Since the left-hand side of (6.35) lies in $\mathcal{M}_{\text{unif}}(\mathbf{R}^3)$, it follows from [12] and [13] (see also the work of Bénilan, Brézis, and Crandall [8] for related issues) that Φ_Λ is bounded in $L_{\text{unif}}^{3/2}(\mathbf{R}^3)$ independently of Λ, and thus, in particular, in $L_{\text{unif}}^1(\mathbf{R}^3)$. Returning to (6.35), we then have that $\triangle\Phi_\Lambda$ is bounded in $\mathcal{M}_{\text{unif}}(\mathbf{R}^3)$ independently of Λ. This latter bound, together with the one for Φ_Λ in $L_{\text{unif}}^1(\mathbf{R}^3)$, yields the following bounds (see, for example, [8] and [13]):

$$\|\Phi_\Lambda\|_{L_{\text{unif}}^{3,\,\infty}(\mathbf{R}^3)} \le C, \tag{6.36}$$

and

$$\|\nabla\Phi_\Lambda\|_{L_{\text{unif}}^{\frac{3}{2},\,\infty}(\mathbf{R}^3)} \le C, \tag{6.37}$$

where C denotes various positive constants that are independent of Λ. In particular, from the embedding properties of the Marcinkiewicz spaces (see (5.40) or the Appendix in [8]), we obtain that Φ_Λ is also bounded in $W_{\text{unif}}^{1,\,p}(\mathbf{R}^3)$, for all $1 \le p < \tfrac{3}{2}$. Thus, extracting a subsequence if necessary and by using Rellich's theorem together with Sobolev's embeddings, we may assume that Φ_Λ converges to $\Phi_\infty \in L_{\text{unif}}^{3,\,\infty}(\mathbf{R}^3)$ weakly in $W_{\text{unif}}^{1,\,p}(\mathbf{R}^3)$, for all $1 \le p < \tfrac{3}{2}$, strongly in $L_{\text{unif}}^p(\mathbf{R}^3)$, for all $1 \le p < 3$ and almost everywhere in $\mathbf{R}^3$. Therefore, passing to the limit in (6.35), we deduce that Φ_∞ coincides with the unique solution to

$$-\triangle\Phi + c\,\Phi^{3/2} = 4\pi\sum_{y\in\mathbf{Z}^3}\delta_y, \tag{6.38}$$

(see Subsection 5.3.3.2 in Chapter 5, and Section 5.3 for some comments on the proof of the existence and uniqueness of the solution of (6.38)). In particular, from the uniqueness and since the right-hand side of (6.38) is periodic, we conclude that Φ_∞ shares the same periodicity as the right-hand side.

Let us now apply these results to obtain some information on the convergence of the sequence of densities. By using the pointwise relationship (6.33) between

ρ_Λ and Φ_Λ and the convergence results we have collected just before for Φ_Λ, we deduce that ρ_Λ converges to some non-negative function $\rho_\infty \in L_{\text{unif}}^{\frac{9}{2}, \infty}(\mathbf{R}^3)$ strongly in $L_{\text{unif}}^p(\mathbf{R}^3)$, for all $1 \leq p < \frac{9}{2}$ and almost everywhere on $\mathbf{R}^3$. In addition, passing to the limit in (6.33) as Λ goes to infinity, we also have

$$\tfrac{5}{3}\rho_\infty^{2/3} = \Phi_\infty; \tag{6.39}$$

hence ρ_∞ is also periodic.

To conclude, we recall from [40] that Φ_{per} defined by (6.31) is another solution to (6.38). Since this solution is unique, we necessarily have $\Phi_\infty = \Phi_{\text{per}}$, and thus $\rho_\infty = \rho_{\text{per}}$, by comparing (6.39) and the Euler–Lagrange equation (6.32) satisfied by ρ_{per}. By the way, let us notice that all the convergences mentioned above for Φ_Λ and ρ_Λ concern the whole sequences (not only subsequences) because of the uniqueness of their limits.

Step 2: uniform convergence on the interior domains. To prove (6.29) and (6.30), we argue as in the proof of Theorem 5.9 in Section 5.5. Indeed, the main two ingredients of this proof are the following.

On the one hand, we have bounds independent of Λ for ρ_Λ and Φ_Λ, respectively, in $L_{\text{unif}}^{\frac{9}{2}, \infty}(\mathbf{R}^3)$ and $L_{\text{unif}}^{3, \infty}(\mathbf{R}^3) \cap W_{\text{unif}}^{1,p}(\mathbf{R}^3)$, for all $1 \leq p < \frac{3}{2}$, which allow us to pass to the limit in (6.35). On the other hand, we have a uniqueness result for the limit equation (6.38). Reproducing the proof of Theorem 5.9, we deduce (6.30) and (6.29).

Step 3: convergence of the energy per unit cell. In order to adapt the proof of Theorem 6.6 to the present case, we essentially have to modify the points where we have used the L^∞ bound on the TFW minimizing density, which no more holds true here. By the way, and since this remark will be useful in the following, let us recall from [40] the following bounds (whose proofs are also sketched in Section 3.3.6 of Chapter 3)

$$\lim_{\Lambda \to \infty} \frac{1}{|\Lambda|} \int_{\Gamma(\Lambda)^c} \Phi_\Lambda^p = 0, \qquad \text{for all } 1 \leq p < +\infty, \tag{6.40}$$

$$\|\Phi_\Lambda\|_{L^\infty(\Gamma(\Lambda)^c)} \leq C, \tag{6.41}$$

which immediately yields, because of (6.33),

$$\lim_{\Lambda \to \infty} \frac{1}{|\Lambda|} \int_{\Gamma(\Lambda)^c} \rho_\Lambda^p = 0, \qquad \text{for all } 1 \leq p < +\infty, \tag{6.42}$$

and

$$\|\rho_\Lambda\|_{L^\infty(\Gamma(\Lambda)^c)} \leq C. \tag{6.43}$$

Let $\Lambda' \subset\subset \Lambda$, according to Definition 1' in Section 5.5. We now follow, step by step, the proof of Theorem 6.6 and indicate where the argument has to be modified.

We first examine the proof of

$$\lim_{\Lambda \to \infty} \frac{1}{|\Lambda|} \int_{\mathbf{R}^3} \rho_\Lambda^{5/3} = \int_{\Gamma_0} \rho_{\text{per}}^{5/3}, \tag{6.44}$$

which corresponds to step (a) in the proof of Theorem 6.6. From (6.42), we already know that

$$\lim_{\Lambda \to \infty} \frac{1}{|\Lambda|} \int_{\Gamma(\Lambda)^c} \rho_\Lambda^{5/3} = 0.$$

To prove that

$$\lim_{\Lambda \to \infty} \frac{1}{|\Lambda|} \int_{\Gamma(\Lambda)\backslash\Gamma(\Lambda')} \rho_\Lambda^{5/3} = 0,$$

we proceed as follows:

$$\frac{1}{|\Lambda|} \int_{\Gamma(\Lambda)\backslash\Gamma(\Lambda')} \rho_\Lambda^{5/3} = \frac{1}{|\Lambda|} \sum_{y \in \Lambda \backslash \Lambda'} \int_{y+\Gamma_0} \rho_\Lambda^{5/3}$$

$$\leq \frac{|\Lambda \backslash \Lambda'|}{|\Lambda|} \|\rho_\Lambda\|_{L_{\text{unif}}^{5/3}(\mathbf{R}^3)}^{5/3}.$$

We conclude, since ρ_Λ is bounded in $L_{\text{unif}}^p(\mathbf{R}^3)$ for every $1 \leq p < \frac{9}{2}$ (including the case when $p = \frac{5}{3}$). It now remains to check that

$$\lim_{\Lambda \to \infty} \frac{1}{|\Lambda|} \int_{\Gamma(\Lambda')} \rho_\Lambda^{5/3} = \int_{\Gamma_0} \rho_{\text{per}}^{5/3}.$$

By virtue of (6.32) and (6.33), and by replicating the proof of (6.11), we thus write

$$\frac{1}{|\Lambda|} \int_{\Gamma(\Lambda')} |\rho_\Lambda^{5/3} - \rho_{\text{per}}^{5/3}|$$

$$= C \frac{1}{|\Lambda|} \int_{\Gamma(\Lambda')} |\Phi_\Lambda^{5/2} - \Phi_{\text{per}}^{5/2}|$$

$$\leq C \frac{1}{|\Lambda|} \int_{\Gamma(\Lambda')} \max\left(|\Phi_\Lambda|, |\Phi_{\text{per}}|\right)^{3/2} |\Phi_\Lambda - \Phi_{\text{per}}|$$

$$\leq C \frac{|\Lambda'|}{|\Lambda|} \max\left(\|\Phi_\Lambda\|_{L_{\text{unif}}^{3/2}}, \|\Phi_{\text{per}}\|_{L_{\text{unif}}^{3/2}}\right)^{3/2} \|\Phi_\Lambda - \Phi_{\text{per}}\|_{L^\infty(\Gamma(\Lambda'))}.$$

Here and below, C denotes various positive constants that are independent of Λ. We reach the conclusion by using the bounds on Φ_Λ and Φ_{per} and (6.30).

Let us now show the analogue of (6.20); that is,

$$\lim_{\Lambda \to \infty} \frac{U_\Lambda^{Cb}}{2|\Lambda|} - \frac{1}{2|\Lambda|} \int_{\mathbf{R}^3} V_\Lambda \rho_\Lambda = \frac{1}{2} \left[M - \int_{\Gamma_0} G(x) \rho_{\text{per}}(x) \, dx + \psi_0 \right]. \qquad (6.45)$$

For this purpose, we rewrite this quantity as

$$\frac{1}{2} \lim_{\Lambda \to \infty} \frac{1}{|\Lambda|} \sum_{y \in \Lambda} \lim_{x \to y} \left[\Phi_\Lambda(x) - \frac{1}{|x-y|} \right].$$

Comparing with the proof of (6.20), we just have to modify the argument which shows that

$$\lim_{\Lambda \to \infty} \frac{1}{|\Lambda|} \sum_{y \in \Lambda \setminus \Lambda'} \lim_{x \to y} \left[\Phi_\Lambda(x) - \frac{1}{|x-y|} \right] = 0. \qquad (6.46)$$

Indeed, for all $0 < \delta < 1$, and for every $y \in \Lambda$, it is easily seen from (6.34) and the bounds on ρ_Λ that $|-\Delta \Phi_\Lambda - 4\pi \delta_y| = \left| -\Delta \left(\Phi_\Lambda - \frac{1}{|x-y|} \right) \right|$ is bounded in $L^p(B(y; \delta))$, for all $1 \le p < \frac{9}{2}$, independently of y. Moreover, $\Phi_\Lambda - \frac{1}{|x-y|}$ is bounded in $L_{\text{unif}}^q(\mathbf{R}^3)$, for all $1 \le q < 3$, independently of Λ and y. Then, for all $0 < \delta' < 1$, $\Phi_\Lambda - \frac{1}{|x-y|}$ is bounded in $W^{2,r}(B(y; \delta'))$, at least for some $r > \frac{3}{2}$. Hence, from Sobolev's embeddings, it is bounded in $L^\infty(B(y; \delta'))$, independently of $y \in \Lambda$ and Λ. Therefore,

$$\lim_{\Lambda \to \infty} \frac{1}{|\Lambda|} \sum_{y \in \Lambda \setminus \Lambda'} \lim_{x \to y} \left| \Phi_\Lambda(x) - \frac{1}{|x-y|} \right| \le \frac{C(|\Lambda| - |\Lambda'|)}{|\Lambda|},$$

and thus goes to zero as Λ goes to infinity.

Our last step consists of showing that

$$\lim_{\Lambda \to \infty} \left[-\frac{1}{2|\Lambda|} \int_{\mathbf{R}^3} V_\Lambda \rho_\Lambda + \frac{1}{2|\Lambda|} D^{Cb}(\rho_\Lambda, \rho_\Lambda) \right]$$

$$= -\frac{1}{2} \int_{\Gamma_0} G(x) \rho_{\text{per}}(x) \, dx + \frac{1}{2} D_G(\rho_{\text{per}}, \rho_{\text{per}}) - \frac{\psi_0}{2}. \qquad (6.47)$$

We first rewrite the quantity on the left-hand side of the above equality as $\frac{1}{2|\Lambda|} \int_{\mathbf{R}^3} \Phi_\Lambda \, \rho_\Lambda$. It follows from (6.41) and (6.42) that

$$\lim_{\Lambda \to \infty} \frac{1}{|\Lambda|} \int_{\Gamma(\Lambda)^c} \Phi_\Lambda \, \rho_\Lambda = 0.$$

On the one hand, using Hölder's inequalities, we obtain

$$\left| \frac{1}{|\Lambda|} \int_{\Gamma(\Lambda) \setminus \Gamma(\Lambda')} \Phi_\Lambda\, \rho_\Lambda \right| \le \frac{1}{|\Lambda|} \sum_{y \in \Lambda \setminus \Lambda'} \int_{y+\Gamma_0} |\Phi_\Lambda|\, |\rho_\Lambda|$$

$$\le \frac{|\Lambda| - |\Lambda'|}{|\Lambda|} \, \|\rho_\Lambda\|_{L^2_{\mathrm{unif}}} \, \|\Phi_\Lambda\|_{L^2_{\mathrm{unif}}}.$$

The left-hand side of the above inequality goes to zero as Λ goes to infinity, by definition of Λ' and from the bounds on ρ_Λ and Φ_Λ.

On the other hand, using (6.30), and then (6.29), we check, as in the proof of Theorem 6.6, that

$$\lim_{\Lambda \to \infty} \frac{1}{|\Lambda|} \int_{\Gamma(\Lambda')} \Phi_\Lambda\, \rho_\Lambda = \int_{\Gamma_0} \Phi_{\mathrm{per}}\, \rho_{\mathrm{per}}.$$

Finally, (6.28) in Theorem 6.8 follows by gathering together (6.44), (6.45), and (6.47). $\diamond$

6.3   A general result for existence and uniqueness

This section, which is the main section of the chapter, is devoted to the statement and proof of a general existence and uniqueness result that extends Lemma 5.5 of Chapter 5, and that has been introduced in Chapter 1 and Section 6.1 above.

Theorem 6.10 (General existence and uniqueness result) *Let $c > 0$, and let m be a non-negative measure on $\mathbf{R}^3$ satisfying*

$$(H1) \quad \sup_{x \in \mathbf{R}^3} m(x + B_1) < \infty,$$

$$(H2) \quad \lim_{R \to +\infty} \inf_{x \in \mathbf{R}^3} \frac{1}{R} \, m(x + B_R) = +\infty,$$

where B_R denotes the ball of radius R centered at 0.

Then, there exists one and only one solution $(u; \Phi)$ on $\mathbf{R}^3$ to the system

$$\begin{cases} -\Delta u + c\, u^{7/3} - \Phi\, u = 0, \\[2mm] u \ge 0, \\[2mm] -\Delta \Phi = 4\pi\, [m - u^2], \end{cases} \tag{6.48}$$

with $u \in L^{7/3}_{\mathrm{loc}} \cap L^2_{\mathrm{unif}}(\mathbf{R}^3)$ and $\Phi \in L^1_{\mathrm{unif}}(\mathbf{R}^3)$.

In addition, $\inf_{\mathbf{R}^3} u > 0$, $u \in L^\infty(\mathbf{R}^3) \cap C^{0,\alpha}(\mathbf{R}^3) \cap W^{2,p}_{\mathrm{unif}}(\mathbf{R}^3)$ for all $0 < \alpha < 1$, $1 \le p < 3$ and $\Phi \in L^{3,\infty}_{\mathrm{unif}}(\mathbf{R}^3)$.

Since Section 6.2, it must be clear to the reader that the uniqueness of the solution to the Euler–Lagrange equation obtained in the thermodynamic limit plays a crucial role in the strategy of proof of the existence of the thermodynamic limit. We have already made use of the result for uniqueness stated in Lemma 5.5 to prove, in Section 6.2, the existence of the thermodynamic limit in cases that we cannot treat, so far as we know, by the direct strategy of Chapters 3 and 5. Likewise, we shall make use of Theorem 6.10 in Section 6.4 to attack very general situations. Furthermore, let us emphasize the following fact. We believe that if such a uniqueness does not hold, then we cannot hope to prove that the density converges in the thermodynamic limit. Indeed, as will be detailed in Subsection 6.3.2 below, we may exhibit a situation where we may build some oscillatory sequence of electronic densities that, to some extent, is not far from the minimizing density (ρ_Λ) and that does not converge. The existence of such an oscillatory sequence comes from the existence of two distinct solutions to the limit equation. Consequently, Theorem 6.10 is likely to be the compulsory tool to study the thermodynamic limit in 'difficult' cases. The next subsection is devoted to its proof.

6.3.1   Proof of Theorem 6.10

We begin this proof by setting some *a priori* estimates on the solutions to the system. Next, we prove the uniqueness and, finally, we present a way to show the existence.

Step 1: a priori estimates. Let us consider a solution $(u; \Phi)$ to the system (6.48) with $u \in L_{loc}^{7/3} \cap L_{unif}^2(\mathbf{R}^3)$ and $\Phi \in L_{unif}^1(\mathbf{R}^3)$.

According to the proof of Lemma 5.5 in Chapter 5, we prove successively that, under the assumptions on $(u; \Phi)$ in (6.48) and because of (H1) and (H2), we have

$$\Phi \in L_{unif}^{3,\infty}(\mathbf{R}^3), \tag{6.49}$$

$$u \in L^\infty(\mathbf{R}^3), \tag{6.50}$$

and

$$\inf_{\mathbf{R}^3} u > 0. \tag{6.51}$$

We have already established the analogues of (6.49) and (6.51) in the course of the proof of Lemma 5.5, in the special case when $m = \sum_{k \in \mathbf{Z}^3} \delta_k$. We refer the reader to this proof for details and just briefly discuss here how our argument has to be modified to adapt to general measures m satisfying both (H1) and (H2).

In order to prove (6.49), we just point out that, because of (H1) and since u is in $L_{unif}^2(\mathbf{R}^3)$, the measure $m - u^2$ lies in

$$\mathcal{M}_{\text{unif}}(\mathbf{R}^3) = \left\{ m \text{ measure on } \mathbf{R}^3, \sup_{x \in \mathbf{R}^3} |m(x + B_1)| < \infty \right\}.$$

Thus, since $\triangle\Phi \in \mathcal{M}_{\text{unif}}(\mathbf{R}^3)$ and $\Phi \in L^1_{\text{unif}}(\mathbf{R}^3)$, we deduce (6.49) as for the proof of (5.41).

We now prove the L^∞ bound (6.50). In Remark 5.6 about the pure Coulomb case in Chapter 5, we have mentioned that the uniqueness of $(u; \Phi)$ in the class $(u \in L^\infty(\mathbf{R}^3); \Phi \in L^1_{\text{unif}}(\mathbf{R}^3))$ may be formulated otherwise. Actually, the L^∞ regularity of u is a by-product of the regularity imposed on m (namely (H1)) and on Φ (namely $\Phi \in L^1_{\text{unif}}(\mathbf{R}^3)$). It is indeed an easy extension of Lemma 4.11 of Chapter 4. In that lemma, we have assumed m to be periodic, which is in fact not necessary, provided that we keep some uniformity on m, as we shall now see. From the equation satisfied by u in (6.48) and from Young's inequality, we may clearly write

$$-\triangle u + \frac{c}{2} u^{7/3} \le C \, \Phi_+^{7/4} \qquad \text{a.e. on } \mathbf{R}^3, \tag{6.52}$$

for some positive constant C. It follows, because of (6.49) and from the continuous embeddings properties recalled by (5.40), that Φ is in $L^p_{\text{unif}}(\mathbf{R}^3)$, for every $1 \le p < 3$. Hence, $\Phi_+^{7/4}$ belongs to $L^q_{\text{unif}}(\mathbf{R}^3)$, at least for one $q > \frac{3}{2}$. Arguing as in the proof of Lemma 4.11, we deduce that $u \in L^\infty(\mathbf{R}^3)$.

It is worth noticing that the same kinds of arguments may also be made if the non-linear term $u^{7/3}$ is replaced by u^{2p-1} with any convenient $p > \frac{3}{2}$. In addition, if the singularities of Φ are like $\frac{1}{|x|}$, the argument we made in Remark 2.6 in Chapter 2 shows that the L^∞ regularity of u also holds for $\frac{3}{2} \ge p > 1$.

We now turn to the proof of (6.51) of which is very much reminiscent of the proof of the same property which we gave in the course of the uniqueness result in the standard Coulomb case (Lemma 5.5). However, we reproduce the argument here, since (6.51) is a key point in the proof of the uniqueness.

Arguing by contradiction, we exhibit a sequence of points $x_n \in \mathbf{R}^3$ such that

$$u(x_n) \longrightarrow 0 \qquad \text{as} \qquad n \to \infty.$$

Since $u^{4/3} - \Phi$ is in $L^p_{\text{unif}}(\mathbf{R}^3)$, for some $p > \frac{3}{2}$, we deduce, from the equation satisfied by u in (6.48) and Harnack's inequality (see [56] or [24]) that the sequence $u(\cdot + x_n)$ converges to zero uniformly on the compact subsets of $\mathbf{R}^3$.

On the one hand, we set $u_n = u(\cdot + x_n)$ and $\Phi_n = \Phi(\cdot + x_n)$. Then, by virtue of the bounds (6.49) and (6.50), we deduce that the sequences u_n and Φ_n are bounded, respectively, in $L^\infty(\mathbf{R}^3) \cap H^1_{\text{unif}}(\mathbf{R}^3)$ and in $L^{3,\infty}_{\text{unif}}(\mathbf{R}^3)$. Thus, extracting subsequences if necessary, we may assume that they converge, respectively, to $\bar{u} \in L^\infty(\mathbf{R}^3) \cap H^1_{\text{unif}}(\mathbf{R}^3)$ and $\bar{\Phi} \in L^{3,\infty}_{\text{unif}}(\mathbf{R}^3)$.

On the other hand, it is clear that the sequence $m(\cdot + x_n)$ converges to some non-negative $\bar{m}$ in the sense of measures. Moreover, $\bar{m}$ still satisfies the assumptions (H1) and (H2). Thus, passing to the limit in (6.48), as n goes to infinity, we obtain a solution $\bar{\Phi}$ to

$$-\triangle\bar{\Phi} = 4\pi\,\bar{m} \quad \text{in } \mathcal{D}'(\mathbf{R}^3), \qquad (6.53)$$

belonging to $L^{3,\infty}_{\text{unif}}(\mathbf{R}^3)$, and thus, in particular, to $L^1_{\text{unif}}(\mathbf{R}^3)$.

We now reach a contradiction.

Indeed, let $\xi \in \mathcal{D}(\mathbf{R}^3)$ be a cut-off function such that $0 \leq \xi \leq 1$, $supp\,\xi \subset B_2$, $\xi \equiv 1$ on B_1 and $|\triangle\xi| \leq C$, for some positive constant C. Let $\xi_R = \xi\left(\frac{\cdot}{R}\right)$. We apply (6.53), to ξ_R, to write

$$\langle -\triangle\bar{\Phi}; \xi_R \rangle_{\mathcal{D}'(\mathbf{R}^3) \times \mathcal{D}(\mathbf{R}^3)} = 4\pi\,\langle \bar{m}; \xi_R \rangle_{\mathcal{D}'(\mathbf{R}^3) \times \mathcal{D}(\mathbf{R}^3)},$$

or, integrating by parts,

$$-\frac{1}{R^2} \int_{R \leq |x| \leq 2R} \bar{\Phi}\,\triangle\xi\left(\frac{\cdot}{R}\right) = 4\pi\,\langle \bar{m}; \xi_R \rangle_{\mathcal{D}'(\mathbf{R}^3) \times \mathcal{D}(\mathbf{R}^3)}. \qquad (6.54)$$

The left-hand side of the equality is bounded as in the proof of Lemma 5.5:

$$\frac{1}{R^2} \int_{R \leq |x| \leq 2R} |\bar{\Phi}|\,|\triangle\xi\left(\frac{\cdot}{R}\right)| \leq C\,R.$$

Besides,

$$\langle \bar{m}; \xi_R \rangle_{\mathcal{D}'(\mathbf{R}^3) \times \mathcal{D}(\mathbf{R}^3)} \geq \bar{m}(B_R).$$

We therefore obtain, for any $R > 0$,

$$\bar{m}(B_R) \leq C\,R.$$

Letting R go to infinity, we reach a contradiction with (H2). Hence, (6.51) follows.

Now that we have proved (6.49), (6.50), and (6.51), we may attempt the proof of the uniqueness.

Step 2: uniqueness. At this stage, we may fairly copy the proof of the uniqueness in the pure Coulomb case to conclude. Indeed, if one looks carefully at the scheme of this proof, one notices that one essentially argues on the differences $u - v$ and $\phi - \psi$ of two solutions to (6.48). Therefore, the form of the measure m has been removed except through the regularity result (6.49) on Φ and the non-degeneracy (6.51) of u that we have precisely checked above.

Let us detail this. Once the *a priori* bounds are obtained, one uses the bound from below (6.51) to prove the inequality (5.45). Then, the next step in the proof of the uniqueness in the pure Coulomb case consists of considering the operator

$$L = -\triangle + \tfrac{1}{2}(u^{4/3} + v^{4/3}) - \frac{\phi + \psi}{2},$$

where $(u; \phi)$ and $(v; \psi)$ are two solutions.

The point in the proof of the estimate (5.51) is to use

$$a = \tfrac{1}{2}(u^{4/3} + v^{4/3}) - \frac{\phi + \psi}{2} \in L^{3,\infty}_{\text{unif}}(\mathbf{R}^3).$$

This holds true here again because of (6.49) and (6.50).

In the same manner, one shows that $F_1 = \dfrac{\phi + \psi}{2\,(u + v)}(u - v)$ also satisfies

$$\int_{\mathbf{R}^3} F_1^2 \xi^2 < \infty,$$

as in (5.64).

The sequel of the argument made in Chapter 5 only makes use of the differences $u - v$ and $\phi - \psi$ and, therefore, can be adapted straightforwardly to the present case.

Step 3: existence. Let $R > 0$ be fixed for the moment and denote by m_R the measure $m \cdot \chi_{B_R}$; that is, the restriction of m on the ball centered at 0 with radius R. Then, because of the assumption (H1) we made on m, we know that m_R is a bounded non-negative measure on $\mathbf{R}^3$. Next, we define the following minimization problem:

$$I_R = \inf\left\{ E_R(\sqrt{\rho});\ \sqrt{\rho} \in H^1(\mathbf{R}^3),\ \int_{\mathbf{R}^3} \rho = m(B_R) \right\}, \tag{6.55}$$

where

$$E_R(\rho) = \int_{\mathbf{R}^3} |\nabla\sqrt{\rho}|^2 + \tfrac{3}{5}c \int_{\mathbf{R}^3} \rho^{5/3} - \int_{\mathbf{R}^3}\left(m_R \star \frac{1}{|x|} \right)\rho(x)\,\mathrm{d}x$$

$$+ \tfrac{1}{2}\iint_{\mathbf{R}^3 \times \mathbf{R}^3} \frac{\rho(x)\rho(y)}{|x - y|}\,\mathrm{d}x\,\mathrm{d}y. \tag{6.56}$$

Admitting for a while that I_R is achieved by ρ_R and writing down the Euler–Lagrange equation satisfied by $u_R = \sqrt{\rho_R}$, we obtain a solution to

$$-\Delta u_R + c\,u_R^{7/3} - \Phi_R\,u_R + \theta_R\,u_R = 0, \tag{6.57}$$

with $u_R \geq 0$, $\theta_R > 0$ and where $\Phi_R = (m_R - u_R^2) \star \dfrac{1}{|x|}$ satisfies

$$-\Delta\Phi_R = 4\pi\,[m_R - u_R^2]. \tag{6.58}$$

The fact that I_R is achieved and that its (unique) minimizer ρ_R satisfies the right constraint has been proved by Lieb in [33] (Theorem 7.19, page 632). Lieb's result applies here because m_R is a bounded measure with compact support in $\mathbf{R}^3$, and because we impose the electrical neutrality through the constraint $\int_{\mathbf{R}^3} \rho = m(B_R) = m_R(\mathbf{R}^3)$. Moreover, the same conclusion holds true if we replace in the definition (6.56) of E_R the term $\int_{\mathbf{R}^3} \rho^{5/3}$ by $\int_{\mathbf{R}^3} \rho^p$, provided that $\frac{3}{2} < p < +\infty$ (see Lions [42] and Lieb [33]).

Our next step now consists of proving some *a priori* bounds on u_R and $\Phi_R - \theta_R$, respectively, in $L^\infty(\mathbf{R}^3)$ and $L^1_{\text{unif}}(\mathbf{R}^3)$ which are independent of R. Once these bounds are obtained, we shall pass to the limit in (6.57) and (6.58) as R goes to infinity (at least in the sense of distributions), and we shall obtain a solution to (6.48).

In order to obtain the bounds on u_R and $\Phi_R - \theta_R$, we argue as we did in Chapter 3 (Propositions 3.10 and 3.12). We first check that the Solovej inequality (3.20) still holds true here (this only makes use of the fact that m is a non-negative measure), and we therefore also obtain here a bound from above for θ_R and a bound from below for Φ_R. Next, we replace, in the proof of Proposition 3.10, Φ_Λ by $\Phi_R - \theta_R$. By the same proof, we then obtain

$$\|\Phi_R - \theta_R\|_{L^{3,\infty}_{\text{unif}}(\mathbf{R}^3)} \leq C,$$

for some constant C that is independent of R. It is worth noticing here that it is the fact that m satisfies (H1) which implies that this bound exists.

Eventually, in view of (3.20), we deduce from this bound a bound in L^∞ for u_R, as we did in the proof of Proposition 3.10. This allows us to pass locally to the limit in (6.57).

At this stage, we have obtained a solution to (6.48) with u in $L^\infty(\mathbf{R}^3)$ and Φ in $L^p_{\text{unif}}(\mathbf{R}^3)$, for all $1 \leq p < 3$. Thus, because of the equation satisfied by u and from elliptic regularity results, it is standard to check that, in fact, u belongs to $W^{2,p}_{\text{unif}}(\mathbf{R}^3) \cap L^\infty(\mathbf{R}^3)$ for all $1 \leq p < 3$ and that ∇u belongs to $W^{1,q}_{\text{unif}}(\mathbf{R}^3) \cap L^p_{\text{unif}}(\mathbf{R}^3)$, for all $1 \leq q < 3$ and $1 \leq p < +\infty$. Hence, from Sobolev embeddings, u also lies in the Hölder spaces $C^{0,\alpha}(\mathbf{R}^3)$, for every $0 < \alpha < 1$.

This concludes the proof.   ◇

6.3.2   *When there is no uniqueness*

In this subsection, we would like to illustrate the following observation: if there is no uniqueness for the Euler–Lagrange equation obtained in the thermodynamic limit, it seems useless to hope that some convergence of the electronic density will hold. Indeed, in this case, we may build some oscillatory sequences of densities that in some sense (that we shall make precise below) approach, for each Λ, the infimum energy, and that have no limit.

As an example of such a situation, let us introduce the following model. For $\Lambda \subset \mathbf{Z}^3$ fixed, we consider the usual set of point nuclei located at the points of Λ, and the non-convex energy functional

$$E_\Lambda(u) = \int_{\mathbf{R}^3} |\nabla u|^2 - \int_{\mathbf{R}^3} V_\Lambda u^2 + \tfrac{3}{5}a \int_{\mathbf{R}^3} |u|^{10/3}$$
$$- \tfrac{3}{4}b \int_{\mathbf{R}^3} |u|^{8/3} + \tfrac{1}{2} \int \int_{\mathbf{R}^3 \times \mathbf{R}^3} u^2(x) V(x-y) u^2(y), \qquad (6.59)$$

where a and b are two positive constants (to be determined below) and V is the short-range interaction potential

$$V(x) = 1 \quad \text{if } x \in \Gamma_0,$$
$$= 0 \quad \text{otherwise,} \tag{6.60}$$

and

$$V_\Lambda(x) = \sum_{y \in \Lambda} V(x - y) = 1_{\Gamma(\Lambda)}.$$

We then define the minimization problem

$$I_\Lambda = \inf \left\{ E_\Lambda(u); u \in H^1(\mathbf{R}^3), \int_{\mathbf{R}^3} |u|^2 = |\Lambda| \right\}. \tag{6.61}$$

Note that in this simplified model with short-range interactions the electrostatic energy modelling the inter-nuclear repulsion is $\frac{1}{2} \sum_{y \neq z \in \Lambda} V(y - z) = 0$.

For Λ large enough, there exists at least one minimizing density u_Λ^2 for this problem. Indeed, any minimizing sequence u_n of (6.61) is clearly bounded in $H^1(\mathbf{R}^3)$. We may therefore assume it weakly converges, to some $u_\Lambda \in H^1(\mathbf{R}^3)$. This function u_Λ, that we may assume to be non-negative, satisfies $\int |u_\Lambda|^2 \leq |\Lambda|$ and, arguing as in the standard TFDW case with Coulomb potential, we see that u_Λ is a minimum for the problem (6.61) with a relaxed constraint (see, for instance, [42]). Next, we may show that necessarily $\int |u_\Lambda|^2 = |\Lambda|$ (at least for Λ large enough): otherwise, by the Harnack inequality, we obtain either that $u_\Lambda \equiv 0$ or that $u_\Lambda > 0$ is a solution to some inequality of the type $-\Delta u_\Lambda + W u_\Lambda \geq 0$ with $[W]_+ \in L^{3/2}(\mathbf{R}^3)$. We then argue as in the proof of Proposition 2.2 of Chapter 2 to reach a contradiction in both cases: in the first case, because we may find, when the domain Ω is large enough, some $\varphi \in \mathcal{D}(\mathbf{R}^3)$ such that $\int \varphi^2 = 1$ and $\int |\nabla \varphi|^2 < \int_\Omega \varphi^2$, while in the second case, we reach a contradiction because of a result by Simon (see [33]).

We do not know if such a minimizing u_Λ is unique (up to a sign), but we may choose one of the convenient u_Λ and wonder whether or not it converges. We now give some ideas that make us believe that it does not converge (but we are not able to prove it). We are going to build a sequence of 'almost minimizers' for a minimization problem that is precisely of the type (6.61) and that does not converge.

We introduce the periodic problem that is to some extent likely to be obtained in the thermodynamic limit for (6.61); namely,

$$I_{\text{per}}(\mu) = \inf \left\{ E_{\text{per}}(u); u \in H^1_{\text{per}}(\Gamma_0), \int_{\Gamma_0} u^2 = \mu \right\}, \tag{6.62}$$

where

$$E_{\text{per}}(u) = \int_{\Gamma_0} |\nabla u|^2 - \int_{\Gamma_0} |u|^2 + \tfrac{3}{5} a \int_{\Gamma_0} |u|^{10/3}$$

$$- \tfrac{3}{4}b \int_{\Gamma_0} |u|^{8/3} + \tfrac{1}{2} \left(\int_{\Gamma_0} |u|^2 \right)^2. \tag{6.63}$$

Indeed, in this case $V_\infty \equiv 1$ over the whole space $\mathbf{R}^3$.

The constant μ in (6.62) is defined by the charge corresponding to the minimum of

$$I_{\mathrm{per}} = \inf_{0 \leq \lambda \leq 1} \inf \left\{ E_{\mathrm{per}}(u); u \in H^1_{\mathrm{per}}(\Gamma_0), \int_{\Gamma_0} u^2 = \lambda \right\}. \tag{6.64}$$

We next show that, when a and b are chosen in a convenient way, no minimizer of (6.62) is constant on Γ_0.

For this purpose, we first remark that when b is large enough (namely, $b \geq a$), the minimum of the energy (6.63) over the set of constant functions on Γ_0 of charge $\lambda \leq 1$ is achieved only for $\lambda = 1$.

Let us now assume that some constant function u is a minimizer of (6.62). The above remark implies that necessarily $u \equiv 1$. We claim that this cannot hold. Indeed, the Euler–Lagrange equation then reads

$$a - b + \theta = 0,$$

where θ is the Lagrange multiplier. In addition, the second order condition reads

$$\forall \varphi \in H^1_{\mathrm{per}}(\Gamma_0) \text{ with } \int_{\Gamma_0} \varphi = 0, \quad \int_{\Gamma_0} |\nabla \varphi|^2 + (\tfrac{7}{3}a - \tfrac{5}{3}b + \theta) \int_{\Gamma_0} \varphi^2 \geq 0;$$

that is

$$\forall \varphi \in H^1_{\mathrm{per}}(\Gamma_0) \text{ with } \int_{\Gamma_0} \varphi = 0, \quad \int_{\Gamma_0} |\nabla \varphi|^2 + (\tfrac{4}{3}a - \tfrac{2}{3}b) \int_{\Gamma_0} \varphi^2 \geq 0. \tag{6.65}$$

Hence, if we choose b large enough in order to have the second eigenvalue of the operator $-\Delta + (\tfrac{4}{3}a - \tfrac{2}{3}b)$ on Γ_0 with periodic boundary conditions negative, we reach a contradiction with (6.65).

Therefore we know that no minimizer of (6.62) is constant. It follows that such a minimizer cannot be unique. Indeed, since the energy functional (6.63) is translation invariant, we may change a minimizer u, for instance, into $u(\cdot + \tfrac{1}{2}e)$, where e is a unit vector in $\mathbf{R}^3$, and obtain another minimizer.

Let us now take two different minimizers, u_1 and u_2, of (6.62). For each Λ, we assume for the sake of simplicity that $\Gamma(\Lambda)$ is the sequence of homothetic cubes centered at 0, and we index this sequence by the integer $n = \tfrac{1}{2}(|\Lambda|^{1/3} + 1)$. We build a function $\bar{u}_n = \bar{u}_\Lambda \in H^1(\mathbf{R}^3)$ satisfying $\int_{\mathbf{R}^3} \bar{u}_\Lambda^2 = \mu |\Lambda|$ by putting together u_1 and u_2 as follows:

$$\bar{u}_{2n}(x) = \begin{cases} u_1(x), & \text{if } x \in \Gamma(\Lambda) \backslash \Lambda^1 \cap \{y \in \mathbf{R}^3; y_1 > \tfrac{1}{2}\}, \\ u_2(x), & \text{if } x \in \Gamma(\Lambda) \backslash \Lambda^1 \cap \{y \in \mathbf{R}^3; y_1 < -\tfrac{1}{2}\}, \\ v(x), & \text{if } x \in \Gamma(\Lambda) \backslash \Lambda^1 \cap \{y \in \mathbf{R}^3; -\tfrac{1}{2} < y_1 < \tfrac{1}{2}\}, \end{cases}$$

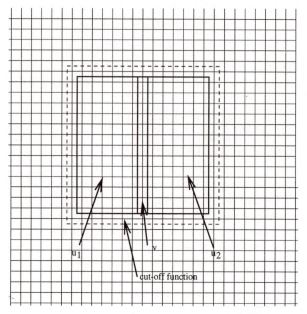

FIG. 6.1. **Construction of $\bar{u}_\Lambda$ in two dimensions.**

$$\bar{u}_{2n+1}(x) = \begin{cases} u_2(x), & \text{if } x \in \Gamma(\Lambda) \backslash \Lambda^1 \cap \{y \in \mathbf{R}^3; y_1 > \frac{1}{2}\}, \\ u_1(x), & \text{if } x \in \Gamma(\Lambda) \backslash \Lambda^1 \cap \{y \in \mathbf{R}^3; y_1 < -\frac{1}{2}\}, \\ v(-x_1, x_2, x_3), & \text{if } x \in \Gamma(\Lambda) \backslash \Lambda^1 \cap \{y \in \mathbf{R}^3; -\frac{1}{2} < y_1 < \frac{1}{2}\}, \end{cases}$$

where v is a function, periodic in the two directions x_2, x_3 that 'glues' u_1 and u_2 and that satisfies $\int_{\Gamma_0} v^2 = \mu$. On the boundary Λ^1 of $\Gamma(\Lambda)$, we use a convenient cut-off function as in Chapter 2 (see Fig. 6.1). One can then see that $E_\Lambda(\bar{u}_\Lambda) = \bar{I}_\Lambda + o(|\Lambda|)$, where, $\bar{I}_\Lambda$ is the minimization problem analogous to (6.61) but with constraint $\mu|\Lambda|$; namely,

$$\bar{I}_\Lambda = \inf \left\{ E_\Lambda(u); u \in H^1(\mathbf{R}^3), \int |u|^2 = \mu|\Lambda| \right\}. \tag{6.66}$$

Therefore, $\bar{u}_\Lambda$ turns out to be a sequence of 'almost minimizers' of a problem of the same family as (6.61). And, of course, $\bar{u}_\Lambda$ oscillates locally between the two functions u_1 and $u_2 \neq u_1$ and therefore does not converge.

Note that, the energy functional not being convex, we cannot claim either that there exists a thermodynamic limit for the energy per unit volume in this setting.

6.4   Geometries which are not that non-periodic

Until Section 6.2 above, we have dealt (in the Coulomb case at least) with cubic periodic lattices. Now that we have Theorem 6.10, we may attempt other situations; this section is devoted to the study of a few cases when the geometry of the lattice is not cubic, or even cases when there is no 'lattice' *sensu stricto*. However, in order to keep some control on the behaviour of the density at infinity, we need to impose some particular property on the measure defining the configuration of nuclei. From the physical viewpoint, we are going to extend our study to the case of general periodic crystals, then to (still periodic) crystals with some impurities (Section 6.4.1) and, finally, we shall make an incursion into the theory of quasicrystals (Section 6.4.2).

6.4.1 *General periodic lattices, impurities.*

Unless otherwise specified, m will denote in the following a positive measure on $\mathbf{R}^3$ satisfying the assumptions, (H1) and (H2) of Theorem 6.10, and is interpreted as the measure defining the configuration of the nuclei, while the positive solution u to (6.48) may be viewed as the square root of the electronic density of the infinite system under study. In the following, we consider various examples of measures satisfying the assumptions (H1) and (H2) and which are relevant from the physical viewpoint, and we study the properties inferred on the solution $(u; \Phi)$ to (6.48) by the uniqueness result of Theorem 6.10 or its proof.

6.4.1.1 *General periodic lattices* Let us first observe that any positive Radon measure which is periodic satisfies both (H1) and (H2). Then, in view of Theorem 6.10, u and Φ are periodic and share the same periodicity as m. By the way, this property holds true whatever the shape and the size of the primitive cell are and, in addition, we need not have one nucleus per cell. Moreover, it is clear that the solution $(u; \Phi)$ does not depend on the definition of the primitive cell we choose.

In particular, the uniform convergence results of the densities on the interior domains stated in Theorem 5.9 in Section 5.5 of Chapter 5, remain valid for any periodic lattice. Therefore, we may apply the strategy of proof developed in Section 6.2 and pass to the thermodynamic limit for the energy per unit cell whatever the size or the shape of the primitive cell is, and whatever the number of nuclei per unit cell is, provided that the (smeared or point) nuclei are periodically distributed in the space.

In addition, it is worth noticing that, as soon as we know that Φ is periodic, we deduce by integrating the equation satisfied by Φ in (6.48) over the primitive cell (say Γ_0) that, necessarily,

$$\int_{\Gamma_0} u^2 = m(\Gamma_0). \tag{6.67}$$

At this point, we want to make the following comment. If we assume that the measure m in Theorem 6.10 is periodic and satisfies (H1), we need not impose that m is a non-negative measure (see also Section 6.4.1.2 below in the non-periodic case). However, since u is positive on Γ_0, (6.67) implies that $m(\Gamma_0) > 0$. Rather to intend to prove this claim in its full generality, let us illustrate with the following example.

We now consider the ground-state energy of a charged crystal; that is, for any $0 \le \mu$, we define the periodic minimization problem

$$I_{\text{per}}^{\text{TFW}}(\mu) = \inf \left\{ E_{\text{per}}^{\text{TFW}}(\rho); \rho \geq 0, \sqrt{\rho} \in H_{\text{per}}^1(\Gamma_0), \int_{\Gamma_0} \rho = \mu \right\}, \tag{6.68}$$

where we recall that

$$E_{\text{per}}^{\text{TFW}}(\rho) = \int_{\Gamma_0} |\nabla \sqrt{\rho}|^2 + \int_{\Gamma_0} \rho^{5/3} - \int_{\Gamma_0} \rho(x) G(x) \mathrm{d}x$$
$$+ \frac{1}{2} \iint_{\Gamma_0 \times \Gamma_0} \rho(x) \rho(y) G(x - y) \mathrm{d}x \, \mathrm{d}y.$$

Then, for every $0 < \mu$, $I_{\text{per}}^{\text{TFW}}(\mu)$ is achieved by a unique minimizer, say ρ, and the Euler–Lagrange equation satisfied by $u = \sqrt{\rho}$ may be written in the following way:

$$\begin{cases} -\Delta u + c\, u^{7/3} - \Phi\, u = 0, \\[2mm] u \geq 0, \\[2mm] -\Delta \Phi = 4\pi \left[m - u^2 \right], \end{cases} \tag{6.69}$$

where

$$m = \sum_{y \in \mathbf{Z}^3} \delta_y - (1 - \mu). \tag{6.70}$$

Let us explain where (6.70) comes from. If we compare (6.69) with the expression (6.68) of $I_{\text{per}}^{\text{TFW}}(\mu)$, it is easy to convince ourselves that

$$\Phi(x) = G(x) - \int_{\Gamma_0} G(x - y)\, \rho(y)\, \mathrm{d}y + \theta,$$

for some Lagrange multiplier θ. Next, we recall that G is, by definition, the periodic solution of

$$-\Delta G = 4\pi \left[\sum_{y \in \mathbf{Z}^3} \delta_y - 1 \right],$$

such that $\int_{\Gamma_0} G = 0$. It is now an easy exercise to deduce the equation satisfied by Φ (and thus (6.70)) from the two properties of G recalled just above, together with the two facts that ρ is periodic and is of mass μ on Γ_0.

It is no more relevant this time to see m as the measure defining the nuclei. However, m may be interpreted as the effective charge to which the electrons are subjected: each cell contains one (point) nucleus of charge 1 together with a uniform density of negative charge $-(1 - \mu)$. Even if m is no longer a non-negative measure, the existence and uniqueness result of Theorem 6.10 remains true in that case. Indeed, it is clear that $|m|$ satisfies (H1) and (H2) but, since m is not signed, we provide a variant of the argument that ensures that any solution u in (6.69) such that $u \in L^\infty(\mathbf{R}^3)$ and $\Phi \in L^{3,\infty}_{\mathrm{unif}}(\mathbf{R}^3)$ satisfies

$$\inf_{\mathbf{R}^3} u > 0.$$

Indeed, we go back to the proof of (6.51) in the course of the proof of Theorem 6.10. We shall only detail the argument in the case of a cubic unit cell for the sake of simplicity, but we should keep in mind that everything remains true with minor modifications for general periodic cells.

For all integer $n \geq 1$, we denote by $\Gamma(2n+1)$ the large cube centered at zero which is the union of $(2n+1)^3$ unit cells, and we choose the cut-off function $\xi_n \in \mathcal{D}(\mathbf{R}^3)$ such that $0 \leq \xi_n \leq 1$, $supp\, \xi_n \subset \Gamma(2n+1)$, $\xi_n \equiv 1$ on $\Gamma(2n-1)$ and $|\triangle \xi_n| \leq C$, for some positive constant C that is independent of n. Following the proof of (6.51), we assume by contradiction that there exists a solution to

$$-\triangle \bar{\Phi} = 4\pi\, m \qquad \text{in } \mathbf{R}^3,$$

that belongs to $L^{3,\infty}_{\mathrm{unif}}(\mathbf{R}^3)$. We multiply this equation by ξ_n and then integrate over $\mathbf{R}^3$. This yields

$$\langle -\triangle \bar{\Phi}; \xi_n \rangle_{\mathcal{D}'(\mathbf{R}^3) \times \mathcal{D}(\mathbf{R}^3)} = 4\pi\, \langle m; \xi_n \rangle_{\mathcal{D}'(\mathbf{R}^3) \times \mathcal{D}(\mathbf{R}^3)},$$

or, integrating by parts,

$$-\int_{\Gamma(2n+1) \backslash \Gamma(2n-1)} \bar{\Phi}\, \triangle \xi_n = 4\pi\, \langle m; \xi_n \rangle_{\mathcal{D}'(\mathbf{R}^3) \times \mathcal{D}(\mathbf{R}^3)}.$$

On the one hand, the left-hand side of the above equality is bounded from above, as in the proof of Lemma 5.5, by

$$\int_{\Gamma(2n+1) \backslash \Gamma(2n-1)} |\bar{\Phi}|\, |\triangle \xi_n| \leq C\, |\Gamma(2n+1) \backslash \Gamma(2n-1)|\, \|\Phi\|_{L^1_{\mathrm{unif}}(\mathbf{R}^3)}$$

$$\leq C\, n^2,$$

where C does not depend on n. On the other hand, we bound from below the right-hand side as follows:

$$|\langle m; \xi_n \rangle|_{\mathcal{D}'(\mathbf{R}^3) \times \mathcal{D}(\mathbf{R}^3)}$$
$$\geq \mu\, |\Gamma(2n-1)| - |\Gamma(2n+1) \backslash \Gamma(2n-1)|\, \sup_{x \in \mathbf{R}^3} |m(x + \Gamma_0)|$$
$$\geq \mu\, (2n-1)^3 - C\, n^2.$$

We reach the desired contradiction by letting n go to infinity.

6.4.1.2 *Impurities* In this section, we choose m in the following form

$$m = m_{\mathrm{per}} + \mu, \tag{6.71}$$

where m_{per} is a periodic non-negative measure satisfying (H1) and (H2) and μ is a signed bounded measure on $\mathbf{R}^3$, which is assumed to describe the measure of the impurities in the periodic lattice defined by m_{per}. Then, it is easy to convince ourselves that the uniqueness result of Theorem 6.10 still holds true in that case. This claim is obvious when μ is non-negative, since in that case m clearly satisfies (H1) and (H2). When μ is non-positive, it is clear that $|m|$ still satisfies (H1). Moreover, (H2) holds true since m_{per} satisfies (H2) and

$$\inf_{x \in \mathbf{R}^3} m(x + B_R) \geq \inf_{x \in \mathbf{R}^3} m_{\mathrm{per}}(x + B_R) - |\mu|(\mathbf{R}^3).$$

As a corollary of Theorem 6.10, we shall then prove some 'periodicity at infinity' for the solution $(u; \Phi)$ of the system (6.48) for the choice (6.71) for m in the sense of the following:

Corollary 6.11 *Let $(u_{\mathrm{per}}; \Phi_{\mathrm{per}})$ be the unique (periodic) solution to (6.48) corresponding to $m = m_{\mathrm{per}}$. Then, for all sequences $(y_n)_{n \geq 1}$, there exists a subsequence such that*

$$u(\cdot + y_n) \longrightarrow u_{\mathrm{per}}(\cdot + k),$$

uniformly on $\mathbf{R}^3$, and, for all $1 \leq p < 3$,

$$\Phi(\cdot + y_n) \longrightarrow \Phi_{\mathrm{per}}(\cdot + k) \qquad \text{strongly in } L^p_{\mathrm{unif}}(\mathbf{R}^3),$$

as n goes to infinity, for some $k \in \bar{\Gamma}_0$.

Proof of Corollary 6.11 Let $(y_n)_{n \geq 1}$ such that $|y_n| \to +\infty$. We set $y_n = k_n + [y_n]$, where $k_n \in \Gamma_0$ and $[y_n] \in \mathbf{Z}^3$. Up to a subsequence, k_n is assumed to converge to some k in $\bar{\Gamma}_0$. We use the notation $m_n = m(\cdot + y_n)$, $u_n = u(\cdot + y_n)$, and $\Phi_n = \Phi(\cdot + y_n)$. From the bounds we have obtained on u and Φ through Theorem 6.10, we know that u_n and Φ_n converge, respectively, to $\bar{u} \geq 0$ belonging to $L^\infty(\mathbf{R}^3)$ and $\bar{\Phi} \in L^{3, \infty}_{\mathrm{unif}}(\mathbf{R}^3)$. In addition, m_n converges to $m_{\mathrm{per}}(\cdot + k)$ for the weak convergence of measures. To conclude, we pass to the limit in the system of equations satisfied by u_n and Φ_n as n goes to infinity, and we appeal to the uniqueness result of Theorem 6.10 to identify $\bar{u}$ with $u_{\mathrm{per}}(\cdot + k)$ and $\bar{\Phi}$ with $\Phi_{\mathrm{per}}(\cdot + k)$. $\diamond$

6.4.2 *Towards quasicrystals: almost periodic measures*

We now turn to a kind of geometry that is not periodic even in the long range, but that still exhibits a convenient long-range behaviour: almost periodicity. It is closely related to a physical problem of some interest; namely, the modelling of quasiperiodic crystals, as we now explain.

It is not clear for physicists, nor for mathematicians, what the definition of a crystal should be, for there seems to be some debate here. If we follow Senechal in [50], we may define a crystal as follows:

A crystal is any solid with an essentially discrete diffraction diagram, that is to say a diffraction diagram with sharp spots.

Consequently, the symmetry of a crystal is defined as simply the symmetry of its diffraction diagram.

It turns out that the diffraction pattern of some solids exhibits a symmetry forbidden by the crystallographic restriction on the ambient space: for instance, symmetry of order 5 is forbidden in two-dimensional space. Crystals giving rise to such forbidden symmetries are called *quasicrystals*. They are of growing importance nowadays in solid state physics. A way to build a quasicrystal mathematically is to make use of almost periodic functions. However, there does not seem to exist a widely accepted rigorous mathematical definition of a quasicrystal. One possible rigorous mathematical definition is given through the so-called Meyer's sets in [3], and a more general one is presented in [18]. In all cases, quasicrystals are particular cases of almost periodic lattices whose definition follows from the notion of almost periodic functions. We therefore turn to the definition and the basic properties of almost periodic functions.

We first introduce the notion of *continuous almost periodic functions*, and we shall generalize this notion to less regular functions afterwards.

Before all this, we shall need the following:

Definition 6.1 A set $E \subset \mathbf{R}^n$ is said to be *relatively dense* if there exists a radius R such that any ball of radius R in $\mathbf{R}^n$ contains at least a point of E.

Now, we may state the following:

Definition 6.2 A continuous function f is called *almost periodic* (or *uniformly almost periodic*) on $\mathbf{R}^n$ if one of the following two equivalent conditions is satisfied:

(i) For any $\varepsilon > 0$, the set of vectors $\tau \in \mathbf{R}^n$ such that

$$\sup_{x \in \mathbf{R}^n} |f(x + \tau) - f(x)| \leq \varepsilon \qquad (6.72)$$

is relatively dense.

(ii) The set of translates of f, namely $\{f(\cdot + \tau); \tau \in \mathbf{R}^n\}$, is relatively compact for the topology of uniform convergence. That is to say: for any arbitrary sequence (y_p), there exists a subsequence $(y_{\alpha(p)})$ such that the sequence of continuous functions $f(\cdot + y_{\alpha(p)})$ converges uniformly.

Remark 6.7 The above definitions may be found in Amerio and Prouse [1], or alternatively in Besicovitch [10], or Bohr [11]. A vector τ satisfying (6.72) may be called an *almost period* or more precisely an ε-*almost period* of the function f. Condition (ii) is called Bochner's criterion, and a function satisfying that criterion is called *normal*. The proof of the equivalence between (i) and (ii) appears, for instance, in the above references.

Let us now give some examples of almost periodic functions.

First of all, it is clear that any periodic function is almost periodic. Indeed, it is obvious that a periodic function satisfies condition (i). To see directly that it satisfies condition (ii), let us observe, in dimension 1 for the sake of simplicity, that if y_n is any arbitrary sequence of points, it is possible to define $[y_n]$ such that for any n, y_n belongs to $\Gamma_0 + [y_n]$, where Γ_0 denotes the periodic cell of f, and $[y_n]$ is therefore the index of the translated cell. Since $|y_n - [y_n]|$ is clearly bounded, we may assume, extracting a subsequence if necessary, that it converges to some $z \in \Gamma_0$. It is then straightforward to check that $f(\cdot + y_n)$ converges uniformly to $f(\cdot + z)$.

Another example is provided by the so-called *quasiperiodic* functions, i.e. functions that are a finite linear combination of periodic functions:

$$f = \sum_{i=1}^{k} f_i,$$

with f_i periodic. The argument we have just made to prove that a periodic function satisfies condition (ii) can be easily adapted to the present situation.

The notion of almost periodic functions is a generalization of quasiperiodic ones in the sense that any almost periodic function is a sum, possibly infinite, of periodic functions. Indeed, it is possible to show that the set of almost periodic continuous functions is the closure for the uniform norm of the vector space generated by the functions $\alpha e^{i\omega \cdot t}$, $\omega \in \mathbf{R}^3$.

An alternative way to see the intimate link between periodic functions and almost periodic ones is the following.

Consider a sequence of periodic functions built as follows: for any integer k, let us take a periodic function $F_k(x_1, ..., x_{m_k})$ of m_k variables. Now suppose that the F_k converge uniformly as k goes to infinity, yielding a so-called *limit periodic function* F of possibly an infinite number of variables. Next, define $f(x) = F(x, x, ..., x, ...)$. It turns out that f is an almost periodic function. Such a construction is generic in the following sense: any uniformly almost periodic function f is the diagonal function of a limit periodic function F of a finite or an infinite number of variables. In fact (see Senechal [50] for the details), this construction has its physical counterpart. A way to build an almost periodic lattice in dimension n is to take the trace on $\mathbf{R}^n$ of a periodic lattice in a dimension larger than n.

We now extend the definition of almost periodicity to L^p_{loc} ($1 \leq p < +\infty$) functions.

Definition 6.3 A function f in $L^p_{\text{loc}}(\mathbf{R}^n)$, $1 \leq p < +\infty$, is called *almost periodic* if, for any $\varepsilon > 0$, the set of vectors $\tau \in \mathbf{R}^n$ such that

$$\sup_{x \in \mathbf{R}^n} \int_{x+B_1} |f(y+\tau) - f(y)|^p dy \leq \varepsilon \qquad (6.73)$$

is relatively dense.

In the same fashion as above, it is worth noticing that the set of almost periodic L^p_{loc} functions is the closure for the L^p_{unif} norm of the vector space generated by the functions $\alpha e^{i\omega \cdot t}$, $\omega \in \mathbf{R}^3$.

Let us now introduce a very useful notion as far as almost periodic functions are concerned, namely the mean value.

For any almost periodic function, we may define its *mean value* by

$$< f > = \lim_{R \longrightarrow \infty} \frac{1}{R^3} \int_{x+B_R} f,$$

the right-hand side being actually uniform in $x \in \mathbf{R}^3$.

Before we turn to another generalization of almost periodicity, we would like to illustrate the link between this notion and the hypotheses (H1) and (H2) appearing in our general result Theorem 6.10.

Let us first show that a non-negative L^1_{loc} almost periodic function f necessarily satisfies (H1).

Let $\varphi \in \mathcal{D}(\mathbf{R}^3)$ such that $0 \le \varphi \le 1$, $supp(\varphi) \subset B_2$, $\varphi = 1$ on B_1. Then $f \star \varphi$ is continuous, almost periodic and thus bounded by some constant $M \in \mathbf{R}_+$ (see [1] or [10]). Therefore,

$$M \ge f \star \varphi(x) \ge \int_{x+B_1} f(y) \mathrm{d}y,$$

and (H1) follows.

Next, we show, under the same assumptions, that if we assume in addition that $f \not\equiv 0$, then it satisfies (H2). More precisely, we are going to show that such a function has a positive mean value. We recall that we denote $\Gamma_0 = [-\frac{1}{2}, \frac{1}{2}[^3$. Since $f \ge 0$ and $f \not\equiv 0$, we may choose R_0 large enough (in particular larger than 2) in order to have

$$\int_{R_0 \Gamma_0} f(x) \mathrm{d}x \ge \delta > 0.$$

Since f is almost periodic, we know there exists a relatively dense set $\{\tau\}$ such that, for any τ,

$$\sup_{x \in \mathbf{R}^3} \int_{x+B_1} |f(y+\tau) - f(y)| \mathrm{d}y \le \frac{\delta}{2R_0^3}.$$

For each $(p, q, r) \in \mathbf{Z}^3$, we may thus find, replacing R_0 by some $R_0' > R_0$ if necessary, $\tau_{(p,q,r)}$ in the cube $R_0\Gamma_0 + 2R_0(p, q, r)$ such that

$$\int_{R_0 \Gamma_0 + \tau_{(p,q,r)}} f(x) \mathrm{d}x \ge \int_{R_0 \Gamma_0} f(x) \mathrm{d}x - \frac{\delta}{2} \ge \frac{\delta}{2} > 0.$$

It is then easy to see that in any cube of the tiling $(4R_0)\mathbf{Z}^3 + 4R_0\Gamma_0$, there exists $(p, q, r) \in \mathbf{Z}^3$ such that

$$R_0 \Gamma_0 + \tau_{(p,q,r)} \subset 4R_0\Gamma_0 + 4R_0(p,q,r).$$

Therefore, there exists a constant $C > 0$ such that

$$\lim_{R \longrightarrow +\infty} \frac{1}{R^3} \int_{B_R} f(x)\mathrm{d}x \geq C\delta.$$

Likewise, with a constant C uniform with respect to x,

$$\lim_{R \longrightarrow +\infty} \frac{1}{R^3} \int_{x+B_R} f(x)\mathrm{d}x \geq C\delta,$$

and (H2) follows.

Conversely, it is worth noticing that there exist functions satisfying (H1) and (H2) that are not almost periodic. Therefore, the scope of Theorem 6.10 is larger than the (soon wide) setting of almost periodic functions. Indeed, we may, for instance, define the function f as follows. Choose $f_0 \geq 0$, a function in $\mathcal{D}(\frac{1}{2}\Gamma_0)$ of total mass 1. In each cube $\Gamma_0 + (p,q,r)$, $\{p,q,r\} \subset \{-1,0,1\}$, we put f_0 at the centre of the cube. In each cube $\Gamma_0 + (p,q,r)$, $\{p,q,r\} \subset \{-2,-1,0,1,2\}$, with at least one of the $|p|, |q|, |r|$ equal to 2, we put $2f_0$ at the centre of the cube. In each cube $\Gamma_0 + (p,q,r)$, $\{p,q,r\} \subset \{-2^2, ..., -1, 0, 1, ..., 2^2\}$, with at least one of the $|p|, |q|, |r|$ larger than or equal to 3, we put $2f_0$ at the centre of the cube. Generically, in each cube $\Gamma_0 + (p,q,r)$, $\{p,q,r\} \subset \{-2^n, ..., -1, 0, 1, ..., 2^n\}$, with at least one of the $|p|, |q|, |r|$ larger than or equal to $2^{n-1} + 1$, we put f_0 at the centre of the cube if n is even, and $2f_0$ if n is odd. This function f obviously satisfies (H1) and (H2), since it is bounded from below by the periodic function built with f_0 and bounded from above by the one built with $2f_0$. However, it is not almost periodic. Indeed, if it was almost periodic, then we could define $< f >$ and obtain $0 << f >< +\infty$. Choosing successively $x_{1,n} = (2^{2n} + 2^{2n-1}, 0, 0)$ and $x_{2,n} = (2^{2n+1} + 2^{2n}, 0, 0)$, we see that $< f >$ is alternatively 1 or 2, and therefore reach a contradiction.

With a view to treating quasicrystals with point nuclei, we now extend the notion of almost periodicity to measures on $\mathbf{R}^3$.

Definition 6.4 A Radon measure m on $\mathbf{R}^3$ is called *almost periodic* if it satisfies the following condition: For any sequence $y_n \in \mathbf{R}^3$, there exist a subsequence $y_{\alpha(n)}$ of y_n and a measure $\bar{m}$, possibly depending on the extraction α and on the sequence y_n, such that the sequence of measures $m_n = m(\cdot + y_{\alpha(n)})$ converges to $\bar{m}$, as n goes to infinity, in the following sense:

$$\sup_{x \in \mathbf{R}^3} \sup \left\{ \int_{\mathbf{R}^3} \Phi(m_n - \bar{m}); \right.$$

$$\left. \Phi \in W^{1,\infty}(\mathbf{R}^3), \|\Phi\|_{W^{1,\infty}} \leq 1, Supp \, \Phi \subset B(x,1) \right\} \overset{n\to\infty}{\longrightarrow} 0.$$

It is straightforward to see that, in the same fashion as above, periodic measures and finite sums of periodic measures, i.e. quasiperiodic measures, are almost periodic. On the contrary, one can check that

$$m = \sum_{i \in \mathbf{Z}^3 \backslash \{0\}} \delta(\cdot - i)$$

is not an almost periodic measure. Indeed, consider $y_n = (n, 0, 0)$: $m(\cdot + y_n)$ converges in $\mathcal{D}'(\mathbf{R}^3)$ to $\bar{m} = \sum_{i \in \mathbf{Z}^3} \delta(\cdot - i)$, but does not converge to $\bar{m}$ in the sense of the above definition, since

$$\sup_{x \in \mathbf{R}^3} \sup \left\{ \int_{\mathbf{R}^3} \Phi(m_n - \bar{m}) = \Phi(y_n); \right.$$

$$\left. \Phi \in W^{1,\infty}(\mathbf{R}^3), \|\Phi\|_{W^{1,\infty}} \leq 1, Supp \ \Phi \subset B(x,1) \right\} = 1.$$

We now prove the following corollary of Theorem 6.10:

Corollary 6.8 *Let m be a non-negative almost periodic measure satisfying conditions (H1) and (H2). Then, the functions u and Φ such that $(u; \Phi)$ is the solution to the system (6.48) are both almost periodic.*

Remark 6.9 Let us first note that if $m \geq 0$ is assumed to be locally bounded and quasiperiodic, then it is clear that (H1) and (H2) are automatically fulfilled. Moreover, if $m \geq 0$ is almost periodic and L^p_{loc}, we have proved above that (H1) and (H2) are also satisfied. This property carries through to almost periodic measures, but we do not want to enter such technicalities here, and that is the reason why we prefer to set (H1) and (H2) as assumptions in the above corollary.

Besides, it is easy to see that if m is almost periodic (and satisfies (H1) and (H2)) then any measure $\bar{m}$ obtained, according to Definition 6.4, as the limit at infinity of translates of m also satisfies conditions (H1) and (H2).

Proof of Corollary 6.8 We consider a sequence y_n going to infinity. Extracting a subsequence if necessary, we may assume, without loss of generality, that $m(\cdot + y_n)$ converges to some measure $\bar{m}$ in the sense of Definition 6.4, and thus in the sense of distributions.

In view of the above remark, $\bar{m}$ satisfies the hypotheses of Theorem 6.10, and we may therefore define $(\bar{u}; \bar{\Phi})$ as the solution to system (6.48) with $\bar{m}$.

We are going to show that $u(\cdot + y_n)$ converges to $\bar{u}$ uniformly on $\mathbf{R}^3$. The convergence of $\Phi(\cdot + y_n)$ to $\bar{\Phi}$ will follow easily, and we shall skip this last part of the proof.

We argue by contradiction and assume there exists some sequence z_n and some positive constant δ such that, up to a subsequence,

$$\forall n, \qquad |u(y_n + z_n) - \bar{u}(z_n)| \geq \delta. \tag{6.74}$$

Since

$$m(\cdot + y_n + z_n) - \bar{m}(\cdot + z_n) \longrightarrow 0$$

in the sense of Definition 6.4, we know, by Theorem 6.10 and by the arguments used in the proof of Theorem 5.9, that $u(\cdot + y_n + z_n)$ and $\bar{u}(\cdot + z_n)$, respectively, converge to the one and only solution $\tilde{u}$ corresponding to the measure $\tilde{m}$, which is the limit (up to an extraction), again in the sense of Definition 6.4, of both sequences of measures $m(\cdot + y_n + z_n)$ and $\bar{m}(\cdot + z_n)$. Consequently, we know that in particular $u(0 + y_n + z_n) - u(0 + z_n)$ converges to zero, which of course contradicts (6.74). $\Diamond$

Remark 6.10 It is easy to see that, any arbitrary triplet of continuous (or L^p_{loc}) almost periodic functions (m, u, Φ) being given, we may find, for any $\varepsilon > 0$, a relatively dense set $\{\tau\}$ that is at once a set of ε-almost period for m, u and Φ. Actually, from the general uniqueness result, Theorem 6.10, and its proof, we may even show in our case (and at least for $m \in L^p_{\text{loc}}$) the following property: for all $\varepsilon > 0$, there exists $\delta > 0$ such that any δ-almost period for m is a ε-almost period for u and Φ.

6.5   Non-neutral systems

We now come back to standard periodic geometries, and are going to consider non-neutral systems.

So far, we have dealt with systems such that, for any Λ, the electronic charge $\int_{\mathbf{R}^3} \rho_\Lambda$ is equal to the total nuclear charge $|\Lambda|$. (Note that, so far again, this total nuclear charge is also the total number of nuclei; since we deal in this section with periodic systems, it will remain so, at least up to some multiplicative constant.) What happens when we consider the sequence of minimizing densities for an ion?

Let us first recall that, in the TFW setting, it is known (see Benguria, Brézis, and Lieb [6]) that, for a given positive charge Z, there exists a minimizing electronic density of prescribed charge provided that the number of electrons is less than or equal to a maximum number, denoted by λ_c, that satisfies $Z < \lambda_c < +\infty$. In other words, when the nuclei are given, any positive ion exists, and so does any negative ion as soon as its negative balance of charge is not too large. Beyond the critical charge λ_c, there is no minimum, which is the mathematical counterpart of the ionization. Let us recall that λ_c may be defined by

$$\lambda_c = \int_{\mathbf{R}^3} \rho_c, \tag{6.75}$$

where ρ_c is the minimizer of

$$\inf\{E^{\text{TFW}}(\rho); \rho \geq 0, \sqrt{\rho} \in H^1(\mathbf{R}^3)\}. \tag{6.76}$$

It is still a challenge to find lower and upper bounds on λ_c with respect to the physical parameters of the system (total nuclear charge Z, number of nuclei K, and so on). In [5], Benguria and Lieb proved a bound of the kind

$$0 < \lambda_c - Z \leq \alpha K, \tag{6.77}$$

where the constant α depends only on the parameters appearing in the definition of the TFW energy. In [53], Solovej improves the value of α, and also gives under some convenient assumptions a positive lower bound for $\lambda_c - Z$. We also refer the reader to Lieb [34] for relative considerations. Let us already mention that we shall improve in some sense the upper bound in (6.77) in the sequel of this section (see Proposition 6.11).

Now that we know which non-neutral systems exist in the TFW setting, we may ask the standard three questions of the thermodynamic limit on such systems.

Before we get to the study of the negative ions first, and then to some positive ions, let us go back to Chapter 3 for a while, and ask the following basic and somewhat formal question: What do we basically need to pass to the thermodynamic limit in a TFW model?

In Section 6.1, when we have summarized the three steps of our new strategy of proof, we have in fact recalled that the basis of our whole argument is some *a priori* estimates. More precisely, we may start our argument as soon as we have the following two conditions:

(i) The energy per unit volume $\frac{1}{|\Lambda|}|I_\Lambda|$ is bounded.

(ii) The Lagrange multiplier θ_Λ is bounded.

We then deduce from these bounds some estimates on ρ_Λ and Φ_Λ, and then proceed further. Next, of course, we need the uniqueness theorem to deal with the limit densities, but these *a priori* estimates clearly are the first step.

We are going to see now that these two conditions are automatically satisfied in the anionic case, which allows us to prove the existence of the thermodynamic limit for such systems.

6.5.1 *Negative ions*

We begin this study of anionic systems with the following estimate:

Proposition 6.11 *We consider Λ, a Van Hove sequence of nuclei of unit charge. If we denote by λ_c the sequence of real numbers that are the maximum number of electrons that may be bound to each Λ (recall that λ_c is defined by (6.75)–(6.76)), then*

$$\lambda_c - |\Lambda| = o(|\Lambda|). \tag{6.78}$$

Proof of Proposition 6.11 In order to show (6.78), we go back to the proof of compactness in Chapter 3, and more precisely to two estimates.

The first one is a consequence of the fact that the energy per unit volume is bounded from above. It is estimate (3.37) of Subsection 3.3.1, valid in the smeared nuclei case as well as in the point nuclei case:

$$\left| 1 - \frac{1}{|\Lambda|} \int_{\Gamma(\Lambda)} \rho_\Lambda(x)\,dx \right| \leq C\left[\left(\frac{|\Lambda^1|}{|\Lambda|} \right)^{\frac{1}{2}} + \left(\frac{|\Lambda^1|}{|\Lambda|} \right)^{\frac{2}{5}} \right], \tag{6.79}$$

where
$$\Lambda^1 = \{x \in \Gamma(\Lambda); d(x; \partial\Gamma(\Lambda)) \leq 1\}.$$

C denotes, here and below, a real constant that depends on the whole sequence (Λ).

The second estimate has been established in Subsection 3.3.5. It is true as soon as the effective potential Φ_Λ is conveniently bounded, which in turn comes from the existence of the two uniform bounds on the absolute energy per unit volume $\frac{1}{|\Lambda|}|I_\Lambda|$ and on the Lagrange multiplier θ_Λ. This estimate is

$$\int_{\Gamma(\Lambda)^c} \rho_\Lambda \leq C \, |\partial\Gamma(\Lambda)|. \tag{6.80}$$

Now we remark that in the maximum ionized case, when the number of electrons is exactly λ_c, we have

$$I_\Lambda^{\mathrm{TF}}(\Lambda) = I_\Lambda^{\mathrm{TF}}(\lambda_c) \leq I_\Lambda(\lambda_c) \leq I_\Lambda(\Lambda),$$

where all minimization problems are set with the nuclei of Λ, and where the number in parentheses denotes the number of electrons. It follows that

$$\frac{|I_\Lambda(\lambda_c)|}{|\Lambda|} \leq C. \tag{6.81}$$

On the other hand, since the number of electrons is maximal, the corresponding Lagrange multiplier is zero:
$$\theta_\Lambda(\lambda_c) = 0. \tag{6.82}$$

From (6.81) and (6.82), we therefore deduce that in our case both (6.79) and (6.80) hold. The claim of Proposition 6.11 follows. $\diamond$

Let us now state and prove an easy extension of the work we have done so far, which allows us to conclude in the case of anionic systems.

Proposition 6.12 *Let us consider a sequence λ_Λ of real numbers such that, for each set Λ, $\lambda_\Lambda \geq |\Lambda|$. Consider, for each Λ, the TFW minimization problem with Λ as the set of nuclei of unit charge and λ_Λ electrons, and denote its ground-state energy by $I_\Lambda(\lambda_\Lambda)$. When $\lambda_\Lambda \leq \lambda_c$, denote its minimizer by ρ_Λ, and if $\lambda_\Lambda > \lambda_c$, then denote by ρ_Λ the minimizer of charge λ_c. Then we have the thermodynamic limits*

$$\lim_{\Lambda \longrightarrow \infty} \frac{1}{|\Lambda|} I_\Lambda(\lambda_\Lambda) = I_{\mathrm{per}} + \frac{M}{2}$$

and

$$\lim_{\Lambda \longrightarrow \infty} \rho_\Lambda = \rho_{\mathrm{per}},$$

with the notation and in the sense of the convergences of Theorem 6.1.

Proof of Proposition 6.12 The proof is a straightforward application of the arguments we have made so far and of the remarks on the proof of Proposition 6.11. All the usual bounds on the energy per unit volume and the Lagrange multiplier are available: let us indeed recall that, in the TFW setting, the infimum energy is a convex function that decreases with respect to the number of electrons. In particular, the Lagrange multiplier is monotonic with respect to this number: since it is bounded in the neutral case, and is zero in the maximum ionized case, it is bounded in between. The rest of the argument is the same as in the proof of Theorem 6.1. ◇

For any anionic system, we therefore know the behaviour in the thermodynamic limit. Let us now turn to cationic systems.

6.5.2 *Positive ions*

First of all, let us recall some notation. When Λ is a Van Hove sequence of nuclei of unit charge, we denote by λ_Λ any arbitrary sequence of 'numbers' of electrons less than or equal to $|\Lambda|$. For each Λ, the minimization problem $I_\Lambda(\lambda_\Lambda)$ with $|\Lambda|$ nuclei at the points of Λ and λ_Λ electrons is well posed (see Chapter 3). There is a unique minimizing density, that we denote by ρ_{λ_Λ}. Of course, when $\lambda_\Lambda = |\Lambda|$, we keep the notation used so far, i.e. I_Λ and ρ_Λ instead of $I_\Lambda(|\Lambda|)$ and $\rho_{|\Lambda|}$.

The first remark we wish to make is the following: if there exists a thermodynamic limit for the energy per unit volume, then necessarily the excess of positive charge must not be too large. More precisely, we have, in the same fashion as Proposition 6.11:

Proposition 6.13 *We consider Λ, a Van Hove sequence of nuclei of unit charge. Suppose that for some sequence $\lambda_\Lambda \leq |\Lambda|$, the sequence $\frac{1}{|\Lambda|} I_\Lambda(\lambda_\Lambda)$ is bounded from above by a constant independent of Λ. Then we have*

$$|\Lambda| - o(|\Lambda|) \leq \lambda_\Lambda \leq |\Lambda|. \tag{6.83}$$

Proof of Proposition 6.13 We recall the estimate (6.79) that holds when the energy per unit volume is bounded from above:

$$\left| 1 - \frac{1}{|\Lambda|} \int_{\Gamma(\Lambda)} \rho_{\lambda_\Lambda}(x) \, dx \right| \leq C \left[\left(\frac{|\Lambda^1|}{|\Lambda|} \right)^{\frac{1}{2}} + \left(\frac{|\Lambda^1|}{|\Lambda|} \right)^{\frac{2}{5}} \right],$$

which yields here

$$\lambda_\Lambda \geq \int_{\Gamma(\Lambda)} \rho_{\lambda_\Lambda}(x) \geq |\Lambda| - C|\Lambda| \left[\left(\frac{|\Lambda^1|}{|\Lambda|} \right)^{\frac{1}{2}} + \left(\frac{|\Lambda^1|}{|\Lambda|} \right)^{\frac{2}{5}} \right]. \tag{6.84}$$

Our claim (6.83) follows. ◇

Remark 6.14 Note that if the sequence $\Gamma(\Lambda)$ is a sequence of large cubes, the estimate (6.83) may be somewhat improved. Indeed, (6.84) and $|\Lambda^1| = O(|\Lambda|^{2/3})$ then yield

$$|\Lambda| - O(|\Lambda|^{13/15}) \leq \lambda_\Lambda \leq |\Lambda|.$$

We now show that, conversely, if the excess of positive charge is small enough, then the energy per unit volume converges.

Proposition 6.15 *Let us consider a Van Hove sequence Λ and a sequence λ_Λ of real numbers such that $|\Lambda| \geq \lambda_\Lambda \geq |\Lambda| - o(|\Lambda|^{2/3})$. Consider, for each Λ, the TFW minimization problem with Λ as the set of nuclei of unit charge and λ_Λ electrons. Then its ground-state energy $I_\Lambda(\lambda_\Lambda)$ has the same thermodynamic limit as the neutral system; namely,*

$$\lim_{\Lambda \longrightarrow \infty} \frac{1}{|\Lambda|} I_\Lambda(\lambda_\Lambda) = I_{\text{per}} + \frac{M}{2}.$$

Proof of Proposition 6.15 We begin with the basic remark

$$I_\Lambda(\lambda_\Lambda) \geq I_\Lambda$$

for each Λ, since $\lambda_\Lambda \leq |\Lambda|$ and the TFW energy $\lambda \longrightarrow I_\Lambda(\lambda)$ is decreasing with respect to the electronic charge. Therefore, passing to the lower limit, we obtain

$$\liminf_{\Lambda \to \infty} \frac{1}{|\Lambda|} I_\Lambda(\lambda_\Lambda) \geq I_{\text{per}} + \frac{M}{2}. \tag{6.85}$$

We now recall that we denote by ρ_Λ the neutral minimizing density. Choosing $\frac{\lambda_\Lambda}{|\Lambda|}\rho_\Lambda$ as a test function for the energy, we have

$$I_\Lambda(\lambda_\Lambda) \leq E_\Lambda\left(\frac{\lambda_\Lambda}{|\Lambda|}\rho_\Lambda\right) + \tfrac{1}{2}U_\Lambda.$$

Now,

$$E_\Lambda\left(\frac{\lambda_\Lambda}{|\Lambda|}\rho_\Lambda\right) + \tfrac{1}{2}U_\Lambda = \frac{\lambda_\Lambda}{|\Lambda|}\int |\nabla\sqrt{\rho_\Lambda}|^2 + \left(\frac{\lambda_\Lambda}{|\Lambda|}\right)^{5/3}\int \rho_\Lambda^{5/3}$$

$$- \frac{\lambda_\Lambda}{|\Lambda|}\int V_\Lambda\rho_\Lambda + \tfrac{1}{2}\left(\frac{\lambda_\Lambda}{|\Lambda|}\right)^2 D(\rho_\Lambda, \rho_\Lambda) + \tfrac{1}{2}U_\Lambda.$$

We write this energy as follows:

$$E_\Lambda\left(\frac{\lambda_\Lambda}{|\Lambda|}\rho_\Lambda\right) + \tfrac{1}{2}U_\Lambda = \left(\frac{\lambda_\Lambda}{|\Lambda|}\right)^2 (E_\Lambda(\rho_\Lambda) + \tfrac{1}{2}U_\Lambda) + \frac{\lambda_\Lambda}{|\Lambda|}\left(1 - \frac{\lambda_\Lambda}{|\Lambda|}\right)\int |\nabla\sqrt{\rho_\Lambda}|^2$$

$$+ \left[\left(\frac{\lambda_\Lambda}{|\Lambda|}\right)^{5/3} - \left(\frac{\lambda_\Lambda}{|\Lambda|}\right)^2\right]\int \rho_\Lambda^{5/3}$$

$$+ \frac{\lambda_\Lambda}{|\Lambda|}\left(1 - \frac{\lambda_\Lambda}{|\Lambda|}\right)\left(U_\Lambda - \int V_\Lambda\rho_\Lambda\right) + \tfrac{1}{2}\left[1 - \frac{\lambda_\Lambda}{|\Lambda|}\right]^2 U_\Lambda. \tag{6.86}$$

We next use $\lambda_\Lambda \leq |\Lambda|$ and the estimates of Propositions 3.4 and 3.8; namely,

$$\int \rho_\Lambda^{5/3} + \int |\nabla\sqrt{\rho_\Lambda}|^2 = O(|\Lambda|)$$

and

$$\left| U_\Lambda - \int V_\Lambda \rho_\Lambda \right| = O(|\Lambda|),$$

to deduce from (6.86) and $E_\Lambda(\rho_\Lambda) + \frac{1}{2}U_\Lambda = I_\Lambda$ that

$$E_\Lambda\left(\frac{\lambda_\Lambda}{|\Lambda|}\rho_\Lambda\right) + \frac{1}{2}U_\Lambda \leq \left(\frac{\lambda_\Lambda}{|\Lambda|}\right)^2 I_\Lambda + \left(1 - \frac{\lambda_\Lambda}{|\Lambda|}\right)O(|\Lambda|)$$

$$+ \frac{1}{2}\left[1 - \frac{\lambda_\Lambda}{|\Lambda|}\right]^2 U_\Lambda. \tag{6.87}$$

We then use the assumption $|\Lambda| \geq \lambda_\Lambda \geq |\Lambda| - o(|\Lambda|^{2/3})$ and the bound $U_\Lambda = O(|\Lambda|^{5/3})$ to obtain, letting Λ go to infinity in (6.87),

$$\limsup_{\Lambda\to\infty} \frac{1}{|\Lambda|}I_\Lambda(\lambda_\Lambda) \leq \limsup_{\Lambda\to\infty} \frac{1}{|\Lambda|}I_\Lambda = I_{\text{per}} + \frac{M}{2}. \tag{6.88}$$

Comparing with (6.85), we conclude the proof of Proposition 6.15. $\diamondsuit$

Remark 6.16 There might be room for improvement in the above proof by choosing some more convenient test function.

Remark 6.17 (On the TF model) It is worth noticing that the above proof, and therefore the result of Proposition 6.15, also holds true in the TF case. Indeed, for each Λ, we have on the one hand $I_\Lambda(\lambda_\Lambda)^{\text{TF}} \leq I_\Lambda(\lambda_\Lambda)$, while on the other hand $I_\Lambda(\lambda_\Lambda)^{\text{TF}} \geq I_\Lambda^{\text{TF}}$, since the TF energy is also decreasing with respect to the number of electrons. It then remains to pass to the limit in both inequalities as Λ goes to infinity.

We now deal with the density. The optimal result we are able to prove is the following:

Proposition 6.18 *Let us consider a Van Hove sequence Λ and a sequence λ_Λ of real numbers such that $|\Lambda| \geq \lambda_\Lambda \geq |\Lambda| - O(|\Lambda|^{1/3})$. Consider, for each Λ, the TFW minimization problem with Λ as the set of nuclei of unit charge and λ_Λ electrons. Then its ground-state density has the same thermodynamic limit as the neutral density; namely,*

$$\lim_{\Lambda\to\infty} \rho_{\lambda_\Lambda} = \rho_{\text{per}},$$

with the notation and in the sense of the convergences of Theorem 6.1.

Proof of Proposition 6.18 Of course, since $|\Lambda| \geq \lambda_\Lambda \geq |\Lambda| - O(|\Lambda|^{1/3}) \geq |\Lambda| - o(|\Lambda|^{2/3})$, the result of Proposition 6.15 holds, i.e. the energy per unit volume has a limit. In order to deal with the sequence of densities, we just have to show that the Lagrange multiplier θ_{λ_Λ} is bounded. Then the usual bounds on ρ_{λ_Λ} and Φ_{λ_Λ} will follow, and the rest of the argument will be the same as in the neutral case. For this purpose, we argue as follows.

Let C be a constant such that

$$|\Lambda| \geq \lambda_\Lambda \geq |\Lambda| - C|\Lambda|^{1/3}. \tag{6.89}$$

We go back to inequality (6.87), from which we deduce, using that we have here (6.89), a better estimate than in the course of Proposition 6.15; namely,

$$
\begin{aligned}
I_\Lambda(\lambda_\Lambda) &\leq E_\Lambda\left(\frac{\lambda_\Lambda}{|\Lambda|}\rho_\Lambda\right) + \tfrac{1}{2}U_\Lambda \\
&\leq \left(\frac{\lambda_\Lambda}{|\Lambda|}\right)^2 I_\Lambda + \left(1 - \frac{\lambda_\Lambda}{|\Lambda|}\right)O(|\Lambda|) + \tfrac{1}{2}\left[1 - \frac{\lambda_\Lambda}{|\Lambda|}\right]^2 U_\Lambda. \\
&\leq I_\Lambda + O(|\Lambda|^{1/3}). \tag{6.90}
\end{aligned}
$$

We therefore have

$$\forall C \geq 0, \qquad I_\Lambda(|\Lambda| - C|\Lambda|^{1/3}) \leq I_\Lambda + O(|\Lambda|^{1/3}). \tag{6.91}$$

Next, we use the fact that the Lagrange multiplier is (up to a sign) the first derivative of the energy with respect to the electronic charge. Therefore, for any C, we have

$$I_\Lambda(|\Lambda| - (C+1)|\Lambda|^{1/3}) = I_\Lambda(|\Lambda| - C|\Lambda|^{1/3}) + \int_{|\Lambda|-(C+1)|\Lambda|^{1/3}}^{|\Lambda|-C|\Lambda|^{1/3}} \theta_\lambda d\lambda.$$

Now, the energy being convex, the Lagrange multiplier decreases when the electronic charge increases; thus

$$
\begin{aligned}
I_\Lambda(|\Lambda| - (C+1)|\Lambda|^{1/3}) &\geq I_\Lambda(|\Lambda| - C|\Lambda|^{1/3}) + |\Lambda|^{1/3}\theta_{|\Lambda|-(C+1)|\Lambda|^{1/3}} \\
&\geq I_\Lambda + |\Lambda|^{1/3}\theta_{|\Lambda|-(C+1)|\Lambda|^{1/3}}.
\end{aligned}
$$

Then using (6.91) to bound the left-hand side from above, we obtain

$$|\Lambda|^{1/3}\theta_{|\Lambda|-(C+1)|\Lambda|^{1/3}} = O(|\Lambda|^{1/3}). \tag{6.92}$$

Using the monotonicity of the Lagrange multiplier again this yields the bound on the Lagrange multiplier for any sequence λ_Λ satisfying $|\Lambda| \geq \lambda_\Lambda \geq |\Lambda| - O(|\Lambda|^{1/3})$, and the proposition follows. $\diamond$

As a conclusion to this section, let us mention that there exist some open questions as far as non-neutral systems are concerned. In particular, we do not know whether or not the Lagrange multiplier is bounded when the excess of positive charge is larger than $|\Lambda|^{1/3}$ but is still $o(|\Lambda|)$ for a sequence of cubes, for instance. In addition, the limit cases may be of some interest, since it is likely that the shape of the domain $\Gamma(\Lambda)$ plays a role. We hope to come back to such questions in the near future.

REFERENCES

[1] Amerio, L. and Prouse, G. (1971). *Almost periodic functions and functional equations*. Van Nostrand Reinhold, New York.

[2] Ashcroft, N. W. and Mermin, N. D. (1976). *Solid-state physics*, Saunders College Publishing.

[3] Axel, F. and Gratias , D. (ed.,) (1995). *Beyond Quasicrystals*. Centre de Physique Les Houches, Les Editions de Physique. Springer-Verlag.

[4] Balian, R. (1991). *From microphysics to macrophysics; methods and applications of Statistical Physics*, I and II. Springer-Verlag.

[5] Benguria, R. and Lieb, E. H. (1985). The most negative ion in the Thomas–Fermi–von Weizsäcker theory of atoms and molecules, J. Phys. B, 18, 1985, 1045–1059.

[6] Brézis, H., Benguria, R., and Lieb, E. H. (1981). The Thomas-Fermi-von Weizsäcker theory of atoms and molecules. *Comm. Math. Phys.*, 79, 167–180.

[7] Bénilan, Ph. and Brézis, H. *Nonlinear problems related to the Thomas–Fermi equation*, unpublished work.

[8] Bénilan, Ph., Brézis, H., and Crandall, M. G. (1975). A semilinear equation in $L^1(\mathbf{R}^N)$. *Ann. Scuola. Norm. Pisa*, 2, 523–555.

[9] Bergh, J. and Löfström, J. (1976). *Interpolation spaces, an introduction*. Springer-Verlag, 223.

[10] Besicovitch, A. S. (1954). *Almost periodic functions*. Dover, New York.

[11] Bohr, H. (1947). *Almost periodic functions*. Chelsea, New York.

[12] Brézis, H. (1984). Semilinear equations in $\mathbf{R}^N$ without condition at infinity. *Appl. Math. Optim.*, 12, 271–282.

[13] Brézis, H. (1980). Some variational problems of the Thomas–Fermi type. In (ed. Cottle, Giannessi, and J.-L. Lions), *Variational inequalities and complementary problems*). Wiley, New York, 53–73.

[14] Callaway, J. (1974). *Quantum theory of the solid state*. Academic Press.

[15] Catto, I. and Lions, P.-L. (1992, 1993). Binding of atoms and stability of molecules in Hartree and Thomas-Fermi type theories, parts 1, 2, 3, 4. *Commun. Part. Diff. Eq.*, 17 and 18.

[16] Catto, I., Le Bris, C., and Lions, P.-L. (1996). Limite thermodynamique pour des modèles de type Thomas–Fermi. *C. R. Acad. Sci. Paris*, t. **322**, Série I, 357–364.

[17] Catto, I., Le Bris, C., and Lions, P.-L. *On the Thermodynamic limit for Hartree–Fock type models*, work in preparation.

[18] Chen, L., Moody, R.V., and Patera, J. Non-crystallographic root systems. In Proceedings of the Workshop on Aperiodic Order, Fields Institute Commu-

nications, American Mathematical Society, 1996, to appear.

[19] Dreizler, R.M. and Gross, E.K.U. (1990). *Density functional theory.* Springer-Verlag.

[20] Ekeland, I. (1979). Nonconvex minimization problems. Bull. A.M.S., 1 (3), 443–474.

[21] Fefferman, C. (1985). The thermodynamic limit for a crystal. *Commun. Math. Phys.*, 98, 289–311.

[22] Fefferman, C. (1985). The atomic and molecular nature of matter. *Rev. Mat. Iberoam.*, **1**, (1).

[23] Gérard, P. (1991). Mesures semi-classiques et ondes de Bloch. Exposé XVI, Séminaire EDP, Ecole Polytechnique.

[24] Gilbarg, D. and Trudinger, N. S. (1983). *Elliptic partial differential equations of second order* (2nd ed.). Springer-Verlag.

[25] Gregg, J. N. Jr. (1989). The existence of the thermodynamic limit in Coulomb-like systems. *Commun. Math. Phys.*, 123, 255–276.

[26] Hohenberg, P. and Kohn, W. (1964). Inhomogeneous electron gas. *Phys. Rev.*, 136, B864–B871.

[27] Kittel, C. (1986). *Introduction to solid-state physics* (6th ed.). Wiley.

[28] Le Bris, C. (1993). Some results on the Thomas–Fermi–Dirac–von Weizsäcker model. *Diff. Int. Eq.*, 6 (2), pp 337–353.

[29] Lebowitz, J. L. and Lieb, E. H. (1969). Existence of thermodynamics for real matter with Coulomb forces. *Phys. Rev. Lett.*, 22 (13), 631–634.

[30] Léon, J.-F. (1988). Existence and uniqueness of positive solutions for semi-linear elliptic equations on unbounded domains. *Commun. Part. Diff. Eq.*, 13 (10), 1223–1234.

[31] Léon, J.-F. (1987). Existence et unicité de la solution positive de l'équation TFW sans répulsion électronique. *Math. Mod. Num. Anal.*, 21 (4), 641–654.

[32] Lieb, E. H. (1976). The stability of matter. *Rev. Mod. Phys.*, 48, 553–569.

[33] Lieb, E. H. (1981). Thomas–Fermi and related theories of atoms and molecules. *Rev. Mod. Phys.*, 53 (4), 603–641.

[34] Lieb, E. H. (1984). Bound on the maximum negative ionization energy of atoms and molecules. *Phys. Rev. A*, 29 (6), 3018–3028.

[35] Lieb, E. H. (1990). The stability of matter: from atoms to stars. *Bull. A.M.S.*, 22 (1), 1–49.

[36] Lieb, E. H. and Lebowitz, J. L. (1972). The constitution of matter: existence of thermodynamics for systems composed of electrons and nuclei. *Adv. Math.*, 9, 316–398.

[37] Lieb, E. H. and Lebowitz, J. L. (1973). *Lectures on the thermodynamic limit for Coulomb systems.* Lecture Notes in Physics, 20, 136–161. Springer-Verlag.

[38] Lieb, E. H. and Narnhofer, N. (1975). The thermodynamic limit for Jellium. *J. Stat. Phys.*, 12, 291–310.

[39] Lieb, E. H. and Simon, B. (1977). The Hartree–Fock theory for Coulomb systems. *Communs. Math. Phys.*, 53, 185–194.

[40] Lieb, E. H. and Simon, B. (1977). The Thomas-Fermi theory of atoms, molecules and solids. *Adv. Math.*, 23, 22–116.

[41] Lieb, E. H. and Thirring, W. (1975). Bounds for the kinetic energy of fermions which prove the stability of matter. *Phys. Rev. Lett.*, 35, 687–689. Errata ibid. 35, 1116.

[42] Lions, P.-L. (1987). Solutions of Hartree-Fock equations for Coulomb systems. *Communs. Math. Phys.*, 109, 33–97.

[43] Madelung, O. (1981). *Introduction to solid-state theory.* Springer-Verlag, Solid State Sciences, 2.

[44] Nakano, F. (1996). The thermodynamic limit of the magnetic Thomas-Fermi energy. *J. Math. Sci. Univ. Tokyo*, 3, 713–722.

[45] Parr, R. G. and Yang, W. (1989). *Density-functional theory of atoms and molecules.* Oxford University Press, Oxford.

[46] Pisani, C. (1996). *Quantum mechanical* ab initio *calculation of the properties of crystalline materials.* Lecture Notes in Chemistry, 67. Springer-Verlag.

[47] Quinn, Ch. M. (1973). *An introduction to the quantum theory of solids.* Clarendon Press, Oxford.

[48] Ruelle, D. (1969). *Statistical Mechanics: rigorous results.* Benjamin, New York. (Also Advanced Books Classics, Addison-Wesley, 1989.)

[49] Reed, M. and Simon, B. (1978). *Methods of modern mathematical physics*, IV. Academic Press, New York.

[50] Senechal, M. (1995). *Quasicrystals and geometry.* Cambridge University Press, Cambridge.

[51] Slater, J. C. (1963). *Quantum theory of molecules and solids.* McGraw Hill, New York.

[52] Slater, J. C. (1972). *Symmetry and energy bands in crystals.* Dover, New York.

[53] Solovej, J. P. (1990). *Universality in the Thomas–Fermi–von Weizsäcker model of atoms and molecules. Communs. Math. Phys.*, **129**, 561–598.

[54] Solovej, J. P. (1994). *An improvement on stability of matter in mean field theory.* Proceedings of the Conference on PDEs and Mathematical Physics, University of Alabama. International Press.

[55] Tolman, R. C. (1962). *The principles of statistical mechanics.* Oxford University Press, Oxford.

[56] Trudinger, N. S. (1973). Linear elliptic operators with measurable coefficients. *Ann. Scuola Norm. Sup. Pisa*, 27, 265–308.

[57] Véron, L. (1992). Semilinear elliptic equations with uniform blow-up on the boundary. *J. Anal. Math.*, 59, 231–250.

[58] Ziman, J. (1972). *Principles of the theory of solids* (2nd ed.). Cambridge University Press, Cambridge.

INDEX